Smart Surface Design for Efficient Ice Protection and Control

Online at: https://doi.org/10.1088/978-0-7503-6009-8

Smart Surface Design for Efficient Ice Protection and Control

Edited by
Carlo Antonini
Department of Materials Science, University of Milano-Bicocca, Milan, Italy

Irene Tagliaro
Department of Materials Science, University of Milano-Bicocca, Milan, Italy

IOP Publishing, Bristol, UK

ISBN 978-0-7503-6009-8 (ebook)
ISBN 978-0-7503-6007-4 (print)
ISBN 978-0-7503-6010-4 (myPrint)
ISBN 978-0-7503-6008-1 (mobi)

DOI 10.1088/978-0-7503-6009-8

Version: 20250601

IOP ebooks

British Library Cataloguing-in-Publication Data: A catalogue record for this book is available from the British Library.

Published by IOP Publishing, wholly owned by The Institute of Physics, London

IOP Publishing, No.2 The Distillery, Glassfields, Avon Street, Bristol, BS2 0GR, UK

US Office: IOP Publishing, Inc., 190 North Independence Mall West, Suite 601, Philadelphia, PA 19106, USA

Contents

Foreword

Back in 2017, when we started our first brainstorming on the topics of icing and ice protection technologies, Carlo Antonini, Ilia Roisman and I did not anticipate that a Marie Skłodowska-Curie project would follow and that a very dedicated and skilled team of PhD students would advance several aspects in this research area so considerably. As sometimes happens, it came differently, and now you have in front of you the latest findings in the form of this book.

As an Airbus expert who has worked for many years in this research field, I am truly amazed how effective a team of young and passionate researchers guided by some very dedicated academic supervisors could be in addressing the icing challenge and in delivering results that are scientifically relevant and industrially valuable. Indeed, icing and ice protection are very interdisciplinary topics. Atmospheric physics, thermodynamics, mechanical engineering, chemistry and materials science must come together, willing to learn each other's languages and to contribute experimentally and through theoretical and numerical models and tools. That such disciplines would find common ground and ways of interfacing and collaborating was not clear in the beginning, but it is exactly what happened within the SURFICE consortium.

Most supervisors had previous experience in one or more aspects of icing, such as physical modeling of ice formation, mechanical modeling of ice adhesion, the design and development of icephobic coatings, the design and prototyping of mechanical devices, or icing wind tunnel testing. However, as you will realize reading this book, these separate disciplines have successfully 'fusioned' and proposed next-generation technological solutions. These are achievements of which the young researchers who contributed this breadth of new knowledge should be especially proud!

The common thread connecting the chapters of this book addresses three critical aspects of icing: (i) understanding the physics of ice formation in most diverse atmospheric conditions and understanding what chemistry can do to control ice formation and ice removal from surfaces; (ii) understanding the physico-chemical mechanisms controlling ice adhesion and designing functional coatings and electro-mechanical technologies to efficiently crack ice off surfaces; and (iii) making use of the knowledge gained in (i) and (ii) for designing and developing novel icephobic coatings that are not only effective, but also durable enough to be adopted in industrial applications.

Indeed, the surfaces of aircraft wings, wind turbine rotor blades, high-voltage line cables, or antennas for telecommunications must resist unaffected most atmospheric weather conditions, from heavy rain to sand storms to UV radiation to snow and ice storms. You will find the latest scientific outcomes and technological solutions addressed in this book—so enjoy the read!

Munich, 31 January 2025
Elmar Bonaccurso
Airbus Central R&T

Preface

The context—Icing is a natural phenomenon affecting our daily life and safe operations in diverse areas, such as aeronautics and ground transportation, operations in critical environment, power systems (such as power lines, wind turbines or solar panels), communication systems and infrastructures. In particular, human activities at high latitudes or at high altitudes require energy-efficient and environmentally sustainable ice protection solutions.

The challenge(s)—The existing solutions for prevention or delay of icing, for reduction of the icing rate and for ice removal facilitation include both active and passive systems. Active systems are sophisticated mechanical or energy-demanding thermal systems, whereas passive systems include specialized materials or morphological modifications of surfaces or coatings. The passive systems are often based on the use of hydrophilic polymers and utilization of colligative properties of solutions for lowering of the freezing point. Also, superhydrophobic and lubricant-infused surfaces may reduce the time of water contact with the surface or reduce adhesion strength. However, at present, technological problems are associated with the design of coating-based anti-icing systems: as examples, substrate erosion and corrosion, liquid impalement and consequent increase in ice adhesion to the solid surface due to high-speed drop impact. These factors have so far represented a bottleneck for the transition of coating-based technologies to industrial applications.

The book's genesis—In November 2017, an initial core team of three scientists (Carlo Antonini, Ilia Roisman, Elmar Bonaccurso) initiated the discussion on how to tackle the above challenges and conceived a project within the framework of Marie Skłodowska-Curie European Training Networks, a unique scheme to train a cohort of interdisciplinary scientists. The core team gathered a consortium comprising universities, research institutions, enterprises and partner organizations in eight European countries (Italy, Germany, France, the UK, the Netherlands, Belgium, Austria, Switzerland). A few years down the line came this book, as a result of that initial vision.

The scientific objectives: why read the book?—The project SURFICE (smart surface design for efficient ice protection and control), that gives the title to the book, has the goal to formulate a rational framework for the design, fabrication and testing of icephobic surfaces for industrially relevant applications. The three major research objectives of SURFICE, tackling the existing knowledge gaps, have been: (i) the investigation of the physics of icing of morphologically complex microstructured surfaces; (ii) the rational design of new icephobic materials and coatings; and (iii) the development of new technologies and systems for efficient ice prevention and control. Consequently, the objective of this book is to integrate information about the modeling, coating and material preparation, characterization and applications of icephobic surfaces in one volume. By combining the discussion of all aspects together, our hope is that the reader will benefit from a single reference book to understand the fundamentals of coating strategies based on icephobicity and the possible implementation into industrial systems.

The book's audience—This book has been designed as useful reading for materials scientists, physicists, chemists and engineers, interested in any aspect of surface, colloid and polymer science and thermodynamics relevant to icing phenomena and related coating strategies.

How to read the book—The book is structured in two sections. The first part covers the fundamentals, including the theory of physics of icing (chapter 1) and the state-of-the-art on existing surface-based strategies for icephobicity (chapter 2). This first part should help readers familiarize themselves with the terminology, the relevant physical mechanisms and surface chemistry concepts, especially to guide those who approach the topic for the first time or are looking for a comprehensive systematic overview of the research field. The second part of the book discusses the most recent significant scientific advances, spanning from the understanding of ice adhesion and fracture mechanisms (chapters 3–5) to strategies for effective icephobic coatings (chapters 6–10).

Milan, 31 January 2025
Carlo Antonini, Irene Tagliaro

Acknowledgements

All the book authors and contributors acknowledge funding through SURFICE (smart surface design for efficient ice protection and control) (www.surfice-itn.eu), a European Training Network funded by the European Union's Horizon 2020 research and innovation programme under the Marie Skłodowska-Curie grant agreement No. 956703. The project started in January 2021 and was completed in January 2025.

A personal acknowledgment from Carlo Antonini: I am particularly grateful to Ilia Roisman and Elmar Bonaccurso, the other two members of the initial core team, for kindly nudging me, giving me the opportunity to take the lead as a project coordinator, despite my initial self-doubt about being fit for the role. I hope I will be able to pay it forward.

Editor biographies

Carlo Antonini

Carlo Antonini received a BSc in Aerospace Engineering (2004) and MSc in Aeronautical Engineering (2007) from Politecnico di Milano, and a PhD from University of Bergamo (2011), Italy. From 2012 to 2018 he worked in Zurich, first at ETH as a Marie Sklodowska-Curie post-doc, then at EMPA as a scientist. In September 2018 he joined the Department of Materials Science at the University of Milano-Bicocca (Italy) as assistant professor, where he created and currently leads the 'Surface Engineering and Fluid Interfaces Laboratory' (SEFI Lab, sefilab.mater.unimib.it). In 2021 he received tenure as associate professor. From 2021 to 2025 Antonini has been the coordinator of SURFICE, an MSCA-ITN project supported by the EU, focusing on surface based-strategies to contrast icing.

Irene Tagliaro

Irene Tagliaro completed her MSc in chemistry science and technology at the University Milano-Bicocca (2015), Italy, where she also obtained a PhD in materials science in 2019. Later, she worked as consultant for small and medium enterprises (SMEs), until she returned as a post-doc to the University Milano-Bicocca at the 'Surface Engineering and Fluid Interfaces Laboratory' (SEFI Lab, sefilab.mater.unimib.it), led by Professor Antonini, working as project manager of SURFICE (smart surface design for efficient ice protection and control) an ITN-MSC project, where she coordinated the work of 13 PhD students located in different European countries and a Consortium of 10 institutes and several partners. In 2025, she became an assistant professor, developing her research interests in the fields of coatings, biopolymers and sustainable materials.

List of contributors

Carlo Antonini
Department of Materials Science, University of Milano-Bicocca, Milan, Italy

Alexandros A Atzemoglou
SuSoS AG, Dübendorf, Switzerland
and
Macromolecular Engineering Laboratory, Department of Mechanical and Process Engineering, ETH Zurich, Zurich, Switzerland

Simrandeep Bahal
Nanoengineered Systems Laboratory, Mechanical Engineering, University College London, London, UK
and
UCL Hawkes Institute, University College London, London, UK

Niccolò Bartalucci
SuSoS AG, Dübendorf, Switzerland

Marc Budinger
Institut Clement Ader, University of Toulouse, UPS, INSA, ISAE-SUPAERO, MINES-ALBI, CNRS, Toulouse, France

Anna Maria Coclite
Institute of Solid State Physics, NAWI Graz, Graz University of Technology, Graz, Austria
and
Department of Physics, University of Bari, Bari, Italy

Theodoros Dimitriadis
Department of Engineering, FT Technologies (UK) Ltd, Sunbury-on-Thames, UK
and
Institute of Solid State Physics, Graz University of Technology, Graz, Austria

Mingyue Ding
Institute for Fluid Mechanics and Aerodynamics, Technical University of Darmstadt, Darmstadt, Germany

Santiago J Garcia
Department of Aerospace Structures and Materials, Faculty of Aerospace Engineering, Delft University of Technology, Delft, Netherlands

Giulia Gastaldo
Fédération ENAC ISAE-SUPAERO ONERA, University of Toulouse, Toulouse, France

Pablo F Ibáñez Ibáñez
Department of Applied Physics, University of Granada, Granada, Spain

Prasenjit Kabi
Nanoengineered Systems Laboratory, Mechanical Engineering, University College London, London, UK
and
UCL Hawkes Institute, University College London, London, UK

Priya Mandal
Nanoengineered Systems Laboratory, Mechanical Engineering, University College London, London, UK
and
UCL Hawkes Institute, University College London, London, UK

Andrea Maslov
Department of Materials Engineering, KU Leuven, Leuven, Belgium

Navid Mostofi Sarkari
Department of Materials Engineering, KU Leuven, Leuven, Belgium

Anny Catalina Ospina Patino
Department of Materials Science, University of Milano-Bicocca, Milan, Italy

Valérie Pommier-Budinger
Fédération ENAC ISAE-SUPAERO ONERA, University of Toulouse, Toulouse, France

Gabriel Hernandez Rodriguez
Institute of Solid State Physics, Graz University of Technology, Graz, Austria

Ilia V Roisman
Institute for Fluid Mechanics and Aerodynamics, Technical University of Darmstadt, Darmstadt, Germany

David Seveno
Department of Materials Engineering, KU Leuven, Leuven, Belgium

Ali Raza Shaikh
Institute for Fluid Mechanics and Aerodynamics, Technical University of Darmstadt, Darmstadt, Germany

Vikramjeet Singh
Nanoengineered Systems Laboratory, Mechanical Engineering, University College London, London, UK
and
Manufacturing Futures Lab, Mechanical Engineering, University College London, London, UK

Luca Stendardo
Department of Materials Science, University of Milano-Bicocca, Milan, Italy

Irene Tagliaro
Department of Materials Science, University of Milano-Bicocca, Milan, Italy

Miisa Tavatstjerna
Department of Aerospace Structures and Materials, Faculty of Aerospace Engineering, Delft University of Technology, Delft, The Netherlands

Mark W Tibbitt
Macromolecular Engineering Laboratory, Department of Mechanical and Process Engineering, ETH Zurich, Zurich, Switzerland

Manish K Tiwari
Nanoengineered Systems Laboratory, Mechanical Engineering, University College London, London, UK
and
UCL Hawkes Institute, University College London, London, UK
and
Manufacturing Futures Lab, Mechanical Engineering, University College London, London, UK

Samuele G P Tosatti
SuSoS AG, Dübendorf, Switzerland

Jianhui Zhang
Nanoengineered Systems Laboratory, Mechanical Engineering, University College London, London, UK
and
UCL Hawkes Institute, University College London, London, UK

Stefan Zürcher
SuSoS AG, Dübendorf, Switzerland

Chapter 1

The physics of icing

Ilia V Roisman, Mingyue Ding, Andrea Maslov, Luca Stendardo and Ali Raza Shaikh

1.1 Phenomena leading to ice accretion

Parts of this section have been reproduced with permission from [31]. Copyright (2021) Elsevier.

Ice accretion can be dangerous in various situations and industries. Some fields where ice accretion poses significant risks include the aerospace industry, power transmission, automotive or maritime transportation, infrastructure and building, wind turbines, and many other industries exposed to low-temperature conditions (figure 1.1).

Ice accretion on aircraft surfaces [1–3], such as wings, tail, and engine components, can lead to reduced aerodynamic performance, increased drag, and altered weight distribution. This compromises the safety and efficiency of flight, potentially leading to accidents. Ice accumulation on power lines and transmission equipment [4–7] can lead to increased weight, sagging of lines, and potential damage to infrastructure. This poses a risk of power outages and can compromise the reliability of electrical grids. Ice accretion on roads and highways [8, 9] can create hazardous driving conditions. It leads to reduced traction, slippery surfaces, and increased braking distances, contributing to road accidents and traffic disruptions. Ice accretion on the hulls of ships [10] can affect their stability and maneuverability. It can also lead to additional weight, affecting fuel efficiency and potentially causing accidents, especially in icy waters. Ice accumulation on structures, such as bridges [11], towers, and buildings, can lead to structural damage. The added weight of ice can compromise the integrity of these structures, increasing the risk of collapse or failure. Ice accretion on wind turbine blades [12–15] can reduce their efficiency and lead to imbalance issues. This can result in decreased power generation and potential damage to the turbine components.

Various types of atmospheric ice accretion on cold surfaces are currently recognized, such as precipitation icing, in-cloud icing (resulting from the impact

doi:10.1088/978-0-7503-6009-8ch1

Figure 1.1. Various industrial applications influenced by ice accretion. (Created by Bing AI Image Generator, powered by OpenAI's DALL-E 3 model.)

of supercooled drops), and hoar frost icing caused by vapor sublimation to ice. Ice accretion due to liquid drop impact is further categorized into glaze ice and rime ice, the latter being divided into hard or soft rime. Rime ice forms rapidly through the immediate freezing of a supercooled drop after impact [1, 16–20], while glaze ice involves the creation of a liquid film that, upon stable solidification, smoothens surface roughness and fills gaps.

Various computational models have been formulated to predict ice accretion on cold surfaces. A notable example is a quasi-one-dimensional solution addressing thin liquid flow over a thin ice layer on an arbitrary cold solid substrate, leading to film solidification [21]. In this model, rime icing is addressed without a liquid film, and approximate balance equations for the thermal problem and liquid film flow are incorporated. The film is propelled by viscous stresses and aerodynamic pressure in the air flow. Similar methodologies are employed in computational codes such as [22–25], relying on assumed collision efficiency, sticking efficiency, and accretion efficiency. While not universally grounded in a complete understanding of microphysical processes during drop–surface collisions, these models, backed by extensive experimental tests, generally exhibit accurate predictions of ice accretion rates.

Enhanced comprehension of icing processes is imperative for modeling liquid freezing in microfluidic devices [26] or in scenarios involving water imbibition in porous and granular media, accompanied by solidification or melting [27].

Recent theoretical models addressing water freezing in aerodynamically driven liquid films [28] or rivulets [29] have emerged. These models hold potential applications in simulating icing in the runback flow on aircraft surfaces.

As an intermediary stage between models of single-drop impacts with subsequent freezing and the ice accretion process resulting from the impact of multiple drops, the scenario of successive impacts of several drops in a cloud has been explored, as detailed in [30]. However, the current understanding of the interactions between freezing drops remains insufficient, with research primarily confined to observations rather than comprehensive physical theories.

Icing can also be a result of partially melted ice crystal impacts. This type of icing is relevant to the icing of aircraft engines. The deposition of snowflakes is another possible reason for the icing of surfaces. The artistic images of the main types of ice accretion are shown in figure 1.2.

Figure 1.2. The main impact phenomena leading to ice accretion on solid surfaces: impact of supercooled drops (left), hailstones or ice crystals (middle image), and snowflakes (right). (Created by Bing AI Image Generator, powered by OpenAI's DALL-E 3 model.)

Understanding the ice accretion processes and their modeling is a rather challenging task since the phase-change phenomena are combined with the liquid flows associated with wetting and dewetting of solid surfaces [31].

1.2 Hydrodynamics and phase change in an impacting drop

1.2.1 Drop impact onto a wetted wall

Parts of this section have been reproduced from [263]. CC BY 4.0.

For reliable modeling of icing resulting from drop impacts, it is necessary to understand the hydrodynamics of drop collisions with a wall. Comprehensive reviews on these phenomena can be found in [32–35]. Two main phenomena are usually considered: drop impact onto a dry or wetted substrate.

Different outcomes of drop impact are shown in figure 1.3. The impact of a drop onto a wall film is affected by both the characteristics of the liquid, including surface tension σ, viscosity ν, and density ρ, as well as the impact parameters, which encompass the droplet diameter D, the height of the liquid film before impact H_{f}, and the velocity of the droplet upon impact U. These variables can be consolidated into the following dimensionless groups:

$$\mathrm{Re}=\frac{UD}{\nu}, \quad \mathrm{We}=\frac{\rho U^2 D}{\sigma}, \quad \delta=\frac{H_{\mathrm{f}}}{D}, \quad \kappa=\frac{\nu_{\mathrm{f}}}{\nu_{\mathrm{d}}}. \tag{1.1}$$

The subscripts d and f denote that the property either corresponds to the drop or the film liquid.

When a droplet collides with a solid substrate, whether it is dry or wetted, it generates a flow that expands radially, forming a relatively thin lamella. A remote asymptotic solution for the flow in the lamella, valid for times $\tilde{t}\equiv tU/D > 1$, is obtained in [37] in the form

$$\tilde{u}_{\mathrm{r}}=\frac{\tilde{r}}{\tilde{t}+\tau}, \quad \tilde{u}_{\mathrm{z}}=-\frac{\tilde{2}z}{\tilde{t}+\tau}, \quad \tilde{h}=\frac{\eta}{(\tilde{t}+\tau)^2}, \tag{1.2}$$

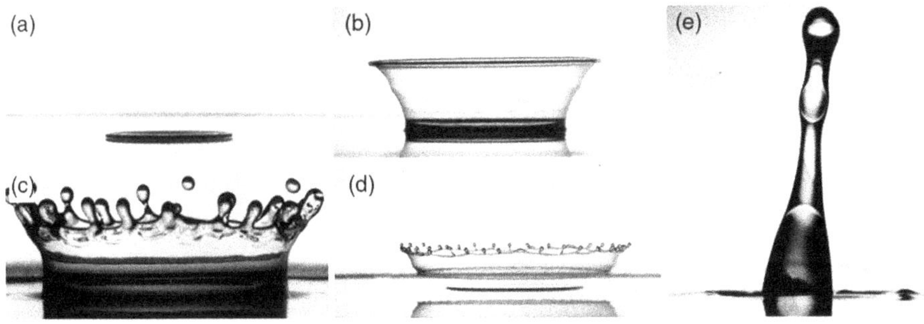

Figure 1.3. Different outcomes of drop impact onto a wetted wall, captured using a high-speed video system: (a) deposition, (b) corona (without splash), (c) corona splash, (d) corona detachment, and (e) partial rebound. (Reproduced with permission from [36]. Copyright (2018) the American Physical Society.)

where $\tilde{u}_r$ and $\tilde{u}_z$ are, respectively, the dimensionless radial and axial velocity components in the lamella, $\tilde{h}$ is the dimensionless uniform lamella thickness, and η and τ are dimensionless constants determined from experiments. This solution satisfies the mass and momentum balance equations for inviscid flow. All length variables are normalized by the initial drop diameter D, velocities are normalized by the impact velocity U, and time is normalized by the characteristic time D/U. The linear profile of the radial velocity (1.2) has been validated in previous studies such as [38, 39], employing particle velocimetry within the spreading drop.

The evolution of the lamella thickness h_{inviscid} from the inviscid solution, valid for long times, is obtained in [40] in the form

$$\tilde{h}_{\text{inviscid}} = \frac{\eta}{(\tilde{t}+\tau)^2} \exp\left[-\frac{6\eta\tilde{r}^2}{(\tilde{t}+\tau)^2}\right], \quad \eta = 0.39, \quad \tau = 0.25. \tag{1.3}$$

The values of η and τ are obtained from the computational fluid dynamics (CFD) computations of drop impact onto a dry substrate with $\text{We} > 70$ and $\text{Re}>30$.

The applicability of solutions (1.2) and (1.3) is restricted to an interval where the lamella thickness exceeds the thickness of the viscous boundary layer expanding in proximity to the wall. In [41], an analytical solution of the Navier–Stokes equations for a thinly spreading viscous film is derived. The resulting expression for the lamella thickness h incorporates the influence of viscous effects

$$\tilde{h} = \tilde{h}_{\text{inviscid}} + \frac{4}{5}\gamma\tilde{t}^{1/2}\text{Re}^{-1/2}, \quad \tilde{t}<\tilde{t}_\nu \approx \frac{\eta^{2/5}}{(1.88 - 0.8\gamma)^{2/5}}\text{Re}^{1/5}. \tag{1.4}$$

where t_ν is the instant at which the thicknesses of the lamella and the viscous boundary layer are equal. The value of the coefficient $\gamma = 0.6$ is predicted theoretically in [41] from the solution for drop impact onto a smooth planar dry substrate.

At times $t > t_\nu$ the dynamics of the lamella are governed by the balance of the inertial and viscous terms. At large times $\tilde{t} \gg \tilde{t}_\nu$ the lamella thickness approaches an asymptotic value

$$\tilde{h}_{\text{res}} = 0.79\,\text{Re}^{-2/5}, \quad \text{for} \quad \delta = 0, \tag{1.5}$$

the thickness of the residual film after impact. The theoretical prediction for $\tilde{h}_{\text{res}}$ agrees very well with the experimental data [42] obtained for drop impact onto a spherical target.

Moreover, investigations into the evolution of the lamella at the impact axis $r = 0$ [43] reveal that the residual film thickness resulting from the impact of a drop onto a liquid film of finite thickness can be expressed as follows:

$$\tilde{h}_{\text{res}} = A(\delta)\text{Re}^{-2/5}, \quad A \approx 0.79 + 0.098\delta^{4.04}, \tag{1.6}$$

Here, A represents a dimensionless constant contingent on the initial dimensionless film thickness δ. Notably, as the initial film thickness decreases ($\delta < 0.2$), the value of A converges toward the solution for drop impact onto a dry substrate ($A \to 0.79$).

If the Reynolds and Weber numbers reach sufficiently high values, the impact results in the formation of a crown-like liquid jet. Described in [37] as a kinematic discontinuity, this jet arises from the interaction between the flow in the expanding lamella and the outer wall film. The advancement of the corona's base with a radius R_{corona} is primarily governed by inertia. The expression for R_{corona} is derived in [37] as follows:

$$\tilde{R}_{\text{corona}} = \beta(\tilde{t} + \tau)^{1/2}, \tag{1.7}$$

where $\beta(\delta)$ represents a dimensionless coefficient. This expression aligns well with experimental observations, particularly for relatively small time intervals. The expression can be modified by accounting for the capillary and gravity effects [44]

$$\widetilde{R}_{\text{corona}} = \sqrt{\beta T - \left(\frac{4G\delta^{1/2}}{\beta G^{1/2}\text{We} - 2\delta^{1/2}} + \frac{4}{\text{We}} + \frac{\delta^2}{\text{Fr}}\right)\frac{T^2}{\delta}}, \tag{1.8}$$

$$T = \tilde{t} + \tau, \quad G = \frac{4}{\text{We}} + \frac{\delta^2}{\text{Fr}}, \tag{1.9}$$

$$\text{Fr} \equiv \frac{U^2}{gD}, \tag{1.10}$$

where Fr is Froude number. Expression (1.8) predicts corona expansion and its receding. The empirical relations for the dimensionless parameters β and τ are obtained

$$\tau = -0.8\delta^{1.7}, \quad \beta = 0.62\delta^{0.33}, \quad \text{at} \quad 0.5 < \delta < 2. \tag{1.11}$$

Another crucial parameter that characterizes the outcome of drop impact is the splashing threshold, representing the minimum impact velocity at which the generation of secondary drops occurs. The empirical relation for the splashing threshold, applicable to drops impacting a wetted wall, is expressed as follows [45, 46]:

$$K_{\text{splash}}^{8/5} = 2100 + 5880\delta^{1.44}, \qquad K \equiv \text{We}^{1/2}\,\text{Re}^{1/4}. \tag{1.12}$$

Various empirical correlations of the splashing threshold are presented in several experimental studies [47–53]. In a comprehensive experimental and theoretical study [54], a model for the splashing threshold on stationary and flowing wall liquid films is developed based on an energy balance approach.

For modeling of the spray deposition and its further solidification, it is important to determine the deposited mass ratio $\dot{m}_{\text{deposited}}/\dot{m}_{\text{impacted}}$ and the parameters of the secondary drops. For the angles of the impact velocity with the normal to the wall direction smaller than 45°, these values are determined only by the normal component of the impact velocity [55, 56]:

$$\frac{\dot{m}_{\text{deposited}}}{\dot{m}_{\text{impacted}}} = 0.5 + 0.616\exp[-K] \tag{1.13}$$

$$\frac{\dot{m}_{\text{splashed}}}{\dot{m}_{\text{impacted}}} = 1 - \frac{\dot{m}_{\text{deposited}}}{\dot{m}_{\text{impacted}}} \tag{1.14}$$

$$D_{\text{secondary}} \approx 24.\, D_{\text{impacted}}\text{Re}^{-1/2}, \tag{1.15}$$

where $D_{\text{secondary}}$ and D_{impacted} are the averaged diameters of the primary and secondary sprays.

1.2.2 Drop impact onto a dry substrate

Dry wall impact is relevant only to the early stages of ice accretion before the formation of a thin liquid wall film by a deposited liquid from the drops spray. The main outcomes of drop impact onto a dry substrate [57, 58], shown in figure 1.4, include deposition on a hydrophilic substrate, prompt splash on a rough surface,

Figure 1.4. Different outcomes of drop impact onto a dry substrate, captured using a high-speed video system: (a) deposition, (b) prompt splash, (c) corona splash, (d) receding break-up detachment, (e) partial rebound and (f) rebound. (Adapted with permission from [57]. Copyright (2015) Elsevier.)

corona splash if the impact velocity is above a splashing threshold, receding break-up on a hydrophobic rough substrate, and partial and complete rebound.

A drop impacting onto a solid substrate leads to the generation of an expanding lamella bounded by a Taylor rim [59, 60]. The dynamics of the rim determines the rate of the drop spreading and possible receding. Drop impact onto a dry substrate is determined not only by the material properties of the liquid and the impact parameters but also by the wetting properties and morphology of the substrate (figure 1.5).

The deposition of the drops on a dry substrate without splash is observed if the capillary and viscous effects are sufficient to balance the inertia of the spreading liquid. The condition for deposition on a smooth solid substrate is determined in [61]

$$\mathrm{We}^{1/2}\,\mathrm{Re}^{1/4} < 84, \quad \text{or} \quad \mathrm{We}\mathrm{Re}^{-2/5} < 12. \tag{1.16}$$

It should be noted that the drop splashing on a solid substrate is influenced also by the aerodynamic effects [62] and surface roughness [57].

At high impact velocities, the prompt splash appears if the height of the corona, determined by its break-up length, is comparable with the lamella thickness [63]. The analysis yields that the prompt splash is limited by the following conditions

$$\mathrm{Oh} < 0.004 \quad \text{and} \quad \mathrm{We} > 34\mathrm{Re}^{1/4}. \tag{1.17}$$

The average diameter of secondary drops in the case of prompt splash on a dry substrate is expressed as

$$D_{\mathrm{secondary}} \approx 2.\ D_{\mathrm{impacted}}\mathrm{Re}^{-1/2}. \tag{1.18}$$

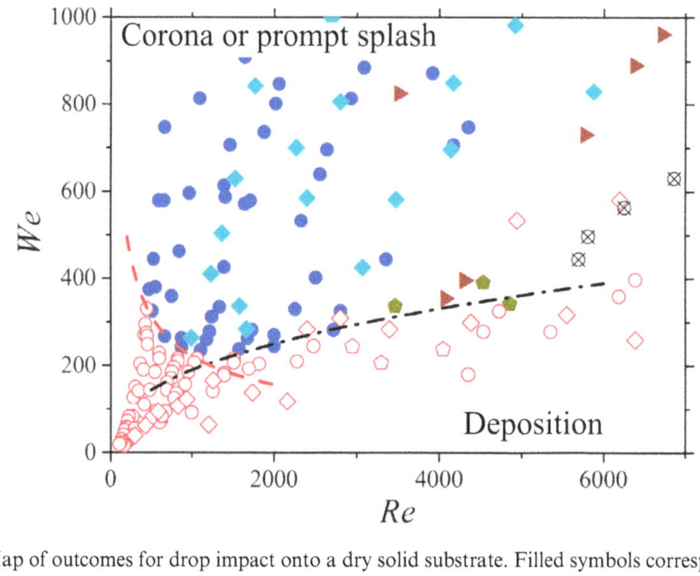

Figure 1.5. Map of outcomes for drop impact onto a dry solid substrate. Filled symbols correspond to corona splash, empty symbols correspond to deposition, and a symbol ⊗ corresponds to prompt splash. The curves correspond to the threshold conditions $\mathrm{We}^{1/2}\,\mathrm{Re}^{1/4} = 84$ and $\mathrm{We}\mathrm{Re}^{-2/5} = 12$. (Adapted with permission from [61]. Copyright (2018) Elsevier.)

1.2.3 Solidification in a liquid at a cold solid substrate

Parts of this section have been reproduced with permission from [31]. Copyright (2021) Elsevier.

The dynamic process of substrate wetting combined with icing is inherently intricate. Consequently, gaining insight into the freezing of water is essential to enhance comprehension of how it spreads in conjunction with icing.

The solidification of water is defined by its melting temperature, denoted as T_{m}, which is approximately 0 °C for pure water. On small length scales, the melting temperature also varies with the local curvature κ of the interface. This phenomenon, known as the Gibbs–Thomson effect [64, 65], is expressed in the equilibrium temperature at the melting interface as follows:

$$T_{\mathrm{eq}} = T_{\mathrm{m}}\left(1 - \frac{\sigma_{\mathrm{iw}}\kappa}{\rho L}\right), \tag{1.19}$$

where σ_{iw} is the surface energy of ice in liquid water, and ρ and L are the water density and specific latent heat of solidification, respectively. In the case of pure water, the characteristic temperature increment is denoted as T_{Δ}, with a value of approximately 78 K, and the capillary length is expressed as $d_{\mathrm{iw}} = \sigma_{\mathrm{iw}} T_{\mathrm{m}} c_{\mathrm{p}}/\rho L^2$, where c_{p} represents the specific heat capacity. These parameters were established in [66].

When the liquid temperature exceeds the melting point, the solidification process is stable and can be elucidated by solving the heat transfer equations within both the solid and liquid domains. This solution must adhere to conditions defining the equilibrium temperature at the solidification front and satisfy the energy balance equations, encapsulated in the Stefan boundary conditions [67, 68]:

$$\rho_{\mathrm{i}} v_{\mathrm{n}} L = k_{\mathrm{i}} \frac{\partial T}{\partial n}_{|\mathrm{ice}} - k_{\mathrm{l}} \frac{\partial T}{\partial n}_{|\mathrm{liquid}}, \quad T = T_{\mathrm{eq}}, \tag{1.20}$$

where v_{n} is the solidification rate in the normal direction (directed toward the liquid region), ρ_{i} is the ice density, and k_{l} and k_{i} are the thermal conductivity of the ice and the liquid, respectively.

1.2.4 Drop impact and solidification on a cold substrate

Parts of this section have been reproduced with permission from [31]. Copyright (2021) Elsevier.

Stable solidification typically occurs when the initial temperature of the droplet is above its melting point but the substrate, for example, ice or metal covered by a thin ice layer or ice crystals, serves as the sources of nucleation. In such cases, the freezing process is primarily driven by heat conduction within both the solid wall region and the thin ice layer obtained by water freezing.

For instances where the substrate is sufficiently cold to induce solidification during impact and droplet spreading, several key regions must be considered within both the substrate and the liquid lamella. Immediately after impact, a viscous boundary layer forms near the substrate, while the rapid inviscid outer flow remains unaffected. The thickness of the viscous boundary layer is denoted as $h_{d\nu} \sim \sqrt{\nu t}$, where ν represents the kinematic viscosity of the liquid.

Simultaneously, two thermal boundary layers begin to develop: one within the solid substrate with a thickness $h_{w\theta} \sim \sqrt{\alpha_w t}$ and the other within the spreading lamella with a thickness $h_{d\theta} \sim \sqrt{\alpha_d t}$. Here, α_w and α_d are the thermal diffusivities of the wall and droplet materials. The thermal boundary layer in the liquid water is significantly thinner than the viscous boundary layer, given that the Prandtl number $\mathrm{Pr} = \nu/\alpha_d$ is much greater than unity.

In regions outside the thermal boundary layer, both in the wall far from the solid substrate (where T_{w0} remains nearly unperturbed) and in the lamella outside the thermal boundary layer (where $T = T_{d0}$), the flow can be approximated using the inviscid solution (1.2).

The flow and temperature fields within the spreading droplet must satisfy continuity, momentum, and energy balance equations [69]:

$$\frac{\partial \rho}{\partial t} + \nabla \cdot (\rho \boldsymbol{u}) = 0, \tag{1.21a}$$

$$\rho \frac{\partial \boldsymbol{u}}{\partial t} + \rho(\boldsymbol{u} \cdot \nabla)\boldsymbol{u} = -\nabla p + \nabla \cdot \left(\mu[\nabla \boldsymbol{u} + \nabla \boldsymbol{u}^T] - \frac{2}{3}\mu(\nabla \cdot \boldsymbol{u})\boldsymbol{I} \right), \tag{1.21b}$$

$$\rho c_v \left(\frac{\partial T}{\partial t} + \boldsymbol{u} \cdot \nabla T \right) = \nabla \cdot (k \nabla T). \tag{1.21c}$$

The viscosity, density, specific heat, and thermal conductivity—$\mu = \mu(T)$, $\rho = \rho(T)$, $c_v = c_v(T)$, $k = k(T)$—depending on the local temperature and phase. These equations must be solved subject to initial conditions such as heat flux continuity at the wall interface, melting temperature at the solidification front, and no-slip and no-penetrability conditions for the flow at the ice/liquid interface. Additionally, Stefan boundary conditions are applied to estimate the rate of ice growth.

The similarity solution for this problem has been derived in [70] using the dimensionless similarity variable:

$$\xi = \frac{z}{\sqrt{\nu_0 t}}, \tag{1.22}$$

where ν_0 is the constant characteristic viscosity of the liquid, and $\sqrt{\nu_0 t}$ is the typical viscous length.

Finally, the coordinate of the solidification layer is expressed as $z_i = \Xi^* \sqrt{\nu t}$, with Ξ^* being a dimensionless parameter determined by the thermodynamic properties of the wall and the liquid. An approximate expression for Ξ^* can be obtained for high Prandtl numbers typical for cold water. In this case, the dimensionless ice layer thickness is obtained from the numerical solution of the following equation for Ξ^* [70]:

$$\frac{k_i}{k_d}\sqrt{\mathrm{Pr}_d}(T_m - T_{w0})A - \sqrt{\mathrm{Pr}_i}(T_{d0} - T_m)B = \frac{\mathrm{Pr}_d \sqrt{\pi} L}{c_{pd}} \Xi^*, \tag{1.23a}$$

$$A = \frac{\sqrt{5}\exp\left[\frac{\mathrm{Pr_d}}{5}(\Xi + \gamma)^2 - \frac{\mathrm{Pr_d}\Xi^{*2}}{4}\right]}{\mathrm{erfc}\left[\frac{\mathrm{Pr_d}(\Xi * -4\gamma)}{2\sqrt{5}}\right]}, \tag{1.23b}$$

$$B = \frac{e_w \exp\left[-\frac{\mathrm{Pr_i}\Xi^{*2}}{4}\right]}{e_\mathrm{i} + e_\mathrm{w}\,\mathrm{erf}\left[\frac{\sqrt{\mathrm{Pr_i}}\Xi}{2}\right]}, \tag{1.23c}$$

where the subscripts w, i, and d correspond to the wall, ice, and liquid drop regions, respectively, and e is the thermal effusivity. The Prandtl numbers for all three regions are defined using the value of the liquid viscosity. These equations account for the heat conduction in the wall and ice layer and the heat convection in the spreading drop, leading to compression of the thermal boundary layer by the liquid flow.

It should be noted that at some instant, the viscous boundary layer becomes comparable with the lamella thickness. This instant is approximately

$$t_\nu \sim \frac{\mathrm{Re}^{1/5} D_0}{W_0}, \tag{1.24}$$

where the Reynolds number is determined as $\mathrm{Re}=D_0 W_0/\nu$. At times $t > t_\nu$ the flow is governed by viscosity. The residual lamella thickness is determined in equation (1.5) from the exact solution of the Navier–Stokes equations [71] in the lamella.

Moreover, at some later instant, the thermal boundary layer becomes comparable with the residual lamella thickness. This instant can be estimated as

$$t_\theta \sim \frac{D_0^2}{\alpha_\mathrm{d}\mathrm{Re}^{4/5}}. \tag{1.25}$$

At times $t \gg t_\theta$ the temperature of the lamella can be well approximated by the melting temperature. The solidification rate in this stage is accelerated since no energy is used for the liquid cooling to the melting temperature.

The theoretical predictions [70] for the heat flux in the solid wall have been compared with several direct numerical simulations of drop impact. Very good agreement has been demonstrated in [72, 73] for an impact of a drop onto a heated wall, in [74] for a drop impact onto a cold wall without solidification, or even impact of multiple drops [75].

1.3 Solidification in a wall-bound supercooled liquid

Freezing of supercooled liquids is unstable and is accompanied by the formation and growth of dendrites, leading to the formation of a mushy layer.

1.3.1 The typical phenomena and main stages of solidification in a supercooled liquid

Parts of this section have been reproduced with permission from [31]. Copyright (2021) Elsevier.

The intricate process of solidification following the contact of a supercooled liquid with a solid substrate can be categorized into several discernible stages, as depicted schematically in the illustration presented in figure 1.6.

Heterogeneous nucleation takes place within the supercooled liquid region either at the surface of the substrate or, in certain instances, at the free surface of the liquid drop [76]. In this stage, the flow within the liquid drop remains unaffected by the solidification process. An illustrative example includes the observed dislodging of aerodynamically driven wall-bound drops at temperatures below the freezing point [77, 78].

At this stage, the ice embryos are formed at the substrate, but their sizes are below the critical size, associated with the inception of the solidification. For the phase transition from gas to liquid or from liquid to solid, the embryo must attain sufficient size to surmount the energy barrier represented by the Gibbs free energy change ΔG associated with the creation of an interface between the solid and liquid phases, along with the liberation of latent heat. It represents the initial irreversible formation of a nucleus in the stable phase. This process is influenced by molecular motion, giving rise to temporal and spatial fluctuations in the temperature and density of the liquid phase [79]. Due to the inherently stochastic nature of molecular motion, nucleation itself is a stochastic process. A thorough description of nucleation processes can be found in works such as [66] or other relevant sources.

The stage of nucleus formation plays a crucial role in determining the freezing delay of a wall-bound drop. It is determined by the temperature, the material properties of both the liquid and the wall, as well as the interfacial properties. Furthermore, the onset of solidification can be accelerated by the generation of micro-bubbles at the substrate, a phenomenon that may occur, for instance, during drop impact [80]. Nanostructuring of substrates has been shown to markedly delay nucleation inception and lower the characteristic freezing temperature [81]. An illustrative example depicting nucleation inception and the expansion of the frozen region in a spreading supercooled drop is presented in figure 1.7. A more detailed description of the physics of homogeneous and heterogeneous nucleation can be found in section 1.3.2.

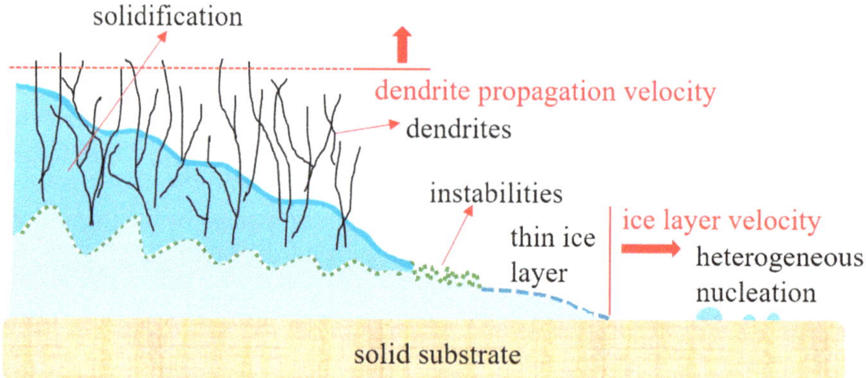

Figure 1.6. Sketch of the stages of solidification of a supercooled liquid on a cold substrate.

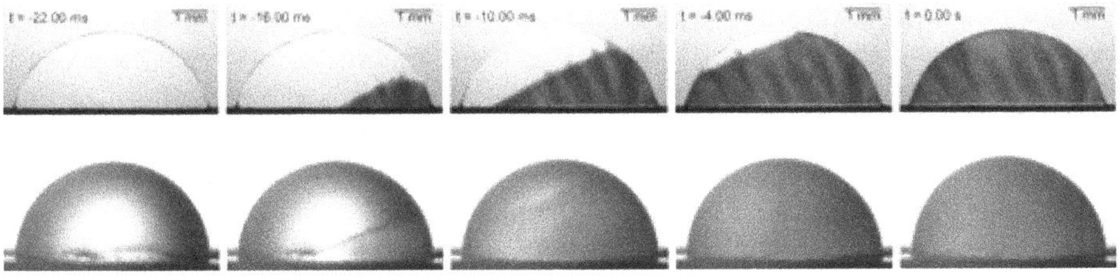

Figure 1.7. First phase of solidification of a water drop, supercooled to approximately −15.8 ° C. Qualitative comparison of a drop within the Hele-Shaw cell (top row) with a sessile drop (bottom row). (Reproduced with permission from [82]. Copyright (2016) the American Physical Society.)

In [80], let J_s represent the immediate nucleation rate per unit area. In the scenario of a droplet impact, this rate may encompass a fixed component linked to self-nucleation at the liquid–solid interface, along with the rate linked to the influx of embryos from the bulk liquid. The total average number $\lambda(t)$ of the nucleation sites per liquid region can be estimated as

$$\lambda(t) = \int_0^{A(t)} \left(\int_{t_w}^{t} J_s \mathrm{d}t \right) \mathrm{d}A, \tag{1.26}$$

where $A(t)$ is the wetted surface area at time t, t_w is the instant when the local area is wetted (if the liquid spreads), and J_s is the instantaneous nucleation rate per unit area. The probability that the liquid region does not include the nucleation site depends on time and can be estimated from the Poisson distribution:

$$p_{\text{liquid}} = \exp(-\lambda). \tag{1.27}$$

In the context of the repetitive experiments with the freezing of sessile drops, the probability $p_{\text{liquid}} = N_{\text{liquid}}/N_0$ represents the ratio of the liquid drops to the total number of experiments. When the freezing delay is comparable to the time of drop spreading and receding, a variety of residual solid drop shapes can be observed [83], including hemispherical or pancake shapes.

Immediately following the initiation of freezing, rapid radial *propagation of a thin ice layer* along the substrate has been observed in experiments [84]. The solidification front of this thin ice layer advances with an almost constant velocity, a parameter significantly influenced by the thermal properties of the substrate. In this study, the shape of the ice layer is approximated by a parabola, and the rate of the ice growth is determined away from the contact edge. To address explosive behavior at the contact edge, an empirical length scale associated with the initial ice layer thickness is introduced. The theoretical predictions [84] for the ice layer propagation velocity align well with the experimental data for various substrate materials. A similar approach has been proposed in [82]. Additionally, an advanced model for ice layer propagation has been developed in a recent study [85], highlighting the clear influence of substrate conductivity and wettability on the speed of ice layer propagation.

The solidification of the ice layer in the supercooled water is unstable. The microscopic structure of the ice layer on various materials has been examined in detail in [86]. Over time, these instabilities result in the emergence and *expansion of the cloud of dendrites*. The temperature within the dendrite cloud closely approaches the melting temperature, given that heat transfer predominantly occurs near the tips of the dendrites, at the advancing front of the dendrite cloud.

The dendrites propagate at a specific velocity dependent on the liquid temperature. Detailed simulations of dendritic growth during the initial stages of drop freezing were conducted in [87]. At lower supercooling, temperature reduction facilitated increased dendrite formation with reduced dimensions. Conversely, at larger supercooling, surface nucleation initiated a planar front, while bulk nucleation resulted in a granular-like dendritic structure. The existing experimental data for the dependence of the velocity of the dendrites on the initial supercooling [88–92] are shown in figure 1.8 in comparison with the theoretical predictions, based on the marginal stability theory (MST) [93]. The agreement is rather good for the supercooling below 10 K. Numerical computations [94, 95] incorporating the Gibbs–Thompson effect (1.19) demonstrate that the deviation can be explained taking into account the kinetic effects, missing in the theory.

The subsequent stage of freezing involves the relatively *slow and stable solidification of the liquid in the dendrite cloud*. This process occurs when the initial temperature of the wall is sufficiently low [94]. This study indicates that the crystallization front grows from the cold boundary, and the velocity of the solidification front is considerably lower than the freezing velocity observed in the earlier stage.

Under increased levels of liquid supercooling, kinetic effects exert a substantial influence on the rate of solidification. In the computations of [94] the solidification velocity influenced by the kinetic effects is assumed to be proportional to ΔT which leads to

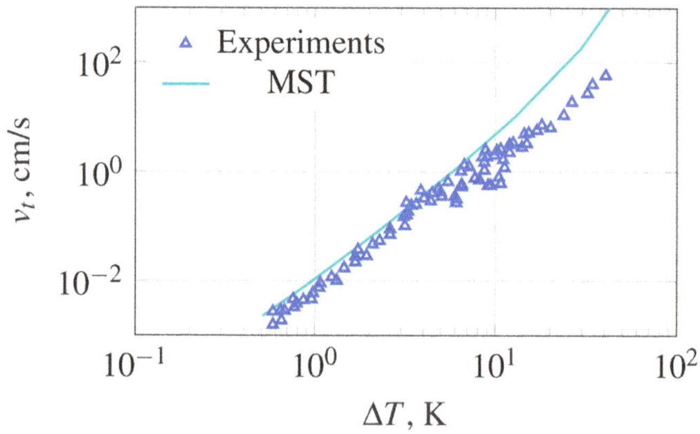

Figure 1.8. Steady tip velocity, v_t of a dendrite in water as a function of the initial supercooling ΔT. Comparison of the theoretical predictions by the MST [93] and experiments [88–92].

$$T_{\mathrm{eq}} = T_{\mathrm{m}} - \frac{v_{\mathrm{n}}}{k_{\mathrm{kin}}}, \tag{1.28}$$

where k_{kin} represents the kinetic coefficient.

As of now, no documented interaction of the dendrite cloud with the liquid flow has been reported in the literature. The rheological and material properties of this region remain unknown. Nevertheless, it can be hypothesized that under certain conditions, the flow may break or interrupt the dendrites, leading to the creation of a dense mixture of ice and water at or near the melting temperature. The excess liquid water from the dendrite cloud region has the potential to generate a runback flow. This phenomenon could explain observations of glaze ice accretion during interactions with supercooled liquid droplets under specific conditions.

1.3.2 The physics of homogeneous and heterogeneous nucleation

Crystallization of a substance when cooled below its melting point begins with the nucleation process. For an ideally pure substance in its liquid phase, the formation of the crystal seeds starts with a process called homogeneous nucleation, however, if this process is in the presence of a foreign agent such as the walls of the container or the presence of solid particles, then it is referred to as heterogeneous nucleation. In this section, the main characteristics of the two different processes will be addressed.

Homogeneous ice nucleation refers to the process of the formation of ice crystals from pure water without any external interference such as a foreign particle or by being in contact with a surface. A certain amount of supercooling, which is the cooling of the system below the melting point, is required for the formation of ice. This is observed because water releases latent heat when the phase-change process occurs, and therefore the probability of the formation of a critical number of ice embryos diminishes. Indeed, the degree of supercooling needed to observe an ice crystal within a system mode of purely water molecules can be substantial. The absence of any impurity or any nucleation seed on a surface drastically increases the free energy barrier that needs to be overcome for the nucleation process to occur. Thus, making the homogeneous ice nucleation a very unlikely event in nature.

Classical nucleation theory (CNT) was first introduced in the 1930s [96] and it served as a starting point for further developments and refinements to study thermodynamics and the kinetics of the nucleation process in supercooled liquids and supersaturated solutions [97]. Although many refinements of the theory have been proposed, it is still a valid methodology for predicting nucleation rates. It is based on the assumption that the atoms of the crystalline cluster can be treated as individual macroscopic objects, regardless of their size. This approximation is also known as the capillary approximation, and it assumes that the crystalline phase is separated by the condensed phase by a thin interface. In this framework, the cooperation of the interfacial free energy, γ_S, and the difference in free energy between liquid and crystal phase, μ_V describe the thermodynamics of the process. The free energy expenditure for an ideal spherical crystal seed with radius r can be written highlighting the dependency of a surface term, $4\pi r^2\gamma_S$, and a volume term, $\frac{4}{3}\pi r^3\Delta\mu_V$, thus resulting in

$$\Delta G = 4\pi r^2 \gamma_S - \frac{4}{3}\pi r^3 \Delta\mu_V. \tag{1.29}$$

From the maximum of this function it is possible to extract the critical nucleus size n^*:

$$n^* = \frac{32\pi\rho_C}{3} \frac{\gamma_S{}^3}{\Delta\mu_V{}^3}. \tag{1.30}$$

The critical nucleus size represents the atoms that have to be included for the free energy difference $\Delta\mu_V$ to overcome the energy cost associated with the formation of the new solid–liquid interface. The ρ_C expression stands for the density of the crystalline phase. To match the energy cost spontaneous and stochastic fluctuations are needed in the liquid in its supercooled state, thus allowing for the crystalline phase to develop. The expression of this free energy barrier for the nucleation is as follows:

$$\Delta G_n^* = \frac{16}{3}\pi \frac{\gamma_S{}^3}{\Delta\mu_V{}^2}. \tag{1.31}$$

Another assumption on which CNT relies is that the formation of a crystal induces a surface tension term, $\sigma_{\text{liq-sol}}$, in the Gibbs free energy that accounts for the formation of the new water–ice interface:

$$\Delta G_{\text{liq-sol}} = n_s(\mu_{\text{sol}} - \mu_{\text{liq}})V + S \cdot \sigma_{\text{liq-sol}}. \tag{1.32}$$

In the equation above, n_s stands for the number of molecules per unit volume of the crystal, V is the volume of the ice with surface S, and μ_{sol}, μ_{liq} are chemical potential for the solid and liquid phase, respectively. It can be derived that the difference in the chemical potential is related to the vapor pressure over the liquid and solid by

$$\mu_{\text{sol}} - \mu_{\text{liq}} = -k_b T \ln\left(\frac{p_{\text{liq}}}{p_{\text{sol}}}\right). \tag{1.33}$$

It is then possible to rewrite the previous equation as a function of the water molecules and two factors, A and B, that are dependent on the surface energy, chemical potential, and geometrical factors:

$$\frac{\Delta G_{\text{liq-sol}}}{k_b T} = -An_i + Bn_i^{2/3}. \tag{1.34}$$

With the increase of the crystal size, the free energy reaches a maximum at a critical size, below that maximum an increase in the crystal size is no longer favorable due to the increase of the surface energy. On the contrary, above that maximum, the spatial term prevails; thus, the ice crystal grows. From that moment the growth is essentially adiabatic and the speed of nucleation is proportional to the degree of supercooling.

Another key aspect of CNT is the ice nucleation rate J ($\text{m}^{-3}\,\text{s}^{-1}$); it estimates the rate at which the ice crystals expand assuming there is no variation in the number of molecules. Another assumption to derive the rate is that the nucleation process is determined only by the most active nucleation site, furthermore, no correlation between successive nucleating events is considered as in a Markov process. Under those circumstances, the rate is given by

$$J \sim \frac{k_b T}{h} \exp\left(-\frac{\Delta G_{\text{liq-sol}}}{k_b T} - \frac{\Delta g_{\text{act}}}{k_b T}\right). \tag{1.35}$$

Here, Δg_{act} stands for the energy barrier related to the transfer of a water molecule across the solid–liquid interface. From this expression, it can also be deducted how the relationship of the nucleation rate is strongly dependent on the temperature. Furthermore, it is clear how there are two energetic barriers to be overcome, namely the Gibbs free energy $\Delta G_{\text{liq-sol}}$ and the Δg_{act} related to the diffusion of the water molecules. The latter takes into account the restricted mobility associated with the motion of the molecules at the liquid–solid interface to join the solid phase. The influence of this term is strongly dependent on the degree of supercooling, for a low supercooling degree its effect is negligible, making the rate solely dependent on $\Delta G_{\text{liq-sol}}$, whilst for a high degree of supercooling the nucleation rate is dependent on both energy barriers. In real liquids, where usually molecular impurities or foreign particles are present in the bulk phase or the system is in contact with a surface, ice nucleation takes place as what is called a heterogeneous process. The heterogeneous nature of the system may highly reduce the nucleation barrier, leading to a faster nucleation rate. This mechanism of freezing is ubiquitous and it has drawn a lot of attention among scientists from different backgrounds [98–101]. The energy barrier that the water molecules have to surpass for the nucleation to proceed is lower than that of the homogeneous case:

$$\Delta G^{\text{het}}_{\text{liq-sol}} < \Delta G^{\text{hom}}_{\text{liq-sol}}. \tag{1.36}$$

Many are the factors that have to be considered, from the chemical composition of the system to the surface characteristics. Here we shall focus on the surface effects to describe the nucleation barrier that can also be expressed as [102]

$$\Delta G^{\text{het}}_{\text{liq-sol}} = f\left(\cos(\theta_{\text{iw}}), \frac{r}{r_{\text{i}}^*}\right)\Delta G^{\text{hom}}_{\text{liq-sol}}. \tag{1.37}$$

The f term accounts for the surface wettability:

$$\begin{aligned} f\left(\cos\theta_{\text{iw}}, \frac{r}{r_{\text{i}}^*}\right) &= 1 + \frac{1 - \cos\theta_{\text{iw}}\frac{r}{r_{\text{i}}^*}}{g} \\ &+ \left(\frac{r}{r_{\text{i}}^*}\right)^3\left[2 - 3\left(\frac{\frac{r}{r_{\text{i}}^*} - \cos\theta_{\text{iw}}}{g}\right) + \left(\frac{\frac{r}{r_{\text{i}}^*} - \cos\theta_{\text{iw}}}{g}\right)\right] \\ &+ 3\cos\theta_{\text{iw}}\left(\frac{r}{r_{\text{i}}^*}\right)^2\left(\frac{\frac{r}{r_{\text{i}}^*} - \cos\theta_{\text{iw}}}{g} - 1\right). \end{aligned} \tag{1.38}$$

Here θ_{iw} stands for the contact angle, r is the surface radius of curvature, and it is divided by r_i^* which represents the ice embryo radius resulting in a dimensionless unit. For the f term to vary between 0 and 1, resulting in a variation in terms of $G_{liq-sol}^{het}$ to be in the interval 0 to $G_{liq-sol}^{hom}$, g is equal to

$$g = \left[1 + \left(\frac{r}{r_i^*}\right)^2 - 2\cos(\theta_{iw})\frac{r}{r_i^*}\right]^{1/2}. \quad (1.39)$$

This model provides a versatile and inclusive description of the different factors that govern both homogeneous and heterogeneous nucleation, additionally, it can be tuned with different values of surface tension and contact angle to estimate the possible icephobic properties of a given surface.

1.3.3 Modeling of icing by molecular dynamics simulations

In this section, a summary of the state-of-the-art methods for studying ice with molecular dynamics (MD) will be presented. Since the implementation of computer simulations, it has been possible to collect many insights about the molecular nature of water in both its liquid and solid forms. Lots of valuable research has been produced since the 1980s and the number of papers is rising year by year. Even though water is ubiquitous, a complete and exhaustive explanation for its peculiar properties remains a challenge for many researchers in the scientific community.

The first works involving the study of ice with MD can be traced down to Weber and Stillinger [103] and Tse *et al* [104]. In the first case the ST2 potential [105] is used to study a cluster of 250 molecules, with no periodic boundary conditions (PBCs), and no cut-off on molecular interactions. They found that for the hexagonal structure of ice, Ih, the melting process starts from the surface of the cluster moving inward at a temperature of 300 K, thus validating Faraday's prediction from 1846 of the existence of a quasi-liquid layer (QLL). In [104] the simple point charge (SPC) potential is used [106] to study the stability of the empty and filled structure of I hydrates and ice Ic, finding similar infrared spectra and heat capacity which they correlate to a similar profile of the phonon density of state (DOS) between the two crystallites. The first research that used the TIP4P water model to investigate the dynamics of the ice–water interface was accomplished in [107]. They studied the interactions between the basal plane of ice Ih and liquid water at 240 K, finding consistent results for the density profile, diffusion constant, and also found the interface ranged from 10 to 15 Å of width. The latter work was followed by another paper in which the SPC model was utilized, showing good agreement with previous findings and when compared to the TIP4P water model, showing no substantial model dependency of the ice–water interface apart from a small change in apparent bulk melting temperature, suggesting that such findings represent actual properties of the real ice–water interfaces.

The first study that implemented a polarizable potential, derived from *ab initio* calculation at the MP24 and Hartree–Fock level, to describe the molecular interactions was conducted in the early 1990s [108]. In his work, through the

Nieser–Corongiu–Clementi potential (NCC-vib) [109] it was possible to describe successfully the intramolecular and intermolecular interactions that are responsible for the hexagonal ice structure and also to reproduce vibrational spectra with fairly good agreement. In [110] the deposition phenomenon of vapor on a cold surface to reproduce amorphous ice has been investigated. In this work, the SPC/E model [111] was utilized, and the substrate was modeled as a two layered surface made up of Lennard-Jones (L-J) spheres. Even though the study had some limitations due to the length of the simulation, it suggested that it was indeed possible to obtain a high-density form of amorphous ice from vapor deposition experiments.

Following the path of studying surface interaction, the ozone depletion mechanism has been studied [112] by addressing the ionization of hydrochloric acid on stratospheric ice. Specifically, the influence of the reactant positions and orientation toward proton-donor water and proton-acceptor water on an ice lattice was investigated. The quasi-liquid layer plays a marginal, if not unnecessary role, to induce ionization in hydrochloride acid. This line of research was followed many years later, in particular in [113], where the Car–Parrinello molecular dynamics was implemented to gain a deeper knowledge of the interaction of hydrochloric acid and ice. The nature of the interaction of the acid molecule with the ice plane is strongly correlated with the density of OH dangling groups on the surface. Indeed, in regions with a high density of dangling OH groups, the ionization of the HCl is barrierless. In [114] the growth kinetics of the basal plane and the prismatic plane are adressed and identified two different kinetic mechanisms of crystal formation. For the basal plane, it was found that the kinetics followed a layer-by-layer mechanism, this means that the nucleation appeared to be two-dimensional. On the other hand, for the prismatic plane, the growth kinetics were found to be completely different from the previous one. In this case, a large number of water molecules cooperatively participate in the crystallization, thus collectively forming the ice crystal structure. In their work, they used the TIP4P with a 2 fs time step for simulation runs of 140 ps in total.

In the TIP4P water model [115] the termination of the basal plane of ice is studied, matching experimental techniques with theoretical work. It was found that the outermost layer of water molecules exhibits an enhanced vibrational motion that challenges the low energy electron diffraction technique (LEED) for the detection of the quasi-liquid layer on top of the well-ordered bulk interlayer. Another unfortunate match between theoretical work and experimental testing was concluded in an attempt to reproduce the inelastic neutron spectrum with the MD technique [116]. Several water models are employed, including the TIP4P, but none of those could successfully reproduce the peaks found using the neutron scattering technique.

A breakthrough in the study of ice behavior using MD [117]. In this pioneer work, the TIP4P model is used with a time step of 1 fs on pure water clusters of different sizes for a simulation time of the order of microseconds. MD trajectories were performed at a different temperature, but crystallization was observed only for the cluster of 512 water molecules at 230 K. In their research they found that the freezing process can be broken down into four stages: a long quiescent period ($t = 0$–256 ns), followed by a short period in which the potential energy slowly decreases

(256–290 ns), and then another short period in which the potential energy decreases more rapidly (290–320 ns) and a final period with relative constant potential energy in which the ice structure forms ($t > 320$ ns). Their findings agree with the predictions of basic nucleation theory [118, 119] the system explores the energy landscape (with a rather flat potential) for a considerable time before entering the nucleation phase. They concluded that ice nucleation occurs only when a sufficient number of long-lived hydrogen bonds develop spontaneously at the same location to form a compact initial nucleus, although individual polyhedral structures are destroyed and reformed continuously. Induction of such nucleation clusters can be achieved by modeling low-density regions to mimic the density fluctuations observed in supercooled water, which have been interpreted to facilitate nucleus formation.

Another study of the ice growth from supercooled water [120] focused on the crystal growth mechanism from an existing water–ice interface. The results are consistent with previous findings [114]. The same tendency was observed as with the TIP4P water model for the ice phase to grow layer by layer, thus forming a smooth interface at all times. Moreover, a theoretical framework was developed to discriminate whether a specific molecule belongs to the solid or condensed phase: a molecule is considered in the liquid state if, throughout 20 ps, the number of hydrogen bonds is smaller than 3.3 on average, otherwise the molecule belongs to the ice structure. They also concluded that the growth rate of ice at the prismatic plane is almost twice as fast as the growth rate for the basal plane, suggesting that the rate-determining step for the growth on the basal plane mechanism is the formation of an in-plane cluster with the proper orientation.

In [121] the transition between amorphous ice is investigated, of low and high density (LDA and HDA, respectively) and ordered crystal structures. They utilized the SPC/E water potential and concluded that the transition from ice (Ih) to HDA ice occurs as a result of mechanical instability in the ice lattice, thus not analogous to the thermodynamic melting process. Furthermore, they found a lack of evidence for connecting the LDA ice to supercooled water, since they found a significant energy difference between those two phases.

Two years later, Shiomi [122] and colleagues applied the same water potential, SPC/E, to study the diameter dependence between ice confined inside carbon nanotubes of different ring sizes. The interactions with the single-walled carbon nanotubes (SWTNs) are modeled through a Lennard-Jones potential and the carbon–carbon interactions with the Brenner potential model. They concluded that the melting temperature of the confined ice structure is in good agreement with experimental values for most of the SWNTs studied, with the limit being for the (10, 10) SWNT, in which the core remains filled with liquid water.

An extensive comparison is made in [123] of polarizable and nonpolarizable water models for simulations of the melting of ice Ih, using the ice slab method: the melting of the ice structure is investigated using an ice slab with two ice–vapor interfaces. The polarizable water models taken into account were the POL3 [124] and SWM4-NDP [125] and were confronted with SPC, SPC/E, TIP3P, TIP4P, TIP5P-Ew, TIP4P/2005, and TIP4P/Ice. It was concluded that adding the polarizability to a water model, regardless of the three or four interaction sites, increases the orientational disorder for hexagonal ice, thus decreasing the ice stability and lowering the melting temperature significantly.

The heterogeneous ice nucleation induced by an electric field was studied [126] using TIP4P/Ice [127] and the six-site Nada potential [128]. It is shown that local electric field, on a narrow surface region, can successfully nucleate cubic ice at a temperature not far below the freezing point.

The behavior of supercooled water on smooth and rough surfaces is addressed in [129]. The monoatomic water (mW) coarse grain model is used to describe water molecules while the interaction with the surface, made up of two atomistic graphene layers, was achieved only by the two-body potential term as described for the water system. They concluded that a linear correlation could be drawn from the reduced work of adhesion and freezing temperature with the roughness parameters for both the Cassie–Baxter and Wenzel states, suggesting the Cassie–Baxter state as a design goal for an icephobic surface.

In [130] the behavior of the four-site model, TIP4P, and the mW are compared in describing ice kinetics. It is concluded that the TIP4P model is to be recommended for quantitative accuracy even though the nature of the heterogeneous crystallization requires the system to be of at least one order of magnitude larger than the crystallizing seed dimension, thus limiting these findings. In [131] the structural properties of liquid water and ice have been assessed with density functional theory (DFT), through evaluation of the performance of the Perdew–Becke–Ernzerhof (PBE) functional with a double zeta as a basis set. This work, however, is limited by the number of molecules in the system and also by the exclusion of the dispersion interaction in the functional, suggesting that a further improvement in the incorporation of the dispersion interaction was needed.

At the same time, the authors of [132] tried to answer the question of what kind of aspects could promote heterogeneous ice nucleation by focusing on the hydrophobicity and the morphology of the surface. The nucleation rates have been obtained by brute force calculations, which means that simulations are carried over until the nucleation event is observed. Due to the long simulation time that is required for the nucleation to initiate, they used the coarse-grained mW model to describe the 4000 water molecules in their system, with a time step of 10 fs. The substrate was modeled as four different crystallographic planes of a generic fcc crystal that exhibited significant differences in terms of roughness and symmetry of the outer crystalline layer. The interaction with the water molecules is achieved by a truncated Lennard-Jones potential at about 7.5 Å, and the position of the surface atoms was kept fixed throughout the simulations. They found the critical nucleus size to be made up of 50 mW molecules at a temperature of 205 K, showing consistency in literature estimates at different degrees of supercooling [28, 133, 134]. Unfortunately, no trivial trend was observed on the ability of an L-J surface to promote ice nucleation with the given experimental set-up, however, it was possible to draw some interesting conclusions: no significant size-effect was observed and more importantly it was detected that the surfaces provided an in-plane template which shows consistency with the different faces of ice that were later formed above the slab. This ability was not found to be an exclusive characteristic of surfaces that already displayed a compatible symmetry.

The reactive force field (ReaxFF) is used to study the dynamics of the collision between ice clusters and amorphous silica, both fully and partially oxidized [135]. The ReaxFF can provide an accurate description of the bond-breaking and bond-formation process because the connectivity is computed from the interatomic distances, which are updated at every MD step. Moreover, long-range interactions are described between every pair of atoms regardless of the connectivity and with a shielding term, any excessive close-range interactions are avoided. In this work, two ice clusters have been considered, the first one being amorphous with 150 water molecules and the second one, in a crystalline form, made up of 128 water molecules. Different velocities are given to the ice clusters in the direction normal to the silica slab to obtain different impact velocities. Due to the intrinsic nature of the chemical and physical properties that might arise from the impact, a very short time step of 0.1 fs is chosen. The initial temperature of the system was set at 150 K and impact velocities were set at 1, 4, and 7 km s^{-1}, the distance between the surface and the ice was set so that the bottom of the ice cluster would be at about 20 Å. It was concluded that with impact velocities greater than 1 km s^{-1} the ice clusters do not accrete on the surface. By computing the sticking coefficients it was determined that the accumulation is greater on the sub-oxidized silica compared to the fully oxidized surface. Lastly, it was assessed that for an impact velocity of 10 km s^{-1} or more, a treatment of the electron excitation needs to be addressed because the temperature of an individual molecule might rise vigorously.

Another attempt to look at the behavior of ice on real surfaces was done in [136] and focused on the ice nucleation on top of the clay mineral kaolinite. In particular, the nucleation of cubic ice, I_c, and hexagonal ice, I_h, on the hydroxylated face of kaolinite is adressed. The simulation protocol was set as follows: TIP4P/Ice and the CLAY-FF [137] were used to model water and kaolinite, respectively, and a time step of 2 fs and different temperatures were examined, from 220 to 240 K. Moreover, three different water–kaolinite interfaces were considered, one in which the surface atoms are kept fixed at their experimental positions of the bulk system (an exception is made for the hydroxyl groups on the outermost layer). The second one also utilizes a constraint for the kaolinite atoms, but the atomic positions are an average of the previously equilibrated system and the last one in which all the kaolinite atoms were unrestrained. The interaction between water and clay was modeled using the Lorentz–Berthelot mixing rules. They concluded that there is a preference toward the hexagonal ice for the surface, and the nuclei of the cubic ice are less stabilized by the surface underneath, thus showing the tendency to shrink back into the liquid phase. The same research group led by Sosso continued to investigate ice at the nanoscale and published an extensive review [138] on the crystal nucleation in molecular dynamics. They highlighted some aspects in which simulations could provide benefits and improve our understanding of nucleation, for example, the evidence of a two-step nucleation mechanism for colloidal particles, how the solvent affects the early stages of the urea crystallization, and the atomistic details of the clathrate nucleation to prevent hydrate formation on gas pipelines. They also addressed some of the shortcomings of the current status of MD simulations, namely, the need to resolve the time scale problem on *ab initio* calculations and the implementation of path sampling techniques to better study rare events, such as nucleation, that intrinsically require longer simulation runs.

In [139], employing molecular dynamics simulations, it has been shown how surfaces with single-walled carbon nanotubes displayed a decrease in ice adhesion, thus opening a new design of passive anti-icing surfaces. They used the TIP4P/Ice model for the Ih ice phase while the interactions with the surfaces (one composed of graphene and the second one as a vertical single-walled carbon nanotube array (CNTA)) were given by a Lennard-Jones potential with the substrate atoms kept electrically neutral and not interacting with each other. Figure 1.6 provides a schematic representation of the simulation procedure. The detachment process is achieved by imposing an increasing acceleration field on the ice cube to obtain the necessary pulling force. In the presence of the CNTA, the ice adhesion strength is decreased by up to 45 m% when compared to the graphene surface. Moreover, they confirmed the hypothesis that a decrease in the substrate–ice interaction energy leads to reduced adhesion strength. They were also able to prove the linear correlation between the wettability and the ice adhesion strength reported in experimental studies.

A comprehensive study [140] on the behavior of kaolinite as a nucleating agent for ice the mW model was used to represent water–water interactions, and hydroxyl groups were modeled as frozen mW molecules. A time step of 10 fs was used to integrate the equations of motion. After an equilibration time of 50 ns simulations were carried at 205 K until a nucleation event happened for the 15 different conformations taken in the experiment. They concluded that the interaction between the surface and the water is not so strong, thus allowing for the liquid layer to rearrange its structure in the contact layer. Furthermore, it was found that the nucleation ability does not relate to the OH pattern on the surface, instead, it depends strongly on the hydroxyl group density. In addition, they found that surfaces that template ice-like contact layers are not more efficient than non-ice-like substrates. To explain this finding, they point to the interaction strength between water and the surface relative to the bulk cohesive energy of ice. If the surfaces do not accommodate water too strongly then the contact layer can rearrange easily to the ice structure, otherwise, if the water adsorbs more strongly then the energy barrier associated with the rearrangement is too high and ice forms less efficiently.

A framework for seeded heterogeneous nucleation has been developed in [141], showing very good agreement with path sampling techniques such as forward flux sampling simulations and metadynamics. Seeded molecular dynamics represent a way to overcome the time limitation of rare events such as nucleation. Even though this technique does not provide some information on the actual nucleation mechanism, it can still provide quantities of interest for experimentalists, such as nucleation rates and the interfacial free energy between water and ice. They used the mW water model and the interactions with two fcc surfaces were obtained through a Lennard-Jones potential. In their work, they validated their approach (HSEED) to study the heterogeneous seeded nucleation also with the fully atomistic force field, TIP4P/Ice, by applying the same methodology on a cholesterol crystal. In both cases, their HSEED algorithm was able to pinpoint the combination of ice polytype and face which is most likely to form on the crystalline surface.

Kling *et al* [142] addressed the dynamical properties of the QLL layer at the ice surface. By using the rigid water model, TIP4P/Ice, with simulations up to 1 μs they were able to characterize the QLL at three different ice surfaces. At low temperature, 200 K, the surface of the ice slab is disordered as in the liquid phase, but the molecular mobility is extremely limited, the same as in an amorphous system well below the glass transition. At 270 K and below only the outermost surface exhibits diffusivity and can therefore be considered to be in the liquid state. Another study that focused on the dynamics and the diffusion coefficient was conducted in [143], only this time focus was given to grain boundaries. They addressed the diffusivity of water in polycrystalline ice and studied the effect of the grain boundaries on a triple junction on the diffusion coefficient, which was found to be negligible. Moreover, they demonstrated that the QLL at the ice–vapor interfaces is similar to high-density liquid water and the diffusion coefficient along the vapor–ice interface is larger than that along the gain boundaries. Water–water interactions were described with the TIP4P/Ice model and good agreement with the experimental result was achieved.

Using the SCAN functional [144], good agreement can be achieved with the predictions of all the properties of interest and, for some cases, the agreement is equal to that for the state-of-the-art semi-empirical water models (mW and TIP4P/Ice). The strategy that they propose is quite an improvement from the current status and implementation of the machine learning (ML) approach. Indeed, the main advantage is to be able to drive the dynamics with an ML potential as a surrogate for the DFT model. This, paired with enhanced sampling [145] and the calculation at the DFT level of the extracted configuration, can be more efficient than the current dynamics performed with forces that are calculated on the fly which represent the standard method in use. Furthermore, all the trends of the properties taken in the study are respected, thus validating the framework in use at least for qualitative comparisons.

1.3.4 Supercooled drop impact

Parts of this section have been reprinted from [31], copyright (2021) with permission from Elsevier.

The impact dynamics of a supercooled droplet introduce a heightened level of complexity to the intricate interplay between various factors. Notably, the internal thermal conditions of the droplet, including its temperature and heat transfer mechanisms, play a crucial role in determining the subsequent events during impact. Additionally, the fluid properties of the supercooled liquid, such as viscosity and surface tension, interact dynamically with the impact surface characteristics, further influencing the outcome of the collision. The delicate balance between these factors contributes to the nuanced behavior exhibited by supercooled droplets upon impact, making it a compelling area of study that unveils essential insights into the thermodynamic and fluid dynamic processes at play during such interactions.

The impact of the drop induces the generation of a radial flow (the right panel in figure 1.9) in the lamella. This flow is disturbed near the substrate due to the

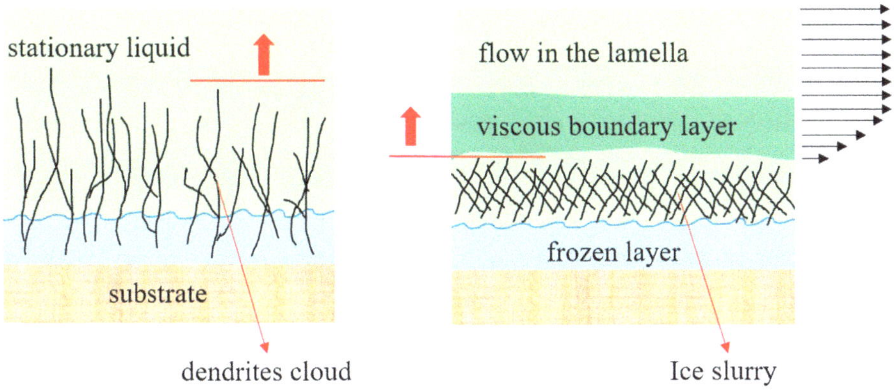

Figure 1.9. Sketch of the solidification in a supercooled liquid lamella. The left panel illustrates a stationary liquid; the right panel illustrates the mechanisms of solidification in the spreading lamella.

formation of the viscous boundary layer and the presence of an ice layer. In the absence of the liquid flow, the ice layer is represented by a dendrite cloud that expands with a constant velocity determined by the liquid initial temperature, as depicted in figure 1.6 and in the left panel of figure 1.9.

The behavior of the dendrite cloud under the stresses induced by the surrounding liquid flow remains unclear. It is conceivable that the dendrites, each a few micrometers in diameter, may fracture, leading to the formation of an ice slurry region. Initially, the temperature of the ice slurry is near the melting point. Similar to many dense suspensions, this material likely exhibits plastic rheological behavior. The yield stress in the ice slurry material can be attributed to the continuous clustering of dendrite fragments. This speculative description relies solely on the physical intuition of the authors of this review. Further modeling efforts would necessitate a precise characterization of the rheological properties derived from the dendrite cloud region.

In [146] the expression for the lamella thickness (1.3) has been extended to incorporate the contribution from the growth of the ice slurry region. The residual film thickness after drop impact is formulated as follows:

$$h_{\text{residual}} = [\xi^* + 3\delta(\xi^*)]\frac{\nu}{u_{\text{i}}}, \tag{1.40}$$

where u_{i} represents a constant rate of growth of the ice slurry region. The dimensionless time ξ^* and the dimensionless thickness of the viscous boundary layer δ at the moment when the viscous effects reach the free interface of the lamella are solely determined by a parameter P calculated as

$$P = \frac{D_0^3 \eta u_{\text{i}}^5}{\nu^3 W_0^2}. \tag{1.41}$$

The measured dimensionless residual lamella thickness, denoted as $h_{\text{residual}} u_{\text{i}}/\nu$, is presented in figure 1.10. The measurements of the residual film thickness are also

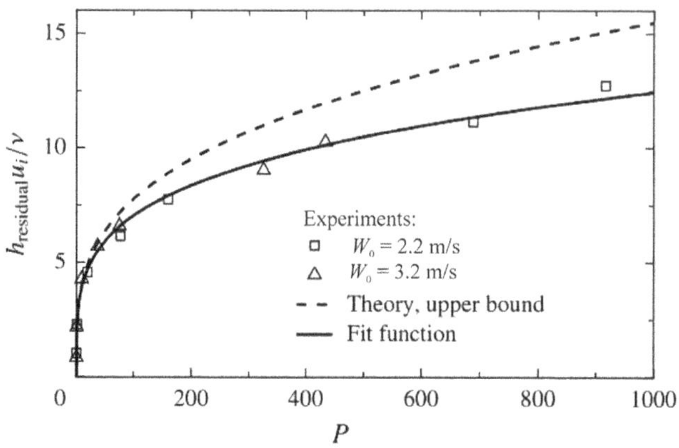

Figure 1.10. Dimensionless residual lamella thickness after supercooled drop impact onto an ice target, as a function of the dimensionless parameter P defined in equation (1.41). The theoretically predicted upper bound for h_{residual}, determined in (1.40), is based on the velocity u_{i} equal to the dendrite propagation speed. (Reproduced with permission from [146]. Copyright (2017) Cambridge University Press.)

compared with the theoretical predictions (1.40), based on the assumption that the ice slurry layer expands with the velocity of the dendrite propagation. It is important to note that this assumption does not consider the deformation and compression of the dendrite cloud caused by the flow in the lamella. Consequently, these predictions can be regarded as an upper bound for h_{residual}. As observed, the predictions align quite well with the measurements.

The optimal fit for the experimental data is achieved with the following expression:

$$h_{\mathrm{residual}} \approx 2.26 \frac{\nu}{u_{\mathrm{d}}(T)} \left[\frac{D_0^3 \eta u_{\mathrm{d}}^5}{\nu^3 W_0^2} \right]^{0.247}, \tag{1.42}$$

where $u_{\mathrm{d}}(T)$ is the known velocity of the dendrite propagation.

It can be demonstrated through the comparison of the theory with the experiments that

$$u_{\mathrm{i}} \approx 0.49 u_{\mathrm{d}}. \tag{1.43}$$

This implies that approximately 51% of the liquid water is squeezed out of the dendrite cloud during drop spreading. This estimation is grounded in the consideration that the solid content of in the dendrite layer is relatively low. The liquid water content of the mushy layer is approximately 70%. It is known [147] that the yield strength of wet sand reaches a maximum at a liquid content in the range $0.15 < c_{\mathrm{l}} < 0.6$. At $c_{\mathrm{l}} = 0.7$ the yield strength is not maximum but already significant.

Hence, in aeronautic applications, particularly in the context of airframe icing resulting from SLD impact, it can be approximately assumed that a mushy layer forms from a fraction η_{mushy} of the deposited drop volume, which can be estimated from the condition $c_l = 0.7$ as $\eta_{\text{mushy}} \approx 3.33c_s$, where the initial concentration of the solid dendrites c_s is determined from the energy balance of the solidification

$$c_s = \frac{\rho_d c_p (T_m - T_{d0})}{L\rho_i}, \tag{1.44}$$

where ρ_d and T_d are the density and the initial temperature of the liquid drop, c_p is the heat capacity, T_m is the melting point, and L is the specific latent heat of solidification. The values of c_s are relatively small. For water, it can be estimated as $c_s \approx 0.013(T_m - T_d)$.

The normal impact of a supercooled drop onto a sandblasted aluminum surface is shown in figure 1.11. Notably, nucleation does not always occur at the impact position. The liquid drop persists in spreading, except in the frozen regions occupied by the cloud of solid dendrites [146]. Additionally, the oblique impact of a supercooled drop was also studies in [80], as shown in figure 1.12. The observed nucleation sites in the drops are distributed uniformly over the wetted area of the target.

The impact of a supercooled drop on a cold substrate was experimentally investigated in [148]. The study found that drop solidification has minimal influence

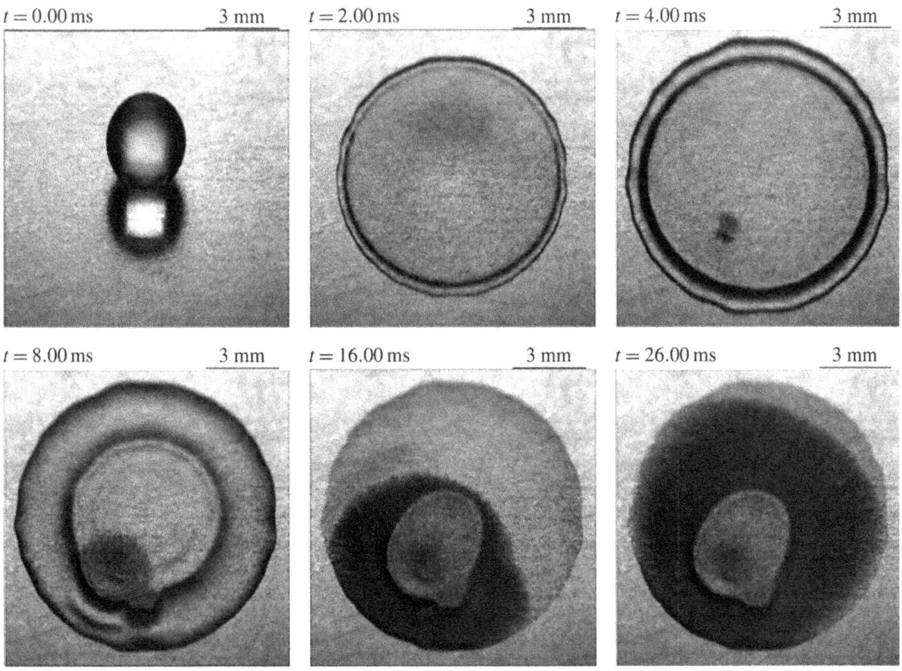

Figure 1.11. Top view observation of a normal impact of a water drop at −16.6 °C onto a sandblasted aluminum surface at the same temperature. Nucleation at a single nucleation site is followed by the radial expansion of a thin ice layer along the solid substrate (not visible in the images) and dendritic freezing of the bulk liquid above the ice layer (dark region in the images), fixing the shape of the deformed drop. (Reproduced with permission from [146]. Copyright 2017 Cambridge University Press.)

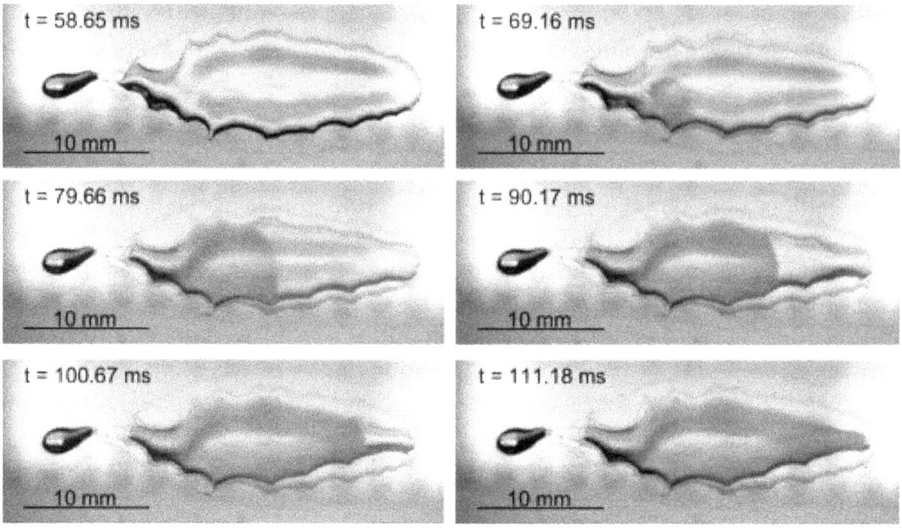

Figure 1.12. Exemplary time sequence of the propagation of a freezing front of dendrites within a receding drop after oblique impact onto polished aluminum. The drop is initially at −14.3 °C and the substrate at −17 °C. To illustrate further receding during freezing, every frame is superposed with the frame at t = 58.65 ms as a reference. (Reproduced with permission from [80]. Copyright (2017) American Physical Society.)

on the kinematics of drop spreading and the maximum spreading diameter. However, solidification significantly affects the minimum receding diameter, primarily due to contact line pinning resulting from the interaction of the ice layer with the substrate surface. The average velocity of the ice layer increases with both substrate and drop temperatures, reaching an estimated velocity of 0.9 m s^{-1} at −20 °C. The influence of the thin ice layer's propagation and freezing delay on the outcome of supercooled drop impact was demonstrated in [149], with computations conducted using a CFD code [150]. The behavior of the liquid drop is characterized by continued spreading, except in regions already frozen and occupied by the solid dendrite cloud.

It has been noted that in the case of the impact of a supercooled drop onto a textured superhydrophobic substrate [151], the maximum spreading diameter is influenced by the temperature. This phenomenon can be attributed to the temperature dependence of liquid viscosity, which becomes more significant during impalement onto surface structures. Furthermore, the significance of the near-wall air flow in determining the outcome of supercooled drop impact on a superhydrophobic substrate has been demonstrated in studies such as [152, 153].

In the experimental and theoretical investigation conducted in [146], the impact of a supercooled drop onto a small ice target has been investigated in detail. The impact onto ice ensures immediate nucleation and generation of dendrites without delay, a phenomenon observed on targets composed of materials other than ice as well.

In a nutshell, researchers have never ceased their exploration of wall-bound supercooled liquids. Commencing in 1724 with Daniel Gabriel Fahrenheit's discovery that fluid motion can potentially aid nucleation [154], there has been an ongoing exploration of diverse factors influencing the nucleation process. Certain researchers have focused on the intrinsic properties of the liquid itself, examining factors such as liquid shear stress [155] and drop diameter [156] as potential influencers on nucleation. Meanwhile, a significant body of work has focused on external influences affecting nucleation, including studies on wettability [157, 158], surface roughness[159, 160], and other triggers such as impact, electric fields, and suspended particles, considering the stochastic nature of the nucleation process. Some researchers investigated the influence of the dynamic parameters of the flow, such as the liquid shear stress [155], or the diameter of the drop [156]. Other studies are interested in external influences on nucleation, such as the wettability [157, 158] and the roughness [159, 160]. In addition, given the inherently random nature of nucleation, researchers have explored ways to enhance nucleation activities. Some studies have specifically investigated the impact of external triggers, including mechanical impact [161, 162], electric fields [163–165], and the presence of suspended particles [166, 167], to increase nucleation efficiency for a better understanding of the physics.

1.4 Ice accretion due to the with density functional theory of ice crystals impact

The undesirable occurrence of ice accretion may occur in the low-pressure compression systems of aircraft [168–175], where walls become wet due to the melting of ice particles in a warm environment or upon their impacts with hot substrates. This ice crystal accretion on internal components of aircraft engines, along with subsequent shedding of ice layers [169, 176], can result in serious consequences such as flameout, mechanical damage, rollback, and more. Additionally, ice crystal accretion poses a significant threat to aircraft sensors and probes.

It is recognized that the onset of accretion is linked to the cooling of an initially hot substrate. When a heated wall is exposed to the flow of cold ice crystals, the impacts cool the wall until it reaches the melting temperature. At this point, ice crystals begin to agglomerate on the wall, initiating the growth of an ice layer through the accumulation of fragments from impacted particles. This phenomenon is driven by the localized freezing of the wall film in the narrow gap between the substrate and the impacting particle.

The interaction of particles with a solid substrate is influenced by various factors, including the properties of both the target and particle materials, as well as the impact parameters. At lower velocities, the impact typically results in elastic rebound. However, beyond a certain threshold velocity, collisions can lead to substantial plastic deformation and even fragmentation of the particles. When the target or particle is wet, the impact may result in particle deposition, leading to agglomeration and accretion when a cloud of such particles impacts the substrate.

Modeling of the dynamics of snowflakes and ice crystals is a challenging task due to the big variety of their shapes and sizes. Some of the shapes of the meteorological

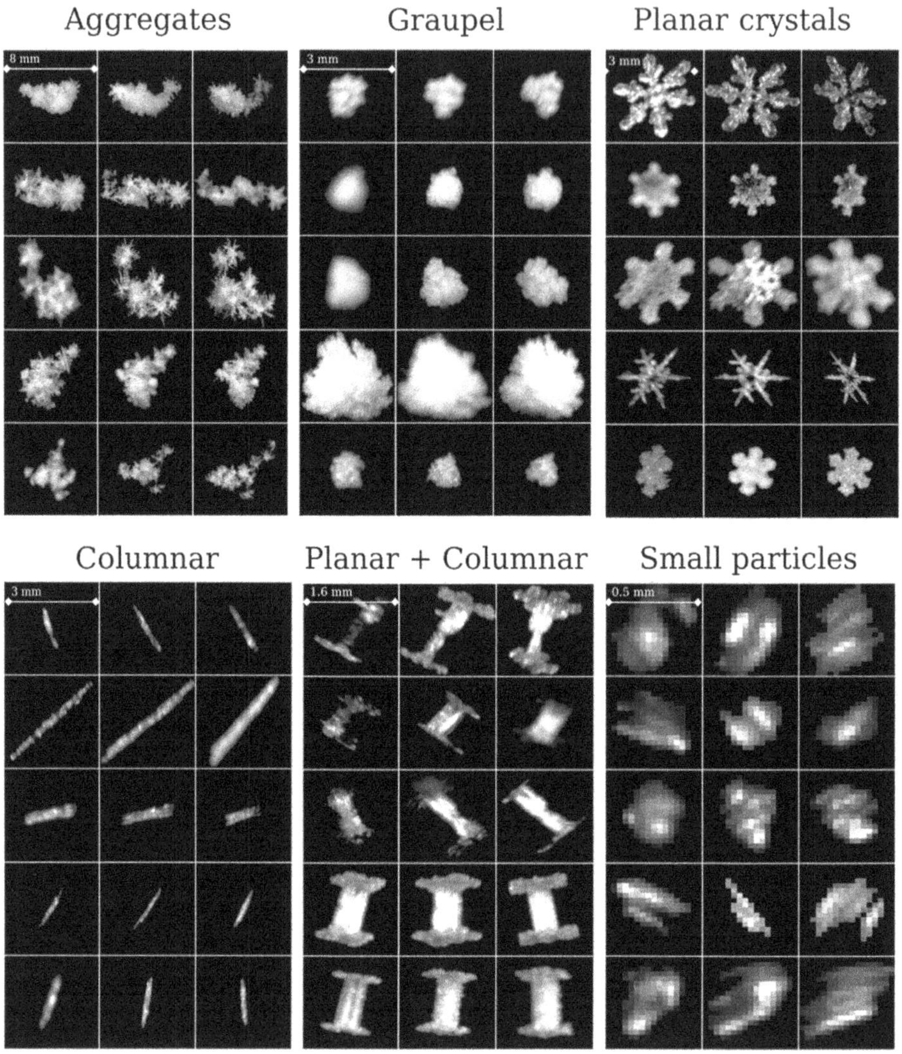

Figure 1.13. Examples of various shapes of natural snowflakes, ice crystals and aggregates of ice particles. (Reproduced from [177]. CC BY 4.0.)

ice particles are shown in figure 1.13. Among the problems is the description of the particles' trajectories [178–180], their melting [181–184] and impact.

1.4.1 Single ice crystal impact

At lower impact velocities, the interaction between a solid particle and a solid substrate is primarily governed by elastic stresses in both media. The elastic deformation of both the particle and the target can be accurately described using Hertzian contact theory [185].

As impact velocities increase, the stress tensor within the particle's contact region may surpass the von Mises yield criterion. It has been demonstrated that the critical interference, denoted as b, between an elastic particle and a perfectly rigid substrate [193] is given by $bR_0 \sim Y^2(1 - \nu^2)^2/E^2$, where Y is the compressive yield strength, E is the elastic modulus, and ν is Poisson's ratio. The maximum interference, $b_{\max}$, is influenced by the particle mass and initial velocity U_0. According to Hertzian contact theory, $b_{\max} \sim \rho^{2/5}U_0^{4/5}/E^{2/5}$, as estimated in [194] or more recently in [195]. Consequently, the critical impact velocity associated with the onset of plastic flow in the impacting particle is $U_{\text{plastic}} \sim Y^{5/2}\rho^{-1/2}E^{-2}$. For typical ice values, such as $Y \sim 10^7$ Pa, $E \sim 10^{10}$ Pa, and $\rho \sim 10^3$ kg m^{-3}, this results in $U_{\text{plastic,ice}} \sim 10^{-4}$ m s^{-1}. This implies that, in most cases relevant to aeronautic applications, the impact of an ice particle on a solid substrate is governed by plastic stresses.

It is widely acknowledged [196] that the yield strength of ice is dependent on the local effective strain rate $\dot{\gamma}$, representing the rate of deformation of a material element. In figure 1.14 existing experimental data are shown for the dependence of the yield strength $Y(\dot{\gamma})$ from the literature [186–191] on the strain rate $\dot{\gamma}$. The value of Y is also significantly influenced by the temperature and purity of the ice [191, 197].

In [192] the uniaxial yield strength is modeled in the form

$$Y = Y_0[1 + \chi - \chi \exp(-\tau\dot{\gamma})], \tag{1.45}$$

where Y_0 is the static yield strength at $\dot{\gamma} \to 0$. The value for the static yield strength of the ice particle $Y_0 = 5.6$ MPa is estimated in [196]. The fitting parameters to the existing data are $\chi = 5.0$ and $\tau = 0.0007$ s.

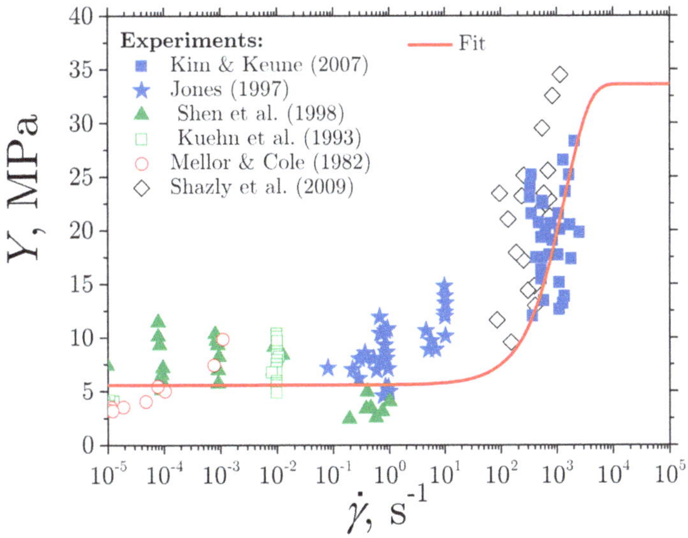

Figure 1.14. Experimental data for the yield strength Y from [186–191] as a function of the strain rate $\dot{\gamma}$ in comparison with the fitting curve in the form defined in (1.45) using the fitting parameters $\chi = 5.0$ and $\tau = 0.0007$ s. (Adapted with permission from [192]. Copyright (2022) Elsevier.)

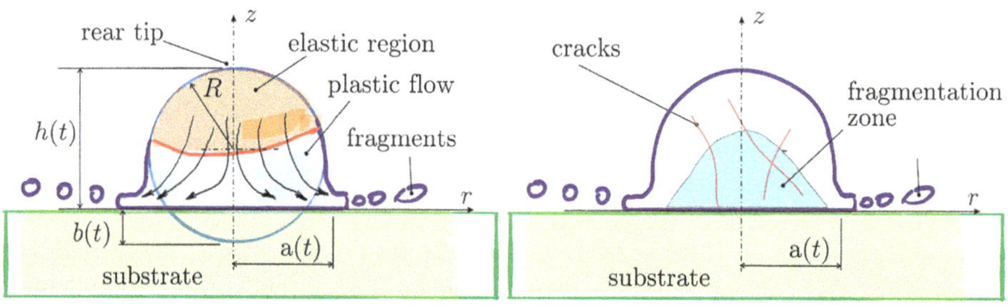

Figure 1.15. Sketch of a particle of the initial radius R impacting onto a perfectly rigid substrate. Definition of the coordinate system, the main geometrical parameters, and the main regions describing particle deformation and fragmentation.

The main geometrical parameters and the regions associated with the particle deformations caused by wall collision, stresses and fragmentation are shown in figure 1.15. At higher impact velocities the plastic region is developed near the wall and expands as soon as the particle material deforms. The impact leads to the formation of expanding cracks. As soon as the crack length is comparable with the particle size the crystal breaks up. The corresponding velocity of the particle splitting [198] is

$$U_{\text{splitting}} \sim \frac{K_c^{4/3}}{\rho^7 Y^{5/6} R_0^{2/3}}. \tag{1.46}$$

The map of ice particle impact events leading to no fragmentation or minor fragmentation outcomes is shown in figure 1.16. The upper boundary for no fragmentation and the upper boundary for minor fragmentation are described by the expressions $U_{\text{nofragment}} = \alpha D_0^{-2/3}$ and $U_{\text{minor}} = \beta D_0^{-2/3}$, respectively. The form of these expressions is determined by the result (1.46). The parameters $\alpha = 0.046$ m$^{5/3}$ s^{-1} and $\beta = 0.14$ m$^{5/3}$ s^{-1} are determined from manual fits to the experimental data.

An advanced theoretical model [192] for ice particle impact onto a rigid substrate takes onto account the inertial effects and plastic stresses in the deforming particles. The model is able to predict the evolution of the particle shape, the final height and diameter of the particle as well as the evolution of the force produced by impact. The particle deformation is characterized by a dimensionless particle dislodging ζ. The value of the maximum dislodging after impact ζ_{max} depends on the initial particle diameter D_0 and impact velocity U_0. The fit of the computations [192] for ζ_{max} yields

$$\zeta_{\text{max}} \approx \left(0.58 + \left[\frac{D_0}{\delta}\right]^{0.77}\right)\left(\exp\left[\frac{U_0}{V}\right] - 1\right), \tag{1.47}$$

where $\delta = 215$ mm and $V = 250$ m s^{-1} are the fitting parameters. The fitting is valid for $D_0 < 30$ mm and $U_0 < 160$ m s^{-1}.

The maximum fragment size D_f correlates rather well with the value of ζ_{max} for a wide range of particle sizes and impact velocities. The empirical correlation for the maximum size, based on the data [199–202], is

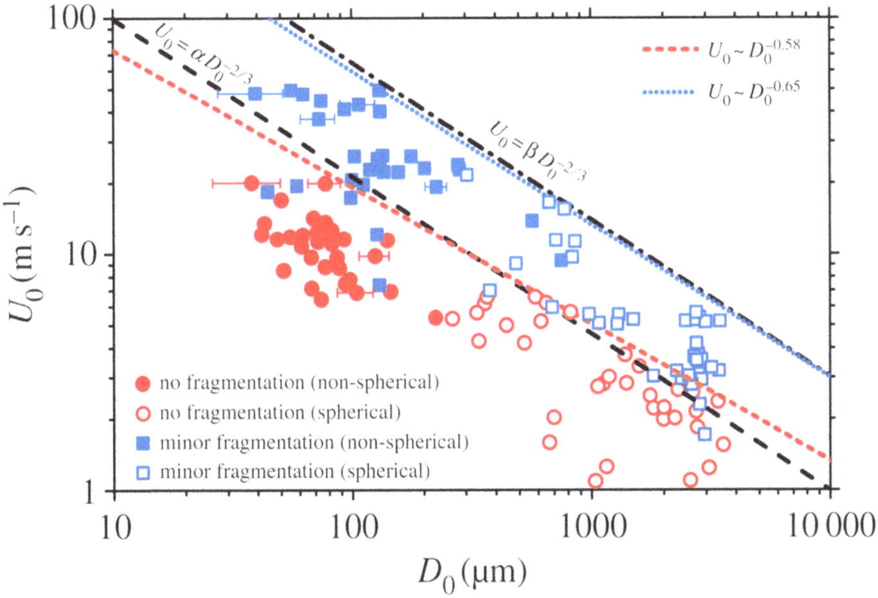

Figure 1.16. Map of no fragmentation and minor fragmentation. The upper bounds of no (long dashed line) and minor (dashed dotted line) fragmentation are shown based on the model for particle splitting (1.46). Best fits to the no fragmentation (short dashed line) and minor fragmentation (dotted line) data are shown. (Reproduced from [198]. Copyright 2015 The Author(s). Published by the Royal Society (UK). All rights reserved.)

$$D_{\mathrm{f}}/D_0 \approx 0.26 + 0.69\exp(-6.11\zeta_{\max}), \qquad \zeta_{\max} < 0.6. \tag{1.48}$$

Certainly, this correlation is valid only for impact velocities exceeding the threshold velocity of particle break-up [198].

For a reliable modeling of ice crystal ice accretion further research and modeling are required for the prediction of the deposited mass after impact onto a solid dry or wetted wall or into ice layer. Current level of knowledge is still not sufficient.

1.4.2 Granular ice layer

Icing occurrences in warm environments, characterized by either warm ambient air or elevated surface temperatures, typically result from the accretion of ice crystals on relevant component surfaces. The ability of solid ice particles to adhere depends on the presence of water, specifically under mixed-phase conditions. Experimental investigations [168, 170, 203, 204] highlight the significant influence of the liquid water content (LWC) to total water content (TWC) ratio in the flowing air on the severity of ice accretion. A typical relationship between relative icing severity and LWC/TWC has been reported in [204]. The critical ratio for the most severe icing ranges from approximately 10% to 25%. Inadequate water content impedes solid particle adhesion, while an upper limit results from insufficient ice particles to adequately cool the surface, leading to rapid removal by liquid water. Therefore, the lower limit for LWC/TWC is dictated by particle impact dynamics, while the upper limit is governed by heat fluxes influencing accretion and the underlying substrate, affecting ice shedding behavior.

Engine icing events typically occur at cruise altitudes, well above the theoretical limit for liquid water presence. Consequently, the liquid water required for significant accretion arises from particle melting during their movement through the warm air within the engine or upon impact with hot surfaces. This liquid water deposits on relevant surfaces in the form of droplets, rivulets, and films. While solid particles rebound from dry surfaces, small particles can adhere to a liquid film. The melting of these particles is accompanied by the cooling of the surface until it reaches the freezing point, allowing the accumulation of ice particles. If the liquid at the surface freezes, adhesion between the ice and surface occurs, creating a firmly bonded connection of the accretion to the wall. Moreover, recent investigation of the mechanical properties of the ice layer in a wind tunnel [205] allow one to suggest that an additional necessary condition for the ice accretion is the sintering of the solid ice granules in the layer.

Ice shedding takes place if the connecting ice at the bottom of the accretion is either broken or melted [27].

1.5 Ice adhesion

Most people have already dealt with the negative consequences of icing in their lives, such as walking or driving on a frozen street. In the industrial sector, icing represents a severe problem for aviation, the nautical sector, offshore structures, TV-towers, overhead power lines, or for meteorological instruments. Insufficient prevention of surface icing causes loss of human lives and huge economic costs in terms of damage and accidents [206].

Icephobic surfaces generally hold three main properties: low ice adhesion, water repellence and ice nucleation suppression. Low ice adhesion surfaces represent a promising strategy to mitigate the negative effects of icing. These surfaces are defined in terms of their adhesion strength. Generally, below 60 kPa of adhesion strength the surfaces can be defined as 'icephobic' [207]. An ice adhesion strength below 10 kPa allows ice to be shed by natural vibrations, the effect of gravity, or wind [208]. These surfaces are defined as 'super-low ice adhesion surfaces'.

The quality of low ice adhesion surfaces is also defined by their durability [209]. Aviation, for example, is one of the most demanding sectors, requiring surfaces that show high durability to erosive and corrosive conditions [209]. Another important aspect that has to be considered is the ice type accreted on the surfaces. The degree of ice adhesion to a surface depends on surface properties, icing conditions, and ice type [209]. Thus, it is important to know exactly which application the surfaces are developed for.

The earliest documentation of test methods to measure ice adhesion date back to the 1930s, where qualitative icing comparisons were made at the National Advisory Committee for Aeronautics (NACA) [210, 211]. In the last 15 years, an impressive increase in research on low ice adhesion surfaces can be observed. Following a search on Scopus, the number of publications that contained the words ‘icephobic’, ‘icephobicity’ or ‘anti-icing surfaces’ from 2008 to 2024 was more than 18700. In the year 2024 alone, 3600 are counted to date (10 April 2025). This shows an impressive expansion of the field and great interest in the subject, both from academia as well as industry.

Even though there is this great interest in the field, the problems and challenges regarding icephobicity are still far from solved. The main challenges concern the better understanding of the physical processes involved in icing that act on length scales that range from nanoscale interactions (10^{-9} m) to large structures (10^0 – 10^1 m). Freezing identical water samples under identical environmental conditions does not always result in identical ice crystal structures. Moreover, depending on the freezing conditions, ice can have different characteristics and properties, such as glaze ice, rime ice, snow, or frost. Each of these ice types has different adhesion properties and might require a different strategy of ice removal.

In recent years, several reviews have tried to provide a better understanding of surface properties against ice formation. Schutzius *et al* [212] have discussed the physics involved in ice formation on surfaces and how these principles can be employed to develop surfaces with extreme resistance to icing. The relationship between hydrophobicity and icephobicity has been investigated by Amirfazli and Antonini [213], while Stone [214] compared the ice adhesion performance of superhydrophobic surfaces to liquid-infused porous surfaces (referred to as SLIPS [215] or LIS [216]), where liquid is generally trapped into a porous material, creating a liquid layer on the surface that strongly reduces ice adhesion.

The multitude of ice types, together with the large number of technological applications where icing can be problematic, has led to a large number of authors using a broad range of test methods. In the last four to five years, more and more attention has been given to the evaluation techniques of ice adhesion, pointing out how discrepancies may arise from different measurement techniques and from different laboratories.

Shen *et al* [217] have highlighted some of the issues and challenges faced by the research community and the fundamental complexity of the environment in which the application of low ice adhesion surfaces takes place. The critical review of Work and Lian [218] analysed a total of 113 papers which deal with the measurement of ice adhesion to solid substrates. The methods used for ice adhesion testing were categorized and data were compared from different set-ups. The authors identified important conceptual problems and key parameters affecting ice adhesion, such as temperature, surface roughness and impact velocity.

Finally, the paper of Rønneberg *et al* [207] pointed out how ice adhesion research could benefit from standard measurement procedures and how there is a lack of comparability between experimentally obtained ice adhesion results. Interestingly, they propose to define a reference test to increase the comparability between results from different research institutes. The exact implementation of this reference test is, however, still to be defined, as the paper does not give further indications.

A direct comparison between different low ice adhesion surfaces may not be straightforward, as the test method and the set of conditions may differ among several studies. This review focuses on the methods of testing and discusses the fundamental and inherent limitations in ice adhesion testing.

1.5.1 Ice adhesion test methods

There is no standard ice adhesion test method to date. In general, each testing method addresses a particular ice accretion or ice loading condition, which is relevant to a specific application. Given the considerable number of applications where ice accretion leads to failures or safety issues, it is no surprise that a multitude of ice adhesion test systems exist in the literature. They can be sorted into three major categories: shear tests, tensile tests, and centrifugal adhesion tests. They are shown schematically in figure 1.17. All methods traditionally calculate the adhesion strength by measuring the detachment force and dividing it by the ice–substrate contact area.

A comparison of the measurement result for ice adhesion strengths measured on aluminum is presented in table 1.1.

Ice type and environmental conditions vary in these tests. As can be seen from the comparison, the values of ice adhesion strength for aluminum vary up to one order of magnitude ($\approx$ 0.1–1 MPa). It is clear that the testing method, ice type and environmental condition play a vital role in measuring adhesion strength.

1.5.1.1 Shear tests

Shear tests represent the biggest sub-group of ice adhesion tests [218]. It is possible to find horizontal shear tests [228–231], vertical shear tests [232–234], shear tensile tests (0° cone test) [224, 235, 236], and rotational shear tests [237, 238].

Horizontal shear tests (figure 1.17(a)) are usually performed by freezing a water sample with defined geometry onto a supercooled surface or by placing it inside a freezer. A force probe is then pushed against the ice sample laterally and the peak force at the exact moment of ice detachment is recorded. Generally, the point of force application on the ice is as close as possible to the substrate, so that the ice sample undergoes relatively pure shear stress.

Vertical shear tests (figure 1.17(b)) are conceptually similar, except that once the ice block has formed, the substate is rotated by 90° and placed vertically. The force is then usually applied from the top to the bottom. In this case, gravity affects the force applied on the ice sample and, therefore, ice adhesion strength depends to a greater extend on the size and geometry of the ice sample [207]. This has direct consequences on the comparison with results that come from horizontal shear tests and generally implies the necessity of correction factors.

In shear tensile test (0° cone test) (figure 1.17(c)) two concentric cylinders are used. They are usually chosen so that a small gap between the inner cylinder and the other cylinder exists. This gap is then filled with water and frozen inside a refrigeration chamber. Once ice has formed, a tensile stress is applied to the inner cylinder until it is pulled out of the outer cylinder. The measured adhesion force is then divided by the contact surface to obtain the maximum adhesion strength.

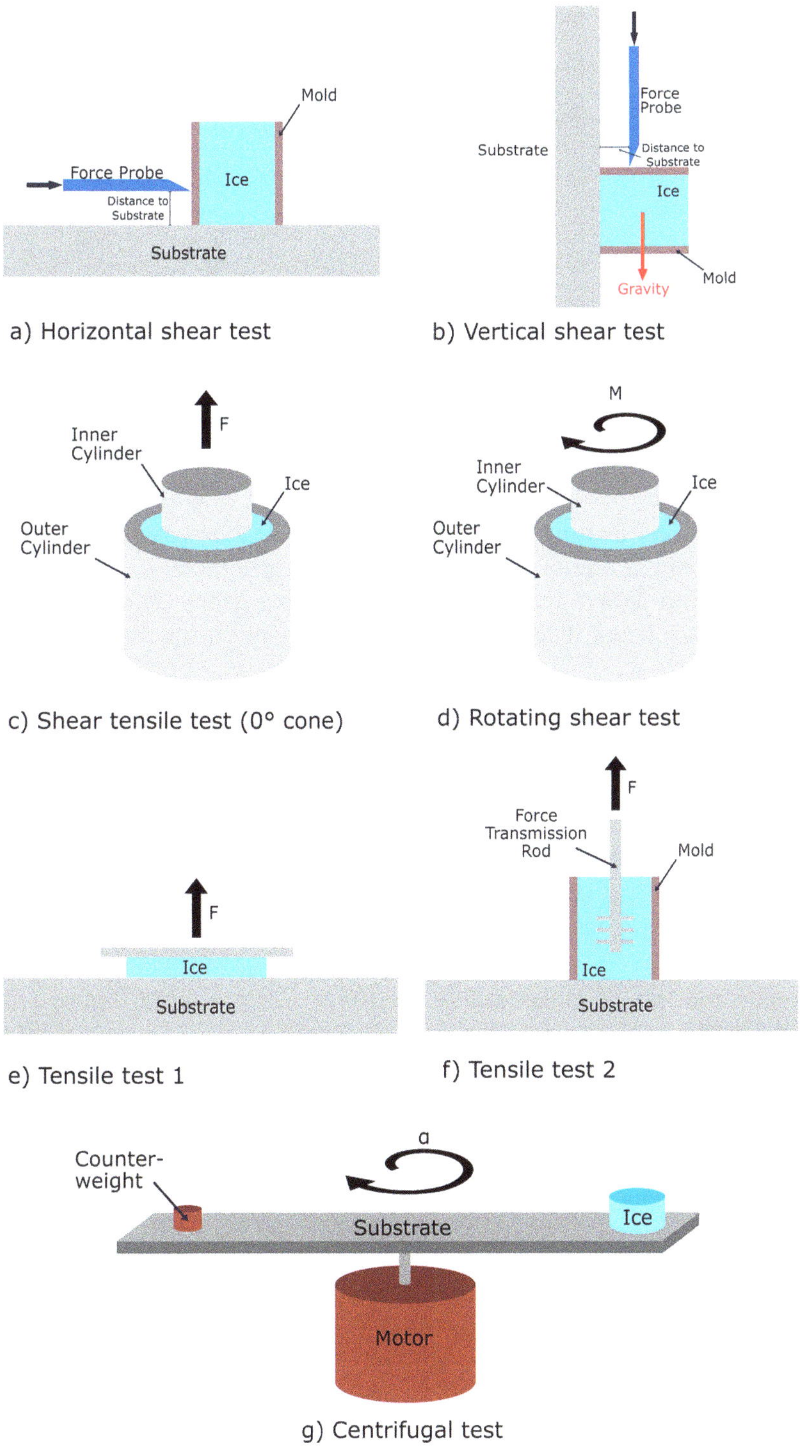

Figure 1.17. Ice adhesion test methods

Table 1.1. Comparison of different ice adhesion strengths on aluminum. (Adapted with permission from [264].)

Test method	Ice adhesion strength (MPa)	References
Horizontal shear test	0.80	Dou *et al* [219]
Horizontal shear test	0.11	Hejazi *et al* [220]
Horizontal shear test	0.7–1.0	Lou *et al* [221]
Horizontal shear test	0.16–0.33	Zou *et al* [222]
Vertical shear test	0.49	He *et al* [223]
Shear tensile test (0° cone)	1.2	Zhu *et al* [224]
Centrifugal adhesion test	0.28–0.78	Rønneberg *et al* [225]
Centrifugal adhesion test	0.19–0.76	Guerin *et al* [226]
Centrifugal adhesion test	0.32	Laforte and Beisswenger [227]

Usually, an icephobic coating is applied on only one of the two cylinders, so that the ice will detach only at one of the two ice–substrate interfaces.

Rotational shear tests (figure 1.17(d)) have a similar geometry to the 0° cone test, the only difference being the application of a rotational torque instead of an axial force.

The disadvantage of shear tests is that the placement of the force probe can have a big influence on the force distribution at the interface [207], while in the case of the 0° cone test, sharp concentrations of shear and tensile stresses near the edge of the ice appear [218].

1.5.1.2 Tension tests

Tension tests are typically designed to produce tension failure under direct mechanical stress [218]. Water is frozen between two surfaces (figure 1.17(e)), which are then pulled apart by an increasing traction force [224]. Alternatively, a force-transmission rod can be embedded in the ice by using a mold (figure 1.17(f)). If engineered properly, the ice detaches from the base surface before it breaks off from the force-transmission rod [209, 239, 240].

Unlike in the shear mode, ice adehsion in the tensile mode had not been explored much in the past few decades. In recent years the tensile ice adhesion approach has received much attention because this approach will help to better understand the adhesion phenonmena between ice and solid substrate. Recently, Mirshahidi *et al* devloped a new method to analyse and calculate the ice adhesion strength in the tensile mode. Furthermore, the study provides the basic understanding of the dependence of ice adhesion on temperture, pull-off speed and material thickness.

1.5.1.3 Centrifugal tests

Centrifuge adhesion tests (figure 1.17(g)) use the centripetal force to measure the adhesion strength. Usually, a small ice sample is placed on a rotor that is accelerated with constant angular acceleration. At the moment of ice detachment, the instantaneous angular velocity is recorded and converted into the centripetal force, which is divided by the contact surface to obtain the maximum adhesion strength ($F_{adh} = F_c = m \cdot V^2/R$).

These test systems are relatively easy to build, cost effective and deliver repeatable results, however, they have a significant number of disadvantages [207]. Most of the papers published to date calculate the shear strength as centripetal force divided by the contact area. Schulz and Sinapius [242] showed, however, that stress concentrations at the interface are located predominantly at the edges of the sample and that peel stresses are present at the rest of the interface. Additionally to that, this type of test system is not able to produce stress–strain curves [218] and the sudden detachment of the ice from the test surface might be damaging for the coatings [243].

1.5.2 Physics of ice adhesion

The comparison in the previous section shows how more work toward standardization must be done. At the current time it is difficult to compare results from different ice adhesion set-ups because of the high number of mechanical and environmental conditions that influence ice adhesion. Moreover, most publications do not report relevant parameters that are crucial to reproducibility (i.e. ice type or density, force application point, stabilization time after freezing or humidity). The reason for this is that defining the relevancy of each parameter on ice adhesion is not trivial and it mostly depends on the chosen testing set-up. The next section attempts to discuss the physics of ice adhesion in different testing modes and define some of the most important physical parameters that influence the ice adhesion.

1.5.2.1 Observation of the ice adhesion process

Ice adhesion is a very spontaneous process. Subramanyam *et al* conducted an experimental study to understand the correlation between the substrate morphology on lubricant impregnated surfaces (LIS) and the strength of ice adhesion by employing complementary macroscale adhesion measurements. Moreover, they examined the fractured ice surface through cryo-SEM imaging, and investigating the ice–LIS interface at the microscale using cryo-FIB/SEM imaging. The results show that the strength of ice adhesion on LIS, having a thermodynamically stable lubricant layer, is surface-texture-dependent, showing a reduction with an increase in texture density.

Xue *et al* experimentally studied the adhesion strength of ice at the point of the separation surface of the aluminum alloy plate and silicone rubber. The results show that the ice adhesion first increases and then decreases by decreasing the ambient temperature. The duration of freezing did not exhibit a notable impact on the adhesion strength of frozen ice to the surfaces of both materials. Notably, the adhesion strength of frozen ice varied based on water quality, with the following descending order: distilled water, lake water, and seawater. The adhesion strength of seawater was observed to be merely 1/22 of that observed in distilled water.

1.5.2.2 Ice type and inner structure

Rønneberg *et al* [225] conducted a systematic analysis of the influence of ice type on ice adhesion. Three types of ice were investigated: precipitation ice, in-cloud ice, and bulk water–ice. All three types of samples were formed on the same aluminum substrate and tested under similar environmental condition. The tests were performed on a centrifugal adhesion test system and showed an adhesion strength of 0.78 ± 0.10 MPa, 0.53 ± 0.12 MPa, and 0.28 ± 0.08 MPa for precipitation ice, in-cloud ice, and bulk water ice, respectively. The authors suggested that ice adhesion strength may inversely correlate with the density of ice [225].

What is even more noticeable, however, are the high standard deviations reported, with values spanning in the range 15%–30% of the mean value. The authors underline the fact that all the experiments were performed under the same conditions for each ice type. Similar scattering has been also reported in other studies [218, 246].

The explanation given by Rønneberg *et al* [207] is that even identical freezing conditions might not give identical crystal structures and shapes of ice. This random nature of ice is difficult to account for in ice adhesion testing and might be the cause of these large standard deviations. It is therefore important to always have a large data set, and studies where only a small number of measurements have been taken might be more inaccurate than expected.

1.5.2.3 Temperature effects

The literature generally converges on the fact that ice adhesion increases with decreasing temperature. The review paper of Work and Lian [218] summarizes this behavior well, which is also confirmed by a set of papers including [246, 247, 248].

On a theoretical level this behavior can be explained by the presence of a liquid-like water film on the ice surface. This is also the same water film that makes us perceive ice as slippery in everyday experience [249].

This so-called QLL was first observed by Faraday [250] in 1859 and is still not fully understood today. Makkonen [206] performed a theoretical estimation of the change in the surface equilibrium melting temperature δT by applying the Lennard-Jones potential of the interatomic forces and the Clausius–Clapeyron equation [206]. The results can be simplified to [206]

$$\delta T = 290\left|\gamma_{\mathrm{i,\,w}} - \gamma_{\mathrm{i,\,s}}\right|,$$

where δT is in °C, $\gamma_{\mathrm{i,\,s}}$ (J m^{-2}) is the interface energy of the ice–solid interface and $\gamma_{\mathrm{i,\,w}}$ (J m^{-2}) (the value of $\gamma_{\mathrm{i,\,w}} = 29 \times 10^{-3}$ J m^{-2} can be used here [206]) represents that of the ice–water interface [206]. This last relation is probably only applicable when the surface energy of the substrate is lower than that of ice [206]. From the paper of Makkonen [206]:

'The theory suggests that the [nanoscale] water film thickness at an interface of ice decreases from infinity at 0 °C to zero at 0 °C $-$ δT, being proportional to $\delta T^{-1/3}$.' (p 420)

The increase in adhesion strength of ice with decreasing temperature is therefore qualitatively explained.

Several studies have tried to make use of this QLL to develop icephobic surfaces. Chen *et al* [251] developed a prototype coating with a self-lubricating liquid water layer (SLWL). The surface makes use of cross-linked hygroscopic polymers inside the micro-pores and shows mechanical robustness against sandpaper abrasion and self-healing capabilities [251].

The ice adhesion strength on the SLWL surface has been measured at different supercooling magnitudes. Interestingly, a sharp increase of ice adhesion was observed at −25 °C and this can be related to the phase transition of the liquid water to ice and, therefore, the disappearing of the liquid water layer [251]. This is consistent with the work of Amirfazli and Antonini [213], who reported other studies that suggested that the QLL exists down to −25 °C in stationary conditions.

At low temperatures and with the disappearance of the QLL, cohesive breaking of the ice sample becomes more likely, as opposed to adhesive breaking close to the melting point [226]. Ice adhesion is therefore influenced as well by the mechanical properties of the ice, which becomes more and more brittle with lower temperatures. Rapid cooling could lead to temperature-induced cracking at the ice–solid interface and an inverse relation between ice adhesion strength and temperature could be assumed [206].

It has been experimentally observed, however, that the increase in adhesion strength due to the disappearing of the QLL is more dominant than the temperature-induced cracking of the ice [206]. Moreover, temperatures close to the freezing point and the presence of a QLL makes the ice–substrate interface ductile. The thermal stresses at the interface can therefore be adjusted and compensated without fracturing [206].

1.5.2.4 Ice sample size

In ice adhesion testing, the size of the ice samples is often not specified. The question whether the size of the sample can play a role in the measured adhesion strength rises spontaneously. The fracture behavior of ice has been investigated by Schulson [252] and is believed to be determined by grain size and thickness. For a test sample to be representative, it should be composed of a minimum of ten ice crystal grains of average size. This results in an ice sample size of at least 1 cm in each dimension [207, 252].

From a fracture mechanical point of view, an ice layer covering a substrate plate can be modeled as a so-called 'bimaterial bonded joint' [253]. According to this model, the delamination of ice can be explained by two competing fracture criteria: the stress condition and the strain energy (or toughness) condition [229, 253–255]. Relatively small interfaces typically fail by critical stress [256], while for larger interface a coupled criterion comes into play [253]. In a stress-dominated interface failure, the crack generally opens instantanously, over the entire interface, without visible crack propagation [229, 257]. The stress value exceeds the critical value in every point of the interface:

$$\tau \geqslant \tau_c, \forall x \in A,$$

or, depending on the stress state,

$$\sigma \geqslant \sigma_c, \forall x \in A,$$

where x is a point of the total interface A [253]. This condition is met for short and thick ice layers, more generally in conditions where the interface stress is uniformly distributed without major stress concentrations.

In contrast, high stress concentrations generally favor fractures controlled by critical strain energy (or interfacial toughness) [257]. The interfacial toughness quantifies the resistance to crack growth along an interface. In this condition, fracture initiation happens in the vicinity of a stress concentration point. The separation of the ice layer from the substrate starts with the opening of a finite crack area ΔA which is smaller than the total interface area [253]. For this first opening of the crack, both the stress as well as the strain energy have to exceed the respective critical values, thus the name 'coupled criterion':

$$\tau \geqslant \tau_c, \forall x \in \Delta A$$

$$-\Delta E \geqslant \Gamma \Delta A,$$

where Γ is the critical strain energy or fracture toughness. After crack initiation, delamination of the ice layer is controlled only by fracture toughness [253, 255].

Recently Golovin *et al* [229] investigated low interfacial toughness (LIT) materials that have been shown to be particularly effective in lowering the forces required to remove large areas of ice. These materials show a particularly low interfacial toughness and, therefore, are very easy to de-ice in all those situations where the fracture is toughness dominated.

The ice blocks in this study had a varying lengths from 0.5 cm up to 1 m, while the width and height were kept constant. The force to remove the ice from these LIT materials increases only up to a critical length of the ice sample size, while above this critical value (for longer ice samples) the force required to detach the samples remains constant [229]. This can be explained by considering again the two different fracture mechanisms explained in the previous paragraph. For an ice block size below the critical length, the detachment is controlled by stress, and therefore the ratio between the force F and area A (which expresses the average shear stress) needs to be constant. If the ice block is longer than the critical length, however, the stress no longer controls the detachment and there is no longer a dependency between F and A. An expression based on the interfacial toughness and the elastic modulus of ice for the critical length L_c has been proposed [229]:

$$L_c = \sqrt{2E_{ice}\Gamma h/\tau^2}.$$

The most important concept of this discussion on ice sample size is that there is not only one fracture mechanism, and so ice adhesion cannot be expressed with only one measure, traditionally $\tau_c = F/A$. Expressing ice adhesion as a stress (in units of Pa), is only meaningful if the ice adhesion test is performed in stress-dominated conditions. This aspect should be taken into account when designing ice adhesion experiments.

1.5.2.5 Stress distributions

Stress concentrations at the ice–substrate interface could lead to adhesion values an order of magnitude lower compared to the true adhesion strength [242]. While it is true that also in real-world applications it is very likely that stress concentrations are present and initiate ice detachment, on a laboratory scale the presence of these stresses makes it very difficult to compare results from different measurement methods or different test systems. It is likely that different stress distributions play an important role in the variations of adhesion strength measured on aluminum in the previous comparison. In fact, from that comparison it is possible to see differences up to one order of magnitude. It is known that the state of stress at the interface is not uniform, but still the majority of studies do not take this into consideration [218].

One of the first papers to account for the stress concentrations on a shear ice adhesion test was the one of Lou *et al* [221]. In this work, the adhesion strength τ_{ad} is expressed as a funcion of the removal force *F*, the adhesion area *A*, and a correction factor θ that accounts for the stress concentrations. The authors also perform numerical finite element analysis (FEA) (static-stress and CZM) to compute the correction factors and to simulate ice detachment, highlighting the complexity of an accurate analysis and assessment of ice adhesion.

Only a few studies make use of FEA simulations, but it is not clear how these stress concentrations should be accounted for [218]. Work and Lian [218] claim that without accounting for stress concentrations it is not possible to obtain quantitative data with better than one order of magnitude accuracy.

A recently published FEA study gives insight into the stress distribution for the horizontal shear test with cylindrical ice blocks [255]. It can be seen how the magnitude of the stress concentrations depends on the geometry of the test system, in particular on the vertical distance between force application point and the substrate. The work also shows how these stress concentrations have to be accounted for, demonstrating that in a stress-dominated fracture regime, the minimum shear stress across the interface should be taken as the characterizing value of the interface strength [255].

Ice accretion speed must not be ignored as well, since residual stresses frozen into the ice might influence the cracking behavior [206, 218]. These frozen residual stresses will also be time dependent, making the timespan between ice accretion and testing another important factor to be considered.

The importance of these stress concentrations is also highlighted by the fact that recent studies were able to use stress concentrations in a positive way, fabricating stress-localized icephobic surfaces [258–260]. The underlying assumption is that stress concentrations can promote micro-cracks at the interface, which then facilitate the propagation of a macroscopic crack.

1.5.2.6 Probe–interface distance and loading rate

When measuring ice adhesion strengths with the horizontal shear configuration, the vertical distance from the interface at which the force probe pushes against the ice has been observed to greatly influence the adhesion value, calculated as $\tau_{\text{c}} = \tau_{\text{ave}} = F/A$ [234].

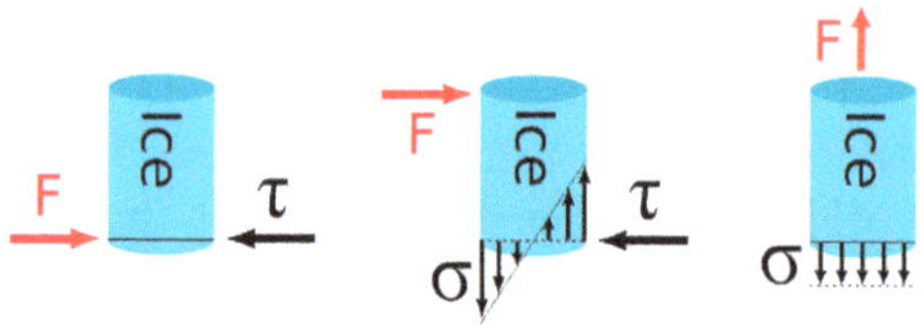

Figure 1.18. Ice sample loading modes

With increasing vertical distance from the interface, the bending moment on the ice sample increases as well. This bending moment produces a tensile component, such that a mixed shear/tensile loading mode is established at the ice–substrate interface (figure 1.18).

The work of Maitra *et al* [261] attempts to take into account the presence of a mixed shear/tensile loading mode and proposes a model to calculate the amount of shear and tensile stress at the interface based on the distance of the force probe from the interface. In their paper, the shear stress τ is calculated based on the entire projected contact area, while only half of the projected area is used to calculate the tensile stress σ, since it is only acting on half of the contact area [261].

The key concept that has to be taken from this paper is that when shear and tensile stresses act simultaneously, the enhanced tensile stress σ on one half of the contact area also increases the contact pressure on the symmetric opposite side. This, in turn, increases τ at the interface. The two loading modes influence one another [261]. With their model, the tensile stress σ has been measured to be always higher than τ in the combined loading mode [261].

Other works demonstrate how τ_{ave} is in fact not a pure surface property, but does also depend on the testing conditions [255]. Even if different pushing heights lead to different values for τ_{ave}, the minimum shear stress was found to be constant. Therefore, to quantify the shear strength of an iced interface, the τ_{ave} can be measured, but then it is essential to correct its value with the stress concentration factor corresponding to the selected pushing height [255]. This way it is possible to find a representative shear strength value τ_c for the tested iced interface.

Also the force probe loading rate has been observed to influence the ice adhesion strength [262]. When the probe rate was increased, the ice adhesion strength increased as well. However, this trend has only been observed for polydimethylsiloxane (PDMS) surfaces and not for other kinds of surfaces [207, 262]. Tests on aluminum have shown that if the force probe loading rate is low enough, then dynamic effects can be neglected, leading to a constant ice adhesion strength for different probe loading rates [255].

1.5.3 Comparison of different test systems

From the discussion above it appears that more effort must be done to standardize or at least to make ice adhesion tests more comparable. At the state-of-the-art, direct comparisons between different papers in literature must be given with great attention. The number of variables involved in ice adhesion is very large, and most of these variables were not reported. Usually, if a paper reports important variables and details, other crucial parameters are left out, making it difficult to properly compare the results across the literature. For example, many authors do not include information about humidity or force probe loading rate. So, before thinking of standardizing ice adhesion test systems, effort must be taken to properly report all the variables and parameters (regarding test system, ice generation, environmental conditions, etc,) that can in any way influence ice adhesion. This is not a trivial task but fundamental to facilitate comparison of ice adhesion measurements on different surfaces and across different institutes and test systems.

The relevant test variables can be classified into two categories: variables that are objectively difficult to obtain but important for the ice adhesion strength measurements and variables that could have been easily reported but, instead, have been neglected. Table 1.2 summarizes the classification of the variables.

Generally, most of the listed variables are reported in ice adhesion testing. However, the missing of only one variable listed under 'necessary' could make a comparison with a different ice adhesion study significantly more difficult. From a technical point of view, the 'necessary variables' would be relatively easy to report.

Until today, no test system has given evidences of being ideal, and it would be unreasonable to think of a test system that could deliver valuable results for all kind of applications. To provide a basis of comparison, it is also not reasonable to think of standardizing ice adhesion test systems completely and only using specific set-ups, with the consensus of the scientific community. Very often, to test the properties of a new surface for a particular kind of application it is necessary to build an entirely new, purpose-built ice adhesion test system.

Trying to calculate the true adhesion strengths, as Work and Lian [218] suggested, free from any type of stress concentrations and environmental influence, would most probably lead to prohibitive costs both for the testing equipment and for the computational power required to perform precise FEA analyses. As discussed above, adhesion strength cannot always be expressed as a stress value, but only in conditions where stress controls the detachment mechanism. It is essential to be aware of the detachment mechanism and how to express the ice adhesion strength correctly.

The other comprehensive review that has been published in the last few years is that of Rønneberg *et al* [207]. Here, a reference test is proposed, with most of the

Table 1.2. Variables to be reported: necessary and desirable. (Adapted with permission from [264].)

Necessary	Desirable
Probe tip surface area	Strain rate
Probe velocity or acceleration	Water resistivity
Probe distance from interface	Ice density
Humidity	Air flow conditions
Freezing time	Water demineralization
Ice sample sizes and shape	
Environmental + interface temperature	
Stabilization time	

experimental details included. As is stated in their paper, this reference test serves as common ground for ice adhesion testing, and after repeating this well-defined test a statistically significant number of times, each experimental parameter is changed individually to understand its effect on the ice adhesion force measured.

It is not clear, however, in which way the variations of the experimental conditions should be made in order to keep consistency between tests from different laboratories. If parameters such as ice type or test methods are changed, it is not straightforward to keep all the other experimental parameters unchanged, given that entirely new test set-ups have to be made. Additionally, even for the proposed standard reference test, not all experimental conditions are defined. For example, the exact geometric shape of the ice sample or the rate of cooling are not defined, factors that might have an influence on ice adhesion. More to that, the conditions chosen for this reference test are also not ideal: it would be easier to control the environmental conditions by using cold plates instead of a freezer and by using a nitrogen environment instead of 80% humidity.

A more realistic approach would be to support an actual application specific test for which the material has been developed by a low-cost reference test. The low-cost reference test could be a horizontal shear test or a centrifuge test with exactly defined geometries, components, and experimental and ice generation conditions. All the variables previously discussed would be reported for this reference test and it would be repeated many times to minimize the effect of the inherent variation of ice structure, as was previously discussed. This test would only be used to establish a comparison with other low ice adhesion materials and then, alongside that, the application specific test would give insight on how this new material behaves under a *specific* set of conditions. These conditions differ from the reference test and are not meant to be compared to other materials or test systems.

References

[1] Lynch F T and Khodadoust A 2001 Effects of ice accretions on aircraft aerodynamics *Prog. Aeosp. Sci.* **37** 669–767

[2] Bragg M, Hutchison T and Merret J 2000 Effect of ice accretion on aircraft flight dynamics *38th Aerospace Sciences Meeting and Exhibit* p 360

[3] Cao Y, Wu Z, Su Y and Xu Z 2015 Aircraft flight characteristics in icing conditions *Prog. Aeosp. Sci.* **74** 62–80

[4] Farzaneh M and Chisholm W A 2009 *Insulators for Icing and Polluted Environments* (New York: Wiley)

[5] Makkonen L 1989 Estimation of wet snow accretion on structures *Cold Reg. Sci. Technol.* **17** 83–8

[6] Nygaard B E K, Ágústsson H and Somfalvi-Tóth K 2013 Modeling wet snow accretion on power lines: improvements to previous methods using 50 years of observations *J. Appl. Meteorol. Climatol.* **52** 2189–203

[7] Bonelli P, Lacavalla M, Marcacci P, Mariani G and Stella G 2011 Wet snow hazard for power lines: a forecast and alert system applied in Italy *Nat. Hazards Earth Syst. Sci.* **11** 2419–31

[8] Ryerson C C and Ramsay A C 2007 Quantitative ice accretion information from the automated surface observing system *J. Appl. Meteorol. Climatol.* **46** 1423–37
[9] Thordarson S and Olafsson B 2008 Weather induced road accidents, winter maintenance and user information *Transp. Res. Arena Eur.* **72** 2008
[10] Panov V V 1978 Icing of ships *Polar Geogr.* **2** 166–86
[11] Kleissl K and Georgakis C T 2010 *Bridge Ice Accretion and De-and Anti-Icing Systems: A Review* (Lyngby: Technical University of Denmark)
[12] Laakso T, Holttinen H, Ronsten G, Tallhaug L, Horbaty R, Baring-Gould I, Lacroix A, Peltola E and Tammelin B 2003 State-of-the-art of wind energy in cold climates *IEA Annex* **24** 53
[13] Parent O and Ilinca A 2011 Anti-icing and de-icing techniques for wind turbines: critical review *Cold Reg. Sci. Technol.* **65** 88–96
[14] Hochart C, Fortin G, Perron J and Ilinca A 2008 Wind turbine performance under icing conditions *Wind Energy: Int. J. Progress Appl. Wind Power Convers. Technol.* **11** 319–33
[15] Jasinski W J, Noe S C, Selig M S and Bragg M B 1998 Wind turbine performance under icing conditions *J. Sol. Energy Eng.* **120** 60–5
[16] Makkonen L 2000 Models for the growth of rime, glaze, icicles and wet snow on structures *Phil. Trans. R. Soc. Lond.* A **358** 2913–39
[17] Gent R W, Dart N P and Cansdale J T 2000 Aircraft icing *Phil. Trans. R. Soc. Lond.* A **358** 2873–911
[18] Charpin J P F 2002 Water flow on accreting ice surfaces *PhD Thesis* Cranfield University, Cranfield
[19] Cebeci T and Kafyeke F 2003 Aircraft icing *Annu. Rev. Fluid Mech.* **35** 11–21
[20] Sultana K R, Dehghani S R, Pope K and Muzychka Y S 2018 A review of numerical modelling techniques for marine icing applications *Cold Reg. Sci. Tech.* **145** 40–51
[21] Myers T G and Charpin J P F 2004 A mathematical model for atmospheric ice accretion and water flow on a cold surface *Int. J. Heat Mass Transf.* **47** 5483–500
[22] Brakel T W, Charpin J P F and Myers T G 2007 One-dimensional ice growth due to incoming supercooled droplets impacting on a thin conducting substrate *Int. J. Heat Mass Transf.* **50** 1694–705
[23] Cao Y, Ma C, Zhang Q and Sheridan J 2012 Numerical simulation of ice accretions on an aircraft wing *Aerosp. Sci. Technol.* **23** 296–304
[24] Janjua Z A, Turnbull B, Hibberd S and Choi K-S 2018 Mixed ice accretion on aircraft wings *Phys. Fluids* **30** 027101
[25] Zhang C, Liu H, Wang F and Kong W 2018 Supercooled large droplet icing accretion and its unsteady aerodynamic characteristics on high-lift devices *Proc. Inst. Mech. Eng.* G **232** 1985–97
[26] Myers T G and Low J 2011 An approximate mathematical model for solidification of a flowing liquid in a microchannel *Microfluid. Nanofluid.* **11** 417–28
[27] Kintea D M, Roisman I V and Tropea C 2016 Transport processes in a wet granular ice layer: model for ice accretion and shedding *Int. J. Heat Mass Transf.* **97** 461–72
[28] Moore M R, Mughal M S and Papageorgiou D T 2017 Ice formation within a thin film flowing over a flat plate *J. Fluid Mech.* **817** 455–89
[29] Monier A, Huerre A, Josserand C and Séon T 2020 Freezing a rivulet *Phys. Rev. Fluids* **5** 062301

[30] Jin Z, Cheng X and Yang Z 2017 Experimental investigation of the successive freezing processes of water droplets on an ice surface *Int. J. Heat Mass Transf.* **107** 906–15
[31] Roisman I V and Tropea C 2021 Wetting and icing of surfaces *Curr. Opin. Colloid Interface Sci.* **53** 101400
[32] Yarin A L 2006 Drop impact dynamics: splashing, spreading, receding, bouncing *Annu. Rev. Fluid Mech.* **38** 159–92
[33] Josserand C and Thoroddsen S T 2016 Drop impact on a solid surface *Annu. Rev. Fluid Mech.* **48** 365–91
[34] Yarin A L, Roisman I V and Tropea C 2017 *Collision Phenomena in Liquids and Solids* (Cambridge: Cambridge University Press)
[35] Cheng X, Sun T-P and Gordillo L 2022 Drop impact dynamics: impact force and stress distributions *Annu. Rev. Fluid Mech.* **54** 57–81
[36] Kittel H M, Roisman I V and Tropea C 2018 Splash of a drop impacting onto a solid substrate wetted by a thin film of another liquid *Phys. Rev. Fluids* **3** 073601
[37] Yarin A L and Weiss D A 1995 Impact of drops on solid surfaces: self-similar capillary waves, and splashing as a new type of kinematic discontinuity *J. Fluid Mech.* **283** 141–73
[38] Smith M I and Bertola V 2011 Particle velocimetry inside Newtonian and non-Newtonian droplets impacting a hydrophobic surface *Exp. Fluids* **50** 1385–91
[39] Gultekin A, Erkan N, Colak U and Suzuki S 2020 PIV measurement inside single and double droplet interaction on a solid surface *Exp. Fluids* **61** 1–18
[40] Roisman I V, Edin Berberović I V and Tropea C 2009 Inertia dominated drop collisions. I. On the universal flow in the lamella *Phys. Fluids* **21** 052103
[41] Roisman I V *et al* 2009 Inertia dominated drop collisions. II. An analytical solution of the Navier–Stokes equations for a spreading viscous film *Phys. Fluids* **21** 052104
[42] Bakshi S, Roisman I V and Tropea C 2007 Investigations on the impact of a drop onto a small spherical target *Phys. Fluids* **19** 032102
[43] van Hinsberg N P, Budakli M, Göhler S, Berberovic E, Roisman I V, Gambaryan-Roisman T, Tropea C and Stephan P 2010 Dynamics of the cavity and the surfacee film for impingements of single drops on liquid films of various thickness *J. Colloid Interface Sci.* **350** 336–43
[44] Roisman I V, Paul van Hinsberg N and Tropea C 2008 Propagation of a kinematic instability in a liquid layer: capillary and gravity effects *Phys. Rev.* **77** 046305
[45] Mundo C H R, Sommerfeld M and Tropea C 1995 Droplet–wall collisions: experimental studies of the deformation and breakup process *Int. J. Multiph. Flow* **21** 151–73
[46] Cossali G E, Coghe A and Marengo M 1997 The impact of a single drop on a wetted solid surface *Exp. Fluids* **22** 463–72
[47] Walzel P 1980 Zerteilgrenze beim Tropfenprall *Chem. Ing. Tech.* **52** 338–9
[48] Rioboo R, Bauthier C, Conti J, Voue M and De Coninck J 2003 Experimental investigation of splash and crown formation during single drop impact on wetted surfaces *Exp. Fluids* **35** 648–52
[49] Okawa T, Shiraishi T and Mori T 2006 Production of secondary drops during the single water drop impact onto a plane water surface *Exp. Fluids* **41** 965–74
[50] Wal R L V, Berger G M and Mozes S D 2006 The splash/non-splash boundary upon a dry surface and thin fluid film *Exp. Fluids* **40** 53–9
[51] Huang Q and Zhang H 2008 A study of different fluid droplets impacting on a liquid film *Pet. Sci.* **5** 62–6

[52] Motzkus C, Gensdarmes F and Géhin E 2011 Study of the coalescence/splash threshold of droplet impact on liquid films and its relevance in assessing airborne particle release *J. Colloid Interface Sci.* **362** 540–52
[53] Okawa T, Kubo K, Kawai K and Kitabayashi S 2021 Experiments on splashing thresholds during single-drop impact onto a quiescent liquid film *Exp. Thermal Fluid Sci.* **121** 110279
[54] Gao X and Li R 2015 Impact of a single drop on a flowing liquid film *Phys. Rev.* **92** 053005
[55] Breitenbach J, Roisman I V and Tropea C 2018 From drop impact physics to spray cooling models: a critical review *Exp. Fluids* **59** 1–21
[56] Roisman I V, Horvat K and Tropea C 2006 Spray impact: rim transverse instability initiating fingering and splash, and description of a secondary spray *Phys. Fluids* **18** 102104
[57] Roisman I V, Lembach A and Tropea C 2015 Drop splashing induced by target roughness and porosity: the size plays no role *Adv. Colloid Interface Sci.* **222** 615–21
[58] Rioboo R and Tropea C 2001 Outcomes from a drop impact on solid surfaces *Atomization Sprays* **11** 155–66
[59] Taylor G I 1959 The dynamics of thin sheets of fluid II. Waves on fluid sheets *Proc. R. Soc. Lond.* A **263** 296–312
[60] Roisman I V, Rioboo R and Tropea C 2002 Normal impact of a liquid drop on a dry surface: model for spreading and receding *Proc. R. Soc. London* A **458** 1411–30
[61] Kittel H M, Alam E, Roisman I V, Tropea C and Gambaryan-Roisman T 2018 Splashing of a Newtonian drop impacted onto a solid substrate coated by a thin soft layer *Colloids Surfaces* A **553** 89–96
[62] Riboux G and Gordillo J 2014 Experiments of drops impacting a smooth solid surface: a model of the critical impact speed for drop splashing *Phys. Rev. Lett.* **113** 024507
[63] Burzynski D A, Roisman I V and Bansmer S E 2020 On the splashing of high-speed drops impacting a dry surface *J. Fluid Mech.* **892** A2
[64] Gibbs J W 1928 *The Collected Works of JW Gibbs* (Green: Longmans)
[65] Thomson W 1872 On the equilibrium of vapour at a curved surface of liquid *Proc. R. Soc. Edinb.* **7** 63–8
[66] Libbrecht K G 2017 Physical dynamics of ice crystal growth *Annu. Rev. Mater. Res.* **47** 271–95
[67] Huppert H E 1990 The fluid mechanics of solidification *J. Fluid Mech.* **212** 209
[68] Rubinšteĭn L I 1971 *The Stefan Problem* (Translation of Mathematical Monographs vol 27). (Providence, RI: American Mathematical Society)
[69] Bird R B, Stewart W E and Lightfoot E N 1960 *Transport Phenomena* (New York: Wiley)
[70] Roisman I V 2010 Fast forced liquid film spreading on a substrate: flow, heat transfer and phase transition *J. Fluid Mech.* **656** 189–204
[71] Roisman I V 2009 Inertia dominated drop collisions. II. An analytical solution of the Navier–Stokes equations for a spreading viscous film *Phys. Fluids* **21** 052104
[72] Berberović E, Roisman I V, Jakirlić S and Tropea C 2011 Inertia dominated flow and heat transfer in liquid drop spreading on a hot substrate *Int. J. Heat Fluid Flow* **32** 785–95
[73] Berberović E, Roisman I V, Jakirlić S and Tropea C 2011 Computational study of hydrodynamics and heat transfer associated with a liquid drop impacting a hot surface *Computational Fluid Dynamics 2010* (Berlin: Springer) pp 543–8
[74] Schremb M, Borchert S, Berberovic E, Jakirlic S, Roisman I V and Tropea C 2017 Computational modelling of flow and conjugate heat transfer of a drop impacting onto a cold wall *Int. J. Heat Mass Transf.* **109** 971–80

[75] Batzdorf S, Breitenbach J, Schlawitschek C, Roisman I V, Tropea C, Stephan P and Gambaryan-Roisman T 2017 Heat transfer during simultaneous impact of two drops onto a hot solid substrate *Int. J. Heat Mass Transf.* **113** 898–907
[76] Jung S, Tiwari M K, Doan N V and Poulikakos D 2012 Mechanism of supercooled droplet freezing on surfaces *Nat. Commun.* **3** 1–8
[77] Roisman I V, Criscione A, Tropea C, Mandal D K and Amirfazli A 2015 Dislodging a sessile drop by a high-Reynolds-number shear flow at subfreezing temperatures *Phys. Rev.* E **92** 023007
[78] Moghtadernejad S, Jadidi M, Tembely M, Esmail N and Dolatabadi A 2015 Concurrent droplet coalescence and solidification on surfaces with various wettabilities *J. Fluids Eng.-Trans. ASME* **137** 071302
[79] Pruppacher H R, Klett J D and Wang P K 1998 *Microphysics of clouds and precipitation* (Springer)
[80] Schremb M, Roisman I V and Tropea C 2017 Transient effects in ice nucleation of a water drop impacting onto a cold substrate *Phys. Rev.* E **95** 022805
[81] Eberle P, Tiwari M K, Maitra T and Poulikakos D 2014 Rational nanostructuring of surfaces for extraordinary icephobicity *Nanoscale* **6** 4874–81
[82] Schremb M and Tropea C 2016 Solidification of supercooled water in the vicinity of a solid wall *Phys. Rev.* E **94** 052804
[83] Wang L, Kong W, Wang F and Liu H 2019 Effect of nucleation time on freezing morphology and type of a water droplet impacting onto cold substrate *Int. J. Heat Mass Transf.* **130** 831–42
[84] Kong W and Liu H 2015 A theory on the icing evolution of supercooled water near solid substrate *Int. J. Heat Mass Transf.* **91** 1217–36
[85] Kong W and Liu H 2018 Unified icing theory based on phase transition of supercooled water on a substrate *Int. J. Heat Mass Transf.* **123** 896–910
[86] Pach E, Rodriguez L and Verdaguer A 2018 Substrate dependence of the freezing dynamics of supercooled water films: a high-speed optical microscope study *J. Phys. Chem.* A **122** 818–26
[87] Staroselsky A, Acharya R and Khain A 2021 Toward a theory of the evolution of drop morphology and splintering by freezing *J. Atmos. Sci.* **78** 3181–204
[88] Shibkov A A, Golovin Y I, Zheltov M A, Korolev A A and Leonov A A 2003 Morphology diagram of nonequilibrium patterns of ice crystals growing in supercooled water *Physica* A **319** 65–79
[89] Shibkov A A, Golovin Y I, Zheltov M A, Korolev A A and Vlasov A A 2001 Kinetics and morphology of nonequilibrium growth of ice in supercooled water *Crystallogr. Rep.* **46** 496–502
[90] Shibkov A A, Zheltov M A, Korolev A A, Kazakov A A and Leonov A A 2005 Crossover from diffusion-limited to kinetics-limited growth of ice crystals *J. Cryst. Growth* **285** 215–27
[91] Langer J S, Sekerka R F and Fujioka T 1978 Evidence for a universal law of dendritic growth rates *J. Cryst. Growth* **44** 414–8
[92] Ohsaka K and Trinh E H 1998 Apparatus for measuring the growth velocity of dendritic ice in undercooled water *J. Cryst. Growth* **194** 138–42
[93] Müller-Krumbhaar H and Langer J S 1978 Theory of dendritic growth. III. Effects of surface tension *Acta Metall.* **26** 1697–708

[94] Criscione A, Roisman I V, Jakirlić S and Tropea C 2015 Towards modelling of initial and final stages of supercooled water solidification *Int. J. Therm. Sci.* **92** 150–61
[95] Criscione A, Kintea D, Tuković Ž, Jakirlić S, Roisman I V and Tropea C 2013 Crystallization of supercooled water: a level-set-based modeling of the dendrite tip velocity *Int. J. Heat Mass Transf.* **66** 830–7
[96] Volmer M and Weber A 1926 Keimbildung in übersättigten Gebilden *Zeitsch. Phys. Chem.* **119** 277–301
[97] Turnbull D and Co Fisher J 1949 Rate of nucleation in condensed systems *J. Chem. Phys.* **17** 71–3
[98] Li C, Liu Z, Goonetilleke E C and Huang X 2021 Temperature-dependent kinetic pathways of heterogeneous ice nucleation competing between classical and non-classical nucleation *Nat. Commun.* **12** 4954
[99] Fitzner M, Sosso G C, Pietrucci F, Pipolo S and Michaelides A 2017 Pre-critical fluctuations and what they disclose about heterogeneous crystal nucleation *Nat. Commun.* **8** 2257
[100] Lupi L, Hudait A and Molinero V 2014 Heterogeneous nucleation of ice on carbon surfaces *J. Am. Chem. Soc.* **136** 3156–64
[101] Wang C, Wu J, Wang H and Zhang Z 2021 Classical nucleation theory of ice nucleation: second-order corrections to thermodynamic parameters *J. Chem. Phys.* **154** 234503
[102] Lamb D and Verlinde J 2011 *Physics and Chemistry of Clouds* (Cambridge: Cambridge University Press)
[103] Weber T A and Stillinger F H 1983 Molecular dynamics study of ice crystallite melting *J. Phys. Chem.* **87** 4277–81
[104] Tse J S, Klein M L and Mcdonald I R 1983 Molecular dynamics studies of ice Ic and the structure I clathrate hydrate of methanet *J. Phys. Chem.* **87** 4198–203
[105] Stillinger F H and Rahman A 1974 Improved simulation of liquid water by molecular dynamics *J. Chem. Phys.* **60** 1528–32
[106] Berendsen H J C, Postma J P M, Van Gunsteren W F and Hermans J 1981 *Intermolecular Forces* (Cambridge: Cambridge University Press)
[107] Karim O A and Haymet A D J 1988 The ice/water interface: a molecular dynamics simulation study *J. Chem. Phys.* **89** 6889–96
[108] Sciortino F and Corongiu G 1993 Structure and dynamics in hexagonal ice: a molecular dynamics simulation with an *ab initio* polarizable and flexible potential *J. Chem. Phys.* **98** 5694–700
[109] Corongiu G and Clementi E 1992 Liquid water with an *ab initio* potential: x-ray and neutron scattering from 238 to 368 K *J. Chem. Phys.* **97** 2030–8
[110] Essmann U and Geiger A 1995 Molecular dynamics simulation of vapor deposited amorphous ice *J. Chem. Phys.* **103** 4678–92
[111] Berendsen H J C, Grigera J R and Straatsma T P 1987 The missing term in effective pair potentialst *J. Phys. Chem.* **91** 6269–71
[112] Gertner B J and Hynes J T 1996 Molecular dynamics simulation of hydrochloric acid ionization at the surface of stratospheric ice *Science* **271** 1563–6
[113] Mantz Y A, Geiger F M, Molina L T, Molina M J and Trout B L 2001 The interaction of HCl with the (0001) face of hexagonal ice studied theoretically via Car–Parrinello molecular dynamics *Chem. Phys. Lett.* **348** 285–92

[114] Nada H and Furukawa Y 1996 Anisotropic growth kinetics of ice crystals from water studied by molecular dynamics simulation *J. Cryst. Growth* **169** 5–7
[115] Materer N, Starke U, Barbieri A, Van Hove M A, Somorjai G A, Kroes G J and Minot C 1997 Molecular surface structure of ice(0001): dynamical low-energy electron diffraction, total-energy calculations and molecular dynamics simulations *Surf. Sci.* **381** 210
[116] Burnham C J, Li J C and Leslie M 1997 Molecular dynamics calculations for ice Ih *J. Phys. Chem.* B **101** 6192–5
[117] Matsumoto M, Shinj S and Iwao O 2002 Molecular dynamics simulation of the ice nucleation and growth process leading to water freezing *Nature* **416** 409–13
[118] Abraham F F 1974 Homogeneous nucleation theory *Phys. Today* **27** 52–3
[119] Krämer B, Hübner O, Vortisch H, Wöste L, Leisner T, Schwell M, Rühl E and Baumgärtel H 1999 Homogeneous nucleation rates of supercooled water measured in single levitated microdroplets *J. Chem. Phys.* **111** 6521–7
[120] Carignano M A, Shepson P B and Szleifer I 2005 Molecular dynamics simulations of ice growth from supercooled water *Mol. Phys.* **103** 2957
[121] Tse J S, Klug D D, Guthrie M, Tulk C A, Benmore C J and Urquidi J 2005 Investigation of the intermediate- and high-density forms of amorphous ice by molecular dynamics calculations and diffraction experiments *Phys. Rev.* B **71** 6
[122] Shiomi J, Kimura T and Maruyama S 2007 Molecular dynamics of ice-nanotube formation inside carbon nanotubes *J. Phys. Chem.* C **111** 12188–93
[123] Gladich I and Roeselová M 2012 Comparison of selected polarizable and nonpolarizable water models in molecular dynamics simulations of ice I_h *Phys. Chem. Chem. Phys.* **14** 11371–85
[124] Caldwell J W and Kollman P A 1995 Structure and properties of neat liquids using nonadditive molecular dynamics: water, methanol, and n-methylacetamide *J. Phys. Chem.* **99** 6208–19
[125] Lamoureux G, Harder E, Vorobyov I V, Roux B and MacKerell A D 2006 A polarizable model of water for molecular dynamics simulations of biomolecules *Chem. Phys. Lett.* **418** 245–9
[126] Yan J Y and Patey G N 2012 Molecular dynamics simulations of ice nucleation by electric fields *J. Phys. Chem.* **116** 7057–64
[127] Abascal J L F and Vega C 2010 Widom line and the liquid–liquid critical point for the TIP4P/2005 water model *J. Chem. Phys.* **133** 12
[128] Nada H and Van Der Eerden J P J M 2003 An intermolecular potential model for the simulation of ice and water near the melting point: a six-site model of H_2O *J. Chem. Phys.* **118** 7401–13
[129] Singh J K and Müller-Plathe F 2014 On the characterization of crystallization and ice adhesion on smooth and rough surfaces using molecular dynamics *Appl. Phys. Lett.* **104** 1
[130] English N J 2014 Massively parallel molecular-dynamics simulation of ice crystallisation and melting: the roles of system size, ensemble, and electrostatics *J. Chem. Phys.* **141** 12
[131] English N J 2015 Structural properties of liquid water and ice I*h* from *ab-initio* molecular dynamics with a non-local correlation functional *Energies* **8** 9383–91
[132] Fitzner M, Sosso G C, Cox S J and Michaelides A 2015 The many faces of heterogeneous ice nucleation: interplay between surface morphology and hydrophobicity *J. Am. Chem. Soc.* **137** 13658–69

[133] Cox S J, Kathmann S M, Slater B and Michaelides A 2015 Molecular simulations of heterogeneous ice nucleation. II. Peeling back the layers *J. Chem. Phys.* **142** 5
[134] Li T, Donadio D, Russo G and Galli G 2011 Homogeneous ice nucleation from supercooled water *Phys. Chem. Chem. Phys.* **13** 19807–13
[135] Rahnamoun A and Van Duin A C T 2016 Study of ice cluster impacts on amorphous silica using the ReaxFF reactive force field molecular dynamics simulation method *J. Appl. Phys.* **119** 3
[136] Sosso G C, Tribello G A, Zen A, Pedevilla P and Michaelides A 2016 Ice formation on kaolinite: insights from molecular dynamics simulations *J. Chem. Phys.* **145** 12
[137] Cygan R T, Liang J J and Kalinichev A G 2004 Molecular models of hydroxide, oxyhydroxide, and clay phases and the development of a general force field *J. Phys. Chem.* **108** 1255–66
[138] Sosso G C, Chen J, Cox S J, Fitzner M, Pedevilla P, Zen A and Michaelides A 2016 Crystal nucleation in liquids: open questions and future challenges in molecular dynamics simulations *Chem. Rev.* **116** 7078–116
[139] Bao L, Huang Z, Priezjev N V, Chen S, Luo K and Hu H 2018 A significant reduction of ice adhesion on nanostructured surfaces that consist of an array of single-walled carbon nanotubes: a molecular dynamics simulation study *Appl. Surf. Sci.* **437** 202–8
[140] Pedevilla P, Fitzner M and Michaelides A 2017 What makes a good descriptor for heterogeneous ice nucleation on OH-patterned surfaces *Phys. Rev.* **96** 9
[141] Pedevilla P, Fitzner M, Sosso G C and Michaelides A 2018 Heterogeneous seeded molecular dynamics as a tool to probe the ice nucleating ability of crystalline surfaces *J. Chem. Phys.* **149** 8
[142] Kling T, Kling F and Donadio D 2018 Structure and dynamics of the quasi-liquid layer at the surface of ice from molecular simulations *J. Phys. Chem.* **122** 24780–7
[143] Yagasaki T, Matsumoto M and Tanaka H 2020 Molecular dynamics study of grain boundaries and triple junctions in ice *J. Chem. Phys.* **153** 9
[144] Piaggi P M, Panagiotopoulos A Z, Debenedetti P G and Car R 2021 Phase equilibrium of water with hexagonal and cubic ice using the scan functional *J. Chem. Theory Comput.* **17** 3065–77
[145] Piaggi P M and Car R 2020 Phase equilibrium of liquid water and hexagonal ice from enhanced sampling molecular dynamics simulations *J. Chem. Phys.* **152** 204116
[146] Schremb M, Roisman I V and Tropea C 2018 Normal impact of supercooled water drops onto a smooth ice surface: experiments and modelling *J. Fluid Mech.* **835** 1087
[147] Takita H and Sumita I 2013 Low-velocity impact cratering experiments in a wet sand target *Phys. Rev.* E **88** 022203
[148] Li H, Roisman I V and Tropea C 2015 a Influence of solidification on the impact of supercooled water drops onto cold surfaces *Exp. Fluids* **56** 133
[149] Sun M, Kong W, Wang F and Liu H 2019 Impact freezing modes of supercooled droplets determined by both nucleation and icing evolution *Int. J. Heat Mass Transf.* **142** 118431
[150] Blake J, Thompson D, Raps D and Strobl T 2015 Simulating the freezing of supercooled water droplets impacting a cooled substrate *AIAA J.* **53** 1725–39
[151] Carlo Antonini M T, Tiwari M K, Mularczyk A, Imeri Z, Schoch P and Poulikakos D 2014 Supercooled water drops impacting superhydrophobic textures *Langmuir* **30** 10855–61

[152] Mohammadi M, Tembely M and Dolatabadi A 2017 Predictive model of supercooled water droplet pinning/repulsion impacting a superhydrophobic surface: the role of the gas–liquid interface temperature *Langmuir* **33** 1816–25

[153] Mohammadi M, Tembely M and Dolatabadi A 2017 Supercooled water droplet impacting superhydrophobic surfaces in the presence of cold air flow *Appl. Sci.* **7** 130

[154] Fahrenheit D G 1724 VIII. experimenta & observationes de congelatione aquæ in vacuo factæ a D G Fahrenheit, R S S Daniel Gabriel Fahrenheit *Phil. Trans. R. Soc. Lond.* **33** 78–84

[155] Srirangam S, Bhendale M and Singh J K 2023 Does supercooled water retain its universal nucleation behavior under shear at high pressure? *Phys. Chem. Chem. Phys.* **25** 21528–37

[156] Heverly J R 1949 Supercooling and crystallization *Eos, Trans. Ame. Geophys. Union* **30** 205–10

[157] Li W, Wang J, Tian L, Zhu C and Zhao N 2023 Numerical investigation on rebound dynamics of supercooled water droplet on cold superhydrophobic surface *Appl. Therm. Eng.* **239** 122007

[158] Tavakoli F, Davis S H and Pirouz Kavehpour H 2015 Freezing of supercooled water drops on cold solid substrates: initiation and mechanism *J. Coat. Technol. Res.* **12** 869–75

[159] Campbell J M, Meldrum F C and Christenson H K 2015 Is ice nucleation from supercooled water insensitive to surface roughness? *J. Phys. Chem.* C **119** 1164–9

[160] Thomas Fernandez K and Coutier-Delgosha O 2023 An advanced aircraft deicing analysis: supercooled liquid dynamics under ultrasonic frequency and surface roughness effects *Bull. Am. Phys. Soc.* **ZC23** 00009

[161] Li H, Roisman I V and Tropea C 2015 Influence of solidification on the impact of supercooled water drops onto cold surfaces *Exp. Fluids* **56** 1–13

[162] Tropea C, Schremb M and Roisman I V 2017 Physics of SLD impact and solidification *Proc. 7th European Conference for Aeronautics and Space Sciences (Milan, Italy. EUCASS)* **vol 332**

[163] Zhou X, Wang H, Zhu X, Chen R and Liao Q 2024 Numerical study of supercooled water droplet impacting on cold superhydrophobic surface under electric field *Int. J. Heat Mass Transfer* **218** 124781

[164] Doolittle J B and Vali G 1975 Heterogeneous freezing nucleation in electric fields *J. Atmos. Sci.* **32** 375–9

[165] Yan J Y, Overduin S D and Patey G N 2014 Understanding electrofreezing in water simulations *J. Chem. Phys.* **141** 074501

[166] Okawa S, Saito A and Minami R 2001 The solidification phenomenon of the supercooled water containing solid particles *Int. J. Refrig.* **24** 108–17

[167] Hallett J 1964 Experimental studies of the crystallization of supercooled water *J. Atmos. Sci.* **21** 671–82

[168] Currie T C, Fuleki D and Mahallati A 2014 Experimental studies of mixed-phase sticking efficiency for ice crystal accretion in jet engines *6th AIAA Atmospheric and Space Environments Conf.* p 3049

[169] Mason J G, Chow P and Fuleki D M 2011 Understanding ice crystal accretion and shedding phenomenon in jet engines using a rig test *J. Eng. Gas Turbines Power* **133** 041201

[170] Currie T, Struk P, Tsao J-C, Fuleki D and Knezevici D 2012 Fundamental study of mixed-phase icing with application to ice crystal accretion in aircraft jet engines *4th AIAA Atmospheric and Space Environments Conf.* p 3035

[171] Bucknell A J, McGilvray M, Gillespie D, Jones G, Reed A and Collier B 2018 Experimental studies of ice crystal accretion on an axisymmetric body at engine-realistic conditions *2018 Atmospheric and Space Environments Conference* p 4223
[172] Saunders C P R and Peck S L 1998 Laboratory studies of the influence of the rime accretion rate on charge transfer during crystal/graupel collisions *J. Geophys. Res. Atmos.* **103** 13949–56
[173] Veres J P and Jorgenson P C 2013 Modeling commercial turbofan engine icing risk with ice crystal ingestion *5th AIAA Atmospheric and Space Environments Conference* p 2679
[174] Ayan E, Ozgen S, Murat C and Tarhan E 2015 Prediction of ice crystal accretion with TAICE *SAE 2015 Int. Conf. on Icing of Aircraft, Engines, and Structures* SAE Technical Paper 2015–01–2148
[175] Veres J, Jorgenson P and Wright W 2012 A model to assess the risk of ice accretion due to ice crystal ingestion in a turbofan engine and its effects on performance *4th AIAA Atmospheric and Space Environments Conf.* p 3038
[176] Mazzawy R S 2007 Modeling of ice accretion and shedding in turbofan engines with mixed phase/glaciated (ice crystal) conditions *2007 SAE Aircraft and Engine Icing International Conference* SAE Technical Paper 2007-01-3288
[177] Grazioli J, Ghiggi G, Billault-Roux A-C and Berne A 2022 MASCDB, a database of images, descriptors and microphysical properties of individual snowflakes in free fall *Sci. Data* **9** 186
[178] Heymsfield A 1972 Ice crystal terminal velocities *J. Atmos. Sci.* **29** 1348–57
[179] Köbschall K, Breitenbach J, Roisman I V, Tropea C and Hussong J 2023 Geometric descriptors for the prediction of snowflake drag *Exp. Fluids* **64** 4
[180] Yan S and Palacios J 2024 A physics-based sphericity, drag coefficient and Nusselt number model of a partially-melted ice crystal *Int. J. Therm. Sci.* **197** 108798
[181] Mason B J 1956 On the melting of hailstones *Q. J. R. Meteorol. Soc.* **82** 209–16
[182] Hauk T, Roisman I V and Tropea C D 2014 Investigation of the melting behaviour of ice particles in an acoustic levitator *11th AIAA/ASME Joint Thermophysics and Heat Transfer Conf.* p 2261
[183] Kintea D M, Hauk T, Roisman I V and Tropea C 2015 Shape evolution of a melting nonspherical particle *Phys. Rev.* E **92** 033012
[184] Köbschall K, Traut B, Roisman I V, Tropea C and Hussong J 2023 Melting of fractal snowflakes: experiments and modeling *Int. J. Heat Mass Transfer* **212** 124254
[185] Hertz H R 1881 Über die Berührung fester elastischer Körper *J. Angew. Math.* **92** 156–71
[186] Kim H and Keune J N 2007 Compressive strength of ice at impact strain rates *J. Mater. Sci.* **42** 2802–6
[187] Jones S J 1997 High strain-rate compression tests on ice *J. Phys. Chem.* **101** 6099–101
[188] Kuehn G A, Schulson E M, Jones D E and Zhang J 1993 The compressive strength of ice cubes of different sizes *ASME. J. Offshore Mech. Arct. Eng.* **115** 142–8
[189] Shen L-T, Zhao S-D, Lu X-N and Shi Y-X 1988 Effects of temperature and strain rate on uniaxial compressive strength of naturally formed fresh-water ice *Proc. 7th Int. Conf. on Offshore Mechanics and Arctic Engineering* pp 19–23
[190] Mellor M and Cole D M 1982 Deformation and failure of ice under constant stress or constant strain-rate *Cold Reg. Sci. Technol.* **5** 201–19
[191] Shazly M, Prakash V and Lerch B A 2009 High strain-rate behavior of ice under uniaxial compression *Int. J. Solids Struct.* **46** 1499–515

[192] Roisman I V 2022 Hydrodynamic model of a collision of a spherical plastic ice particle with a perfectly rigid substrate *Int. J. Impact Eng.* **159** 104019

[193] Jackson R L and Green I 2003 A finite element study of elasto-plastic hemispherical contact *International Joint Tribology Conf.* vol 37068 pp 65–72

[194] Raman C V 1920 On some applications of Hertz's theory of impact *Phys. Rev.* **15** 277–84

[195] Antonyuk S, Heinrich S, Tomas J, Deen N G, Van Buijtenen M S and Kuipers J A M 2010 Energy absorption during compression and impact of dry elastic-plastic spherical granules *Granular Matter* **12** 15–47

[196] Tippmann J D, Kim H and Rhymer J D 2013 Experimentally validated strain rate dependent material model for spherical ice impact simulation *Int. J. Impact Eng.* **57** 43–54

[197] Wu X and Prakash V 2015 Dynamic strength of distill water and lake water ice at high strain rates *Int. J. Impact Eng.* **76** 155–65

[198] Hauk T, Bonaccurso E, Roisman I V and Tropea C 2015 Ice crystal impact onto a dry solid wall. Particle fragmentation *Proc. R. Soc.* A **471** 20150399

[199] Pan H and Render P M 1996 Impact characteristics of hailstones simulating ingestion by turbofan aeroengines *J. Propulsion Power* **12** 457–62

[200] Hauk T 2016 Investigation of the impact and melting process of ice particles *PhD Thesis* Technische Universität Darmstadt, Germany

[201] Reitter L M, Lohmann H, Schremb M, Roisman I V, Hussong J and Tropea C 2022 Impact of an ice particle onto a dry rigid substrate: Dynamic sintering of a residual ice cone *Cold Reg. Sci. Technol.* **194** 103416

[202] Karpen N, Cuco A, Kuenstler D and Bonaccurso E 2021 MUSIC-haic_WP1_T11_ACRT_01_ParticleImpactOnRigidSurface https://zenodo.org/records/5576280

[203] Currie T C, Fuleki D, Knezevici D C and MacLeod J D 2013 Altitude scaling of ice crystal accretion *5th AIAA Atmospheric and Space Environments Conf.* p 2677

[204] Struk P, Bartkus T, Tsao J-C, Currie T and Fuleki D 2015 Ice accretion measurements on an airfoil and wedge in mixed-phase conditions *Technical Report, SAE Technical Paper* 2015-01-2116

[205] Reitter L M, Malik Y A, Jahn A B, Roisman I V and Hussong J 2024 Penetration of a solid spherical particle into artificial and natural wet granular ice layers: dynamic strength characterization *Int. J. Impact Eng.* **183** 104791

[206] Makkonen L 2012 Ice adhesion—theory, measurements and countermeasures *J. Adhes. Sci. Technol.* **26** 413–45

[207] Rønneberg S, He J and Zhang Z 2020 The need for standards in low ice adhesion surface research: a critical review *J. Adhes. Sci. Technol.* **34** 319–47

[208] Beemer D L, Wang W and Kota A K 2016 Durable gels with ultra-low adhesion to ice *J. Mater. Chem.* A **4** 18253–8

[209] Laroche A, Grasso M J, Dolatabadi A and Bonaccurso E 2020 Tensile and shear test methods for quantifying the ice adhesion strength to a surface *Ice Adhesion: Mechanism, Measurement and Mitigation* (Beverley, MA: Scrivener) pp 237–84

[210] Knight M and Clay W C 1930 Refrigerated wind tunnel tests on surface coatings for preventing ice formation *Technical Notes* 339 National Advisory Committee for Aeronautics, Washington

[211] Geer W C and Scott M 1930 The prevention of the ice hazard on airplanes *Technical Notes* 345 National Advisory Committee for Aeronautics, Washington

[212] Schutzius T M, Jung S, Maitra T, Eberle P, Antonini C, Stamatopoulos C and Poulikakos D 2015 Physics of icing and rational design of surfaces with extraordinary icephobicity *Langmuir* **31** 4807–21
[213] Amirfazli A and Antonini C 2016 Fundamentals of anti-icing surfaces *Non-wettable Surfaces: Theory, Preparation and Applications* ed H A Ras, Robin and A Marmur (London: Royal Society of Chemistry) pp 319–46
[214] Stone H A 2012 Ice-phobic surfaces that are wet *ACS Nano* **6** 6536–40
[215] Wilson P W, Lu W, Xu H, Kim P, Kreder M J, Alvarenga J and Aizenberg J 2013 Inhibition of ice nucleation by slippery liquid-infused porous surfaces (slips) *Phys. Chem. Chem. Phys.* **15** 581–5
[216] Rykaczewski K, Anand S, Subramanyam S B and Varanasi K K 2013 Mechanism of frost formation on lubricant-impregnated surfaces *Langmuir* **29** 5230–8
[217] Shen Y, Wu X, Tao J, Zhu C, Lai Y and Chen Z 2019 Icephobic materials: fundamentals, performance evaluation, and applications *Prog. Mater. Sci.* **103** 509–57
[218] Work A and Lian Y 2018 A critical review of the measurement of ice adhesion to solid substrates *Prog. Aerosp. Sci.* **98** 1–26
[219] Dou R, Chen J, Zhang Y, Wang X, Cui D, Song Y, Jiang L and Wang J 2014 Anti-icing coating with an aqueous lubricating layer *ACS Appl. Mater. Interfaces* **6** 6998–7003
[220] Hejazi V, Sobolev K and Nosonovsky M 2013 From superhydrophobicity to icephobicity: forces and interaction analysis *Sci. Rep.* **3** 2194
[221] Lou D, Hammond D and Pervier M-L 2014 Investigation of the adhesive properties of the ice–aluminum interface *J. Aircr.* **51** 1051–6
[222] Zou M, Beckford S, Wei R, Ellis C, Hatton G and Miller M A 2011 Effects of surface roughness and energy on ice adhesion strength *Appl. Surf. Sci.* **257** 3786–92
[223] He Z, Vågenes E T, Delabahan C, He J and Zhang Z 2017 Room temperature characteristics of polymer-based low ice adhesion surfaces *Sci. Rep.* **7** 1–7
[224] Zhu L, Xue J, Wang Y, Chen Q, Ding J and Wang Q 2013 Ice-phobic coatings based on silicon-oil-infused polydimethylsiloxane *ACS Appl. Mater. Interfaces* **5** 4053–62
[225] Rønneberg S, Laforte C, Volat C, He J and Zhang Z 2019 The effect of ice type on ice adhesion *AIP Adv.* **9** 055304
[226] Guerin F, Laforte C, Farinas M-I and Perron J 2016 Analytical model based on experimental data of centrifuge ice adhesion tests with different substrates *Cold Reg. Sci. Technol.* **121** 93–9
[227] Laforte C and Beisswenger A 2005 Icephobic material centrifuge adhesion test *Proceedings of the 11th International Workshop on Atmospheric Icing of Structures (Montreal, QC, Canada)* pp 12–6
[228] Golovin K, Kobaku S P R, Lee D H, DiLoreto E T, Mabry J M and Tuteja A 2016 Designing durable icephobic surfaces *Sci. Adv.* **2** e1501496
[229] Golovin K, Dhyani A, Thouless M D and Tuteja A 2019 Low-interfacial toughness materials for effective large-scale deicing *Science* **364** 371–5
[230] Ling E J Y, Uong V, Renault-Crispo J-S, Kietzig A-M and Servio P 2016 Reducing ice adhesion on nonsmooth metallic surfaces: wettability and topography effects *ACS Appl. Mater. Interfaces* **8** 8789–800
[231] Andersson L-O, Golander C-G and Persson S 1994 Ice adhesion to rubber materials *J. Adhes. Sci. Technol.* **8** 117–32

[232] He Z, Zhuo Y, Wang F, He J and Zhang Z 2020 Design and preparation of icephobic PDMS-based coatings by introducing an aqueous lubricating layer and macro-crack initiators at the ice-substrate interface *Progr. Org. Coatings* **147** 105737

[233] He Z, Xiao S, Gao H, He J and Zhang Z 2017 Multiscale crack initiator promoted super-low ice adhesion surfaces *Soft Matter.* **13** 6562–8

[234] Wang C, Zhang W, Siva A, Tiea D and Wynne K J 2014 Laboratory test for ice adhesion strength using commercial instrumentation *Langmuir* **30** 540–7

[235] Yang S, Xia Q, Zhu L, Xue J, Wang Q and Chen Q 2011 Research on the icephobic properties of fluoropolymer-based materials *Appl. Surf. Sci.* **257** 4956–62

[236] Susoff M, Siegmann K, Pfaffenroth C and Hirayama M 2013 Evaluation of icephobic coatings–screening of different coatings and influence of roughness *Appl. Surf. Sci.* **282** 870–9

[237] Petrenko V F and Peng S 2003 Reduction of ice adhesion to metal by using self-assembling monolayers (sams) *Can. J. Phys.* **81** 387–93

[238] Raraty L E and Tabor D 1958 The adhesion and strength properties of ice *Proc. R. Soc. London* A **245** 184–201

[239] Croutch V K and Hartley R A 1992 Adhesion of ice to coatings and the performance of ice release coatings *JCT J. Coat. Technol.* **64** 41–53

[240] Sarkar D K and Farzaneh M 2009 Superhydrophobic coatings with reduced ice adhesion *J. Adhes. Sci. Technol.* **23** 1215–37

[241] Mirshahidi K, Zarasvand K A, Luo W and Golovin K 2020 A high throughput tensile ice adhesion measurement system *HardwareX* **8** e00146

[242] Schulz M and Sinapius M 2015 Evaluation of different ice adhesion tests for mechanical deicing systems *Technical Report, SAE Technical Paper*

[243] Kasaai M R and Farzaneh M 2004 A critical review of evaluation methods of ice adhesion strength on the surface of materials *Int. Conf. on Offshore Mechanics and Arctic Engineering* vol 37 459 pp 919–26

[244] Subramanyam S B, Rykaczewski K and Varanasi K K 2013 Ice adhesion on lubricant-impregnated textured surfaces *Langmuir* **29** 13414–8

[245] Xue X, Qiang G, Feng Y and Luo T 2022 Experimental study on the adhesion strength of the frozen ice for aircraft moving parts *Aerospace* **9** 589

[246] Scavuzzo R J and Chu M L 1987 Structural properties of impact ICES accreted on aircraft structures *NASA Contractor Report* 179580 National Aeronautics and Space Administration

[247] Soltis J, Palacios J, Eden T and Wolfe D 2015 Evaluation of ice-adhesion strength on erosion-resistant materials *AIAA J.* **53** 1825–35

[248] Druez J, Nguyen D D and Lavoie Y 1986 Mechanical properties of atmospheric ice *Cold Reg. Sci. Technol.* **13** 67–74

[249] Rosenberg R *et al* 2005 Why is ice slippery? *Phys. Today* **58** 50

[250] Faraday P 1859 XXIV. On regelation, and on the conservation of force *Phil. Mag. J. Sci.* **17** 162–9

[251] Chen J, Dou R, Cui D, Zhang Q, Zhang Y, Xu F, Zhou X, Wang J, Song Y and Jiang L 2013 Robust prototypical anti-icing coatings with a self-lubricating liquid water layer between ice and substrate *ACS Appl. Mater. Interfaces* **5** 4026–30

[252] Schulson E M 1999 The structure and mechanical behavior of ice *JOM* **51** 21–7

[253] Martin E, Vandellos T, Leguillon D and Carrère N 2016 Initiation of edge debonding: coupled criterion versus cohesive zone model *Int. J. Fract* **199** 157–68

[254] Leguillon D 2002 Strength or toughness? A criterion for crack onset at a notch *Eur. J. Mech.* A **21** 61–72

[255] Stendardo L, Gastaldo G, Budinger M, Pommier-Budinger V, Tagliaro I, Ibáñez-Ibáñez P F and Antonini C 2023 Reframing ice adhesion mechanisms on a solid surface *Appl. Surf. Sci.* **641** 158462

[256] Yang Q and Cox B 2005 Cohesive models for damage evolution in laminated composites *Int. J. Fract.* **133** 107–37

[257] Huré M, Olivier P and Garcia J 2022 Effect of Cassie–Baxter versus Wenzel states on ice adhesion: a fracture toughness approach *Cold Reg. Sci. Technol.* **194** 103440

[258] Irajizad P, Al-Bayati A, Eslami B, Shafquat T, Nazari M, Jafari P, Kashyap V, Masoudi A, Araya D and Ghasemi H 2019 Stress-localized durable icephobic surfaces *Mater. Horizons* **6** 758–66

[259] Wang J, Wu M, Liu J, Xu F, Hussain T, Scotchford C and Hou X 2021 Metallic skeleton promoted two-phase durable icephobic layers *J. Colloid Interface Sci.* **587** 47–55

[260] Jiang X, Lin Y, Xuan X, Zhuo Y, Wu J, He J, Du X, Zhang Z and Li T 2023 Stiffening surface lowers ice adhesion strength by stress concentration sites *Colloids Surf.* **666** 131334

[261] Maitra T, Jung S, Giger M E, Kandrical V, Ruesch T and Poulikakos D 2015 Superhydrophobicity versus ice adhesion: the quandary of robust icephobic surface design *Adv. Mater. Interfaces* **2** 1500330

[262] Wang C, Gupta M C, Yeong Y H and Wynne K J 2018 Factors affecting the adhesion of ice to polymer substrates *J. Appl. Polym. Sci.* **135** 45734

[263] Stumpf B, Hussong J and Roisman I V 2022 Drop impact onto a substrate wetted by another liquid: flow in the wall film *Colloids Interfaces* **6** 58

[264] Stendardo L 2025 *PhD Thesis* University of Milano—Bicocca

IOP Publishing

Smart Surface Design for Efficient Ice Protection and Control

Carlo Antonini and Irene Tagliaro

Chapter 2

Strategies for icephobicity and the role of surface chemistry

Simrandeep Bahal, Miisa Tavaststjerna, Navid Mostofi Sarkari, Anny Catalina Ospina Patino, Gabriel Hernandez Rodriguez, Alexandros A Atzemoglou, Theodoros Dimitriadis, David Seveno, Irene Tagliaro, Manish K Tiwari and Vikramjeet Singh

2.1 Introduction

Ice accretion on surfaces is one of the fundamental problems presented by nature, and the effects of ice accumulation can be observed on aircraft surfaces, the blades of wind turbines, electrical cables, refrigeration systems, and in many other places [1]. Broadly, this phenomenon of icing has some serious detrimental effects in the aviation and power sectors and on infrastructure such as buildings and bridges [2]. For example, ice build-up on an aircraft's wing disrupts the normal flow of the air over the wing and hence leads to a loss of lift force, which can result in the stalling of the aircraft even at higher speeds [1]. In the case of wind turbine blades, the presence of ice causes a weight disparity and reduces the aerodynamic performance, resulting in lower power generation [2]. In heating, ventilating and air conditioning (HVAC) applications, the formation of ice around the compressor lowers the efficiency of the system and can also clog the system's fans [1]. In household applications, such as in dish antennas, the ice build-up interferes with the electromagnetic signal detection and leads to poor quality signal at the end user output devices. Considering all these detrimental effects of icing in various applications, it would be quite useful to prevent the formation of ice or remove the ice from the various substrates as soon as it builds up. There are various methods of ice removal from a surface, and these methods have been classified into two categories, namely active and passive [3]. Active systems require energy from a source and consist of mechanical, thermal, chemical and electrical methods [1]. On the other hand, passive systems comprise specialized materials or the application of various coatings on the substrate to render icephobic/anti-icing properties to the substrate. Therefore, it is important to understand the theory behind ice nucleation on the substrates to design surfaces that

doi:10.1088/978-0-7503-6009-8ch2

thermodynamically inhibit the formation of ice. The following two sections discuss the mechanism of ice nucleation on a surface and the different types of ice nucleation.

2.1.1 Ice nucleation theory

The phase change of water plays an important role in many natural phenomena such as cloud and fog formation and the precipitation of rain and snow. In industrial and technical applications such as in heat exchangers, condensers and water desalination applications, a high nucleation and condensation rate of water is desired. On the other hand, in the design of anti-icing substrates, the suppression of water and ice nucleation is sought. A phase change occurs when the original or the mother phase has turned into a supersaturated phase. The development of a new phase from the original supersaturated mother phase does not occur in a continuous manner, but it occurs in a spontaneous manner due to fluctuations in density and temperature. This spontaneous process is termed nucleation. Many nucleation theories have been proposed to describe the kinetics of nucleation, but the classical nucleation theory has been the most widely used theory to study the effects of nucleation [4]. This theory is based on two major assumptions. First, the nuclei or the embryos are spherical clusters with macroscopic density and surface tension. Second, the clusters are distributed according to the Boltzmann statistical theory [5].

Nucleation can be further divided into two types—homogeneous and heterogeneous. When the nucleation takes place within the liquid phase itself away from any foreign surfaces, it is termed as homogeneous nucleation. On the other hand, when the nucleation occurs on an external solid substrate, it is known as heterogeneous nucleation.

2.1.1.1 Homogeneous nucleation

In the case of homogeneous nucleation, a growing ice embryo of clustered water molecules is modeled as a sphere of radius r_e. When the temperature of the liquid is reduced below the freezing temperature at a given pressure, the liquid becomes supercooled and is said to be in a metastable state. In liquid water, a cluster of a few molecules due to random agglomeration of molecules is always present [6]. There is a continuous formation and decay of these agglomerates. When the liquid water is in a supercooled, metastable state, the formation of such a cluster is energetically favored due to the difference in the chemical potential. The chemical potential of a molecule within the cluster is reduced compared to the liquid phase. However, the formation of a cluster is also energetically costly because it also leads to the formation of a new interface, namely the ice–liquid interface. Hence, the total change in the free energy of the liquid water due to the formation of a spherical ice embryo is the sum of two competing energy effects. First is the formation of a new lower-energy solid (ice) phase. The second is the creation of a new high-energy ice–liquid interface. The expression for the same is represented as [6]

$$\Delta G\ (T,\, r_e) = -\frac{4}{3}\pi r_e^3 \Delta G_{IW}(T) + 4\pi r_e^2 \gamma_{IW}(T). \tag{2.1}$$

In the above expression, $\Delta G_{\mathrm{IW}}(T)$ denotes the free energy difference between ice and water per unit volume and $\gamma_{\mathrm{IW}}(T)$ represents the interfacial energy (per unit area) of the ice–water interface. The first term in the above equation represents the decrease in the free energy of the volume which is occupied by the ice embryo during the phase change from liquid to solid. On the other hand, the second term represents the increase in the surface energy due to the newly formed ice–liquid interface. At low radii, the interfacial energy term dominates whereas at higher embryo radii, the volume energy term dominates. The curve of the Gibbs free energy ΔG with the embryo radius r_{e} is shown in figure 2.1.

By differentiating the free energy term, ΔG in the equation (2.1) with respect to the ice embryo radius r_{e}, and setting the resulting equation equal to zero, we can find the critical embryo radius r_{e}^*. If the radius of the ice embryo is less than the critical radius, then it is energetically favorable for the ice embryo to decay back into the liquid phase. On the other hand, if the ice embryo has grown to a critical radius, then it is energetically favorable for the embryo to keep on growing into the solid ice phase. The expression for the critical embryo radius comes as

$$r_{\mathrm{e}}^* = \frac{2\gamma_{\mathrm{IW}}}{\Delta G_{\mathrm{IW}}}. \tag{2.2}$$

The critical embryo radius depends upon the free energy of the ice–water interface γ_{IW} and the free energy difference between ice and water per unit volume ΔG_{IW}. To calculate ΔG_{IW}, Gibbs–Helmholtz equation can be used.

At the critical embryo radius r_{e}^*, the free energy barrier ΔG^* can be found by plugging equation (2.2) into equation (2.1):

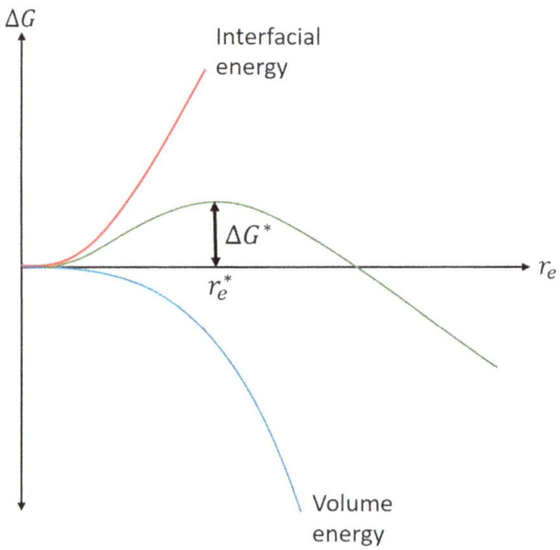

Figure 2.1. Effect of the radius of the ice embryo r_{e} on the Gibbs free energy ΔG. ΔG^* is the free energy barrier for continuous ice growth and r_{e}^* is the critical embryo radius.

$$\Delta G^*(T) = \frac{16\pi\gamma_{IW}(T)^3}{3(\Delta G_{IW}(T))^2}. \tag{2.3}$$

2.1.1.2 Heterogeneous nucleation

In day-to-day life, nucleation of ice embryo is more likely to occur at the liquid–solid interface than within the supercooled liquid itself [7]. Hence, it is important to determine the free energy barrier for the case of heterogeneous nucleation to to design and develop surfaces with anti-icing characterstics [8]. The classical nucleation theory was adapted for the case of heterogeneous nucleation by Fletcher [9]. Fletcher considered a spherical ice embryo of radius r_e growing on a convex substrate with roughness radius R_s within a liquid phase. The free energy of formation of ice embryo for this case can be written as [6]

$$\Delta G(T, r_e, R_s) = \Delta G_{IW} V_I + \gamma_{IW} SA_{IW} + (\gamma_{IS} - \gamma_{SW}) SA_{IS}. \tag{2.4}$$

In the above equation, ΔG_{IW} is the free energy difference between ice and water per unit volume, V_I is the volume of the ice embryo, SA_{IW} is the interfacial area between the ice embryo and the water, and SA_{IS} is the interfacial area between the ice embryo and the substrate. γ_{IS} and γ_{SW} represent the interfacial energies (per unit area) between the ice embryo and solid substrate, and solid substrate and water, respectively. Fletcher noted that a foreign solid substrate creates a low energy interface between the ice embryo and the solid substrate. Hence the third term in the above equation (2.4) would be negative. Hence the magnitude of the Gibbs free energy for heterogeneous nucleation is less than that of homogeneous nucleation. After doing some geometric manipulations, the free energy barrier for the heterogeneous growth of a stable ice embryo is given as

$$\Delta G^*(T) = \frac{16\pi\gamma_{IW}(T)^3}{3(\Delta G_{IW}(T))^2} f\left(\theta_{IW}, R_s\right), \tag{2.5}$$

where f is the geometric factor and depends upon the ice–water contact angle θ_{IW} and the radius of curvature of the surface roughness R_s. It can be seen that equations (2.5) and (2.3) are nearly identical with the only difference being the presence of an additional wetting factor f in equation (2.5). The value of f is always less than 1. Hence $\Delta G^*_{heterogeneous} < \Delta G^*_{homogeneous}$. This implies that ice nucleates with more ease on a foreign solid substrate.

2.1.2 Characteristics of icephobic surfaces

Icephobic surfaces have three main properties. They should be able to repel incoming liquid water droplets, delay the freezing of water droplets and lower the adhesion strength between the ice and the underlying substrate [10].

2.1.2.1 Repelling incoming water droplets

The first property of an icephobic surface is to repel incoming water droplets. In this way, the surface will be free of any liquid water before freezing occurs. For this, it is desired that the surface should be able to repel the incoming water quickly with a short contact time and without losing its water-repellent performance. A superhydrophobic surface was designed with micrometer-sized pillars decorated with a nanometer-sized texture to study the drop impact using supercooled water droplets [3]. The authors found that supercooled droplets even at −30 °C could be repelled by their designed surface.

2.1.2.2 Delaying the freezing of water droplets

In the situations where a sufficiently fast removal of liquid droplets is not possible, there is another option to achieve the icephobic property which is to delay the freezing of water droplets as long as possible. The freezing delay can be measured in terms of ice nucleation temperature [11] and the freezing delay time. The ice nucleation temperature is defined as a temperature at which a sessile water droplet placed on a surface nucleates into ice when the droplet is cooled in a slow and quasi-equilibrium manner. The freezing delay time is the average time taken by a supercooled droplet placed on a surface to nucleate into ice [12].

2.1.2.3 Reducing the ice adhesion strength

In the cases where freezing is inevitable, it is favorable to reduce the adhesion strength between the ice and the underlying substrate so that the ice can be removed easily by external forces such as by the aid of gravity or wind shear. Typically, it is considered that a surface has a low adhesion to ice when the adhesion strength is below 100 kPa and super low ice adhesion when the adhesion strength is below 10 kPa [13].

2.2 Wettability and its relation to icephobic properties

2.2.1 Liquid–solid interaction

The interaction of ice with a surface understandably elicits parallels with liquid–solid interactions and, often, the two sets of interactions are related. However, the relation is not straightforward. Therefore, a quick review of the surface wettability and liquid–solid interaction is in order. The wettability of a surface can be described based on its behavior toward a fluid or as the balance between the surface and interfacial forces. The outcome of an impacting droplet can be characterized either as wetting or anti-wetting and is dependent on the properties of the substrate. This property is of substantial significance when defining the surface chemistry, physics and engineering applications of a material.

The determination of wettability is based on the measurement of the surface contact angle. Contact angle is the main parameter that characterizes the shape of a droplet on a solid surface. The wettability depends on the surface roughness and chemical composition [14]. Changes to these parameters can adjust the value of the contact angle and hence affect the wettability (figure 2.2).

2.2.1.1 Mechanism and definitions of non-wettability

To characterize a surface as non-wettable, it is essential that any drop can be removed with ease, by applying only a small force (a fraction of the drop weight).

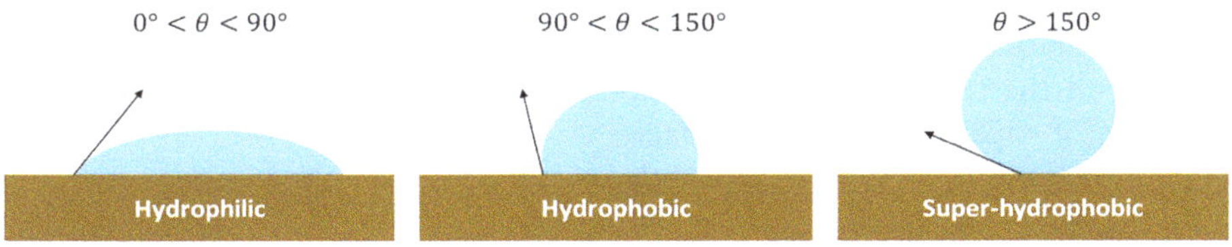

Figure 2.2. Conventional definition of surface wettability: hydrophilic having a contact angle with water smaller than 90°, hydrophobic with an angle greater than 90°, and superhydrophobic with contact angle greater than 150°.

This is typically assessed by tilting the surface, similar to the natural inclination of leaves, and measuring the contact angle at which the droplet rolls off. For a surface to be non-wettable, it is important that the contact angle of a droplet is greater than 150° and the roll-off angle is less than 5°. The easy removal of a drop from a solid surface appears to be dependent on two main factors: (i) the ability of a weak external force to move the drop out of equilibrium and (ii) a high rate of removal from the surface. These two factors characterizing non-wettability translate into the following objectives: (i) minimizing hysteresis by ensuring surface uniformity and (ii) maximizing the contact angle. In theory, both goals can be achieved if the surface interacting with the liquid predominantly consists of gas, such as air trapped within surface roughness grooves [15].

A qualitative definition could suggest that the wetted area has to be sufficiently small. Overall, it is evident that in order to be non-wettable the solid surface must be either rough or porous. The grooves of a rough surface are interconnected and open to the atmosphere. In a porous surface, the pores may be either interconnected or isolated. In the latter case it may be much easier to keep the air in the pores in a stable state, but structural constraints may limit the reduction of the wetted area.

A superhydrophobic surface can even be observed with the bare eye. When a flow of water is applied on the substrate surface, an intuitional wetting behavior and a self-cleaning phenomenon are observed. However, determination of the static water contact angle (WCA) and contact angle hysteresis (CAH), are essential to quantitatively characterize the superhydrophobic property. It is well known that superhydrophobic surfaces can exhibit completely different contact angle hysteresis. Five different wetting states have been described, to differentiate the different superhydrophobic states: totally wetting superhydrophobic surfaces in the Wenzel state, totally air-supporting superhydrophobic surfaces in the Cassie state, the metastable state between the Wenzel and Cassie states (including a 'petal' state), surfaces in the micro/nanostructured two-tier 'lotus' state, and a partially wetting 'gecko' state [16].

2.2.1.2 Artificial super-liquid-repellent surfaces

Natural liquid-repellent surfaces have inspired scientists to develop a range of superhydrophobic and superoleophobic surfaces. Modifying the surface structure and chemical composition are the two key parameters for creating liquid-repellent surfaces. However, fabricating superoleophobic surfaces is significantly more challenging than producing superhydrophobic ones [17]. Therefore, it is essential to control both the surface chemical composition and the surface geometry for the fabrication of superamphiphobic surfaces.

2.2.1.3 Surface structure

Designing the appropriate surface structure constitutes the first step for the fabrication of super-non-wetting surfaces. A suitable roughness can provide a capillary force that prevents liquid from entering into the grooves on the surface. Different methods have been employed for the fabrication of super-non-wettable surfaces, such as grinding, sandblasting, particle coatings, etching, and lithography, which have been used for surface roughening. After surface roughening, the air can be trapped inside the fabricated grooves. This trapped air provides a floating force against the liquid droplets, which is important to achieve a low sliding angle and low contact angle hysteresis.

2.2.1.4 Chemical modification and interfacial energy

Achieving the desired wettability greatly depends on the chemical composition of a material, which plays a crucial role in determining its surface free energy. For a liquid-repellent surface, the contact angle of a liquid droplet should be greater than 90°. Additionally, the surface free energy of the solid substrate should be less than one-quarter that of the liquid. Studies have shown that surface energy levels in monolayer films follow this order: $-CH_2 > -CH_3 > -CF_2 > -CF_2H > -CF_3$. Fluoropolymers, which contain high levels of $-CF_3$ and $-CF_2$ groups, are widely used to reduce surface energy. These materials also offer other benefits, including good chemical and thermal stability and a low friction coefficient [18].

2.2.1.5 Strategies for ice mitigation relying on liquid–solid interactions

The rational design of potential icephobic devices should fulfill requirements in three sequential categories (adapted from [2]) below. Theoretical and experimental evidence suggests that high receding contact angles are beneficial for effective de-icing. However, further development of the equations for ice shedding has revealed that crack size plays a more critical role in determining ice adhesion strength [26].

(i) *Repelling water before it holds on to the surface:* An initial approach to achieving low ice adhesion in materials is based on the idea that preventing surface interactions with liquid water may be crucial. The similar surface free energies of water and ice seem to support this strategy. However, it is important to note that, in the early stages of design, understanding the conditions under which water will approach the surface is essential. For instance, depending on the application, water may contact the surface as vapor, liquid, supercooled droplets, or, under colder conditions, as a solid. Numerous studies have sought to establish a relationship between wetting properties and ice adhesion behavior. While ice adhesion strength generally decreases with increasing hydrophobicity, experimental advances indicate that wettability alone is not a reliable predictor of icephobic behavior in all cases [19–21].

(ii) *Delaying freezing (nucleation) of sessile droplets:* The second requirement addresses the phase change from liquid water to solid ice once water is present on the surface. At this stage, hydrophobic and icephobic behaviors must be compared, highlighting their similarities and differences as explored in several studies. An interesting modeling study investigated the nucleation efficiency during this phase transition. Lupi *et al* designed carbon surfaces with both homogeneous and heterogeneous distributions of hydrophilicity [22]. Their findings suggest that, beyond the hydrophilic character itself, the spatial distribution of hydrophilic areas on the surface significantly influences heterogeneous ice nucleation efficiency by affecting the organization of water molecules at the interface. They observed that the nanoscopic layering in the interfacial region correlates more strongly with ice nucleation efficiency than wettability, according to theoretical molecular dynamics simulations. Consequently, intrinsic wettability, which plays a key role in phase change processes such as boiling and condensation, may not be as effective for controlling ice nucleation behavior. Additional studies have since been conducted to identify surface properties beyond wettability that are critical for understanding the speed of ice nucleation [23].

(iii) *Facilitating ice removal*: Once ice forms on the surface, a solid–solid interaction perspective is essential to understand the ice adhesion strength of the surfaces [24]. As explained by Nosonovsky *et al*, water can withstand compressive or tensile forces but cannot endure shear stress, whereas ice can sustain both shear loads and distributed normal stress [25].

As summarized in figure 2.3, designing icephobic surfaces involves more than simply addressing the adhesive interactions between liquid water and a solid surface. Although water maintains the same chemical composition across different states, its mechanical, thermodynamic, physicochemical, and surface properties differ significantly between liquid and solid phases, making direct extrapolation impossible. Therefore, understanding de-wetting and ice detachment from solid surfaces requires integrating models from liquid, solid, and transitional phases. Given the challenges of controlling all relevant conditions simultaneously, the most practical approach for achieving effective icephobic behavior is to design materials tailored to the specific environmental conditions of the intended application.

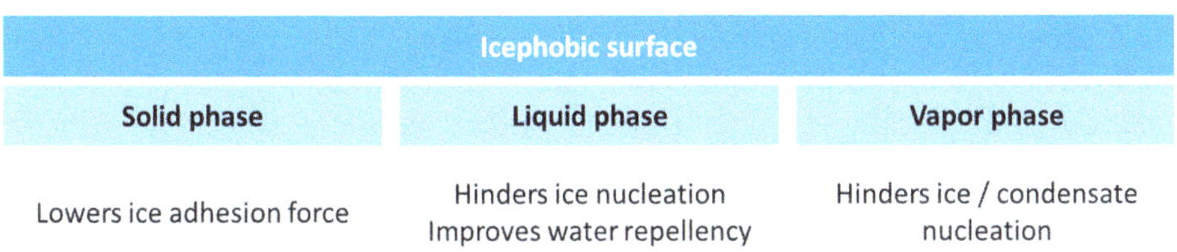

Figure 2.3. Interaction of icephobic surfaces with water in different condensed states.

2.2.1.5.1 Superhydrophobic anti-icing surfaces

Although terms such as 'superhydrophobic' and even 'hydrophobic' remain debated in the literature [27], superhydrophobicity is generally defined as surfaces that ensure water repellence (high contact angles) and high droplet mobility (low contact angle hysteresis) [28]. This effect is typically achieved with rough or hierarchical surfaces, which promote a Cassie–Baxter wetting state. The potential of these surfaces for icephobic design can be assessed based on the previously outlined stages.

Superhydrophobic properties are particularly useful in the initial stage of ice formation, as they can delay water retention by causing liquid or supercooled droplets to rebound or roll off the surface. The air pockets within these surfaces may also act as thermal barriers, reducing freezing rates and lowering ice nucleation efficiency [29, 30]. This makes superhydrophobic surfaces (SHSs) advantageous for developing passive anti-icing systems [26]. However, under high humidity, these textures can promote water droplet condensation, as will be discussed later. Compared with smooth hydrophobic surfaces, superhydrophobic surfaces have shown a significant delay in freezing, a beneficial characteristic in the second stage of ice formation.

A recent review suggests that to control freezing propagation (e.g. through ice bridging, cascade freezing, frost halos, or droplet explosions) [26], it is crucial to manage condensation by precisely designing surface structures. This includes spacing between pillars and optimizing both wettability and roughness to prevent or delay ice propagation under specific environmental conditions. In the final stage, once ice begins to form on the surface, superhydrophobicity can help achieve icephobicity by correlating high receding contact angles and low contact angle hysteresis with reduced ice adhesion strength [31].

Nonetheless, discrepancies persist between expected and actual ice adhesion strength measurements. Theoretical work has shown that microcrack size at the solid interface is a key factor in ice adhesion, indicating that in the third stage of ice testing, the surface should ideally provide sufficient voids at the interface. This characteristic, however, does not always align with the properties of highly superhydrophobic surfaces. As a result, while superhydrophobicity is promising for delaying freezing by enhancing water shedding, it does not always guarantee low ice adhesion. Experiments have shown that this anti-icing effect persists down to −20 °C, regardless of surface effusivity and wettability. Understanding the environmental conditions and wetting mode is crucial for selecting the appropriate approach and material.

2.2.1.6 Delay of icing on superhydrophobic surfaces

Superhydrophobic surfaces employ two main strategies to prevent and delay ice formation: reducing the contact time of water on the surface and hindering heterogeneous nucleation through surface design [32]. For the first approach, superhydrophobic coatings aim to shed most of the liquid water impacting the surface before ice can nucleate and spread. High water repellency of these surfaces is achieved through low surface energy, causing droplets to either rebound directly or be shed by external forces such as gravity, wind, vibrations, or larger droplet sizes

[33]. On superhydrophobic surfaces, water droplets tend to adopt a spherical shape, minimizing contact area with the solid and allowing droplets to roll off with minimal tilting angles [34].

Superhydrophobic surfaces are typically created by increasing surface roughness and/or chemically modifying the surface with low-energy compounds. The combination of roughness and chemical modification has been shown to enhance icephobic properties [35]. Studies have found that materials with low dielectric constants exhibit weak ice adhesion, as they do not contribute to electrostatic attraction with water molecules, making them ideal candidates for icephobic applications. Perfluorinated polymers, for example, are of particular interest due to their low dielectric properties. Poly(tetrafluoroethylene) (PTFE, commercially known as Teflon®) has a low electrical permittivity of about 2.1, making it a strong candidate for minimizing electrostatic interactions with water molecules. When PTFE coatings are applied to aluminum substrates, they form submicron grain structures with gaps smaller than the average water droplet size. These gaps trap air, allowing the surface to achieve a Cassie–Baxter state. The trapped air in these surface defects not only reduces the actual contact angle, thereby lowering ice adhesion, but also provides stress points that promote crack propagation during ice removal. Additionally, the air pockets serve as thermal insulators, decreasing heat transfer between the ice and the coating [36].

2.2.1.7 Ice adhesion

There is considerable debate within the scientific community regarding the mechanisms and forces involved in the icing process. While most researchers recognize three main physical interactions contributing to ice adhesion to a solid—hydrogen bonding, van der Waals forces, and electrostatic interactions—the specific role of each remains a topic of discussion [37].

Some authors propose that electrostatic forces are the primary contributors to ice adhesion, suggesting that materials with low dielectric properties are better suited for achieving low ice adhesion strengths. Conversely, others argue that van der Waals forces primarily determine ice adhesion strength on solid surfaces. Some researchers attribute strong interactions between ice and solids to both van der Waals forces and electrostatic interactions, highlighting the significance of the electrical charge at the ice surface and the induced charge on the solid, making electrostatic interactions a dominant mechanism.

A more integrative perspective published by Farzaneh *et al* demonstrated that ice accumulation and adhesion result from a combination of the three aforementioned forces, along with mechanical adhesion due to the interlocking of water with surface features. Additionally, the chemical composition and environmental conditions are recognized as influential factors in ice adhesion interactions. The size and velocity of droplets, their angle of impact, and surface geometry also play crucial roles in ice adhesion and accretion [31, 38].

To quantitatively assess the ice adhesion of icephobic systems, researchers typically report values for ice adhesion strength to a surface, denoted as τ_{ice}. Icephobic materials are generally defined as those exhibiting $\tau_{ice} < 100$ kPa or $\tau_{ice} <$

20 kPa for passive systems, while materials like bare aluminum demonstrate $\tau_{ice} >$ 1000 kPa. Although the reported ice adhesion values are not directly comparable due to varying measurement conditions and often unique set-ups, they provide insight into material performance. Typically, materials or coatings with $\tau_{ice} <$ 100 kPa are considered to exhibit potential icephobic performance [10].

A noteworthy and contrasting study by Tuteja *et al* focused on developing low interfacial toughness (LIT) materials. Unlike 'classic' icephobic materials, which rely on scaling performance with the iced area and are described primarily by τ_{ice}, LIT materials emphasize toughness. According to this concept, the force required to remove ice is independent of the iced area. Interfacial toughness quantifies the energy required for a crack at the interface to propagate, which is crucial for controlling delamination when the length of the interface is substantial [39, 40]. This distinction is significant for understanding ice adhesion measurements, as evaluating icephobic properties based solely on shear strength may provide an incomplete picture of a material's properties. As noted by the same authors, low interfacial shear strength does not necessarily imply low toughness; thus, the dynamics of removing ice at short interfaces may not directly translate to longer ones [39].

2.2.2 Limitations of superhydrophobic surfaces

The greatest challenge for icephobic materials is their need to perform effectively across various icing scenarios. Ice formation occurs under different temperature conditions, humidity levels, and wind conditions, resulting in diverse types of ice. It is well established that specific modifications to a material's roughness can achieve hydrophobic or superhydrophobic properties. However, there are limitations when these materials are intended for icephobic applications. Excessive roughness can negatively impact hydrophobic properties, potentially transforming the surface into a hydrophilic one. This transformation promotes increased water interaction with the surface and may enhance ice adhesion due to mechanical interlocking.

Mechanical interlocking is not limited to the static presence of water on a very rough surface. When water droplets impact a surface, they lose kinetic energy through spreading and retraction events. During the impact, two downward forces—wetting pressure and effective water hammer pressure—and one upward force, capillary pressure, are at play. If the sum of the downward forces exceeds the upward force, wetting occurs; otherwise, the droplets bounce off the surface. When droplets collide with high kinetic energy, the air pockets in the surface are displaced, leading to a Wenzel state. Under freezing conditions, the displacement of air pockets results in droplets with reduced mobility and an increased contact area, creating an ideal environment for ice nucleation to occur, followed by mechanical interlocking. Therefore, materials intended for icephobic applications should ideally maintain a constant Cassie–Baxter state during droplet collisions [32].

One significant limitation of these strategies is their reliance on wet chemistry for substrate coating. This often requires post-heating treatments, which may alter the polymeric and/or alloy structure or damage the substrate. In the study by Zeng *et al*, a detailed analysis of the contributions of wettability was conducted. Aluminum samples were prepared through an etching method and then coated with PTFE [29]. To examine the effect of heterogeneous coated areas, spray-coated and dip-coated samples were compared. Despite having the same roughness, differences in icephobic properties were directly related to the heterogeneity of the coating technique. The freezing points were approximately −10 °C for the spray-coated sample and −26 °C for the dip-coated sample, highlighting the significant impact of the deposition technique on icephobic performance and underscoring the importance of achieving uniform coverage.

A promising alternative lies in initiated chemical vapor deposition (iCVD), a technique that allows for the synthesis and deposition of complex structures, such as graded, biolayer, or crosslinked coatings, with precise surface coverage and solvent-free synthesis. This method can effectively address the challenges of fluorinated compounds, which are of high interest but often limited by their solubility and compatibility with other molecules. By utilizing iCVD, the low chemical affinity and poor deposition adhesion of fluoropolymers can be managed effectively, leading to the development of new and promising structures for icephobic coatings [41].

2.3 Different functional chemistries

2.3.1 Silicone coated smooth substrates

Soft polysiloxanes are often selected as the base material for icephobic surfaces and coatings [42]. Their popularity in anti-icing applications stems not only from their chemically inert, non-toxic, non-flammable, and easily moldable nature but also from two key properties: softness and low surface energy. Silicones consist of an inorganic silicon-oxygen backbone, with two organic groups attached to each silicon atom in the main chain. The simplest and most common silicone material, polydimethylsiloxane (PDMS), has two methyl groups attached to the silicon atom in its repeating unit, as illustrated in table 2.1. With a Si–O backbone, silicone materials exhibit low intermolecular interactions, which result in low surface tension and surface energy. For instance, PDMS has a surface tension of approximately 20 mN m^{-1} at room temperature, whereas the corresponding surface tension for an aluminum substrate is around 500 mN m^{-1}. This low surface tension contributes to the water repellency and reduced ice adhesion strength of silicone materials. Given that the maximum water contact angle on smooth hydrophobic surfaces is 120°, PDMS, with a contact angle of approximately 110°, is considered one of the most hydrophobic surfaces among smooth substrates [10].

Although the water contact angles on polysiloxanes are not as high as those on superhydrophobic surfaces, silicone-based coatings can demonstrate extremely low ice adhesion strengths below 10 kPa. The icephobicity of polysiloxane is influenced not only by the wettability of the surface but also by the softness of the coating material. The Young's modulus of ice ranges from 0.3 to 3.6 GPa, while that of PDMS is less than 10 MPa. This significant difference in elastic moduli facilitates the removal of ice through DI. As shown in figure 2.4(a), even a small amount of stress can lead to the formation of cavities or cracks at the ice–substrate interface, thereby facilitating passive de-icing on the soft material [43, 44]. However, the DI

Table 2.1. Comparison of ice adhesion strength order and properties of common bulk polymers.

Polymer	Ice adhesion strength order[a]	Surface tension[b] (mN m^{-1})	Water contact[b] angle (°)	T_g[b] (°C)	Chemical structure
Polydimethylsiloxane (PDMS)	1	24	101	−123	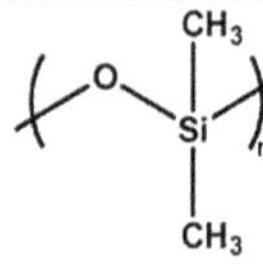
Polytetrafluoroethylene (PTFE)	2	18	108	−73	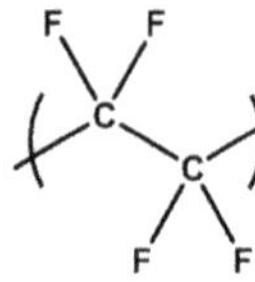
Low-density linear polyethylene (LDPE)	3	33	94	−130	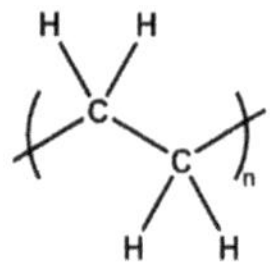
Atactic polypropylene (a-PP)	4	29	116	−10	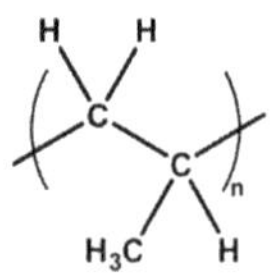
Polystyrene (PS)	5	33	91	95	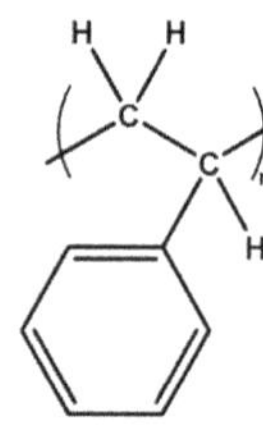
Polyvinyl chloride (PVC)	6	39	87	75	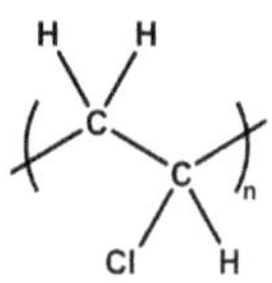

[a] Instead of giving specific values, the polymers are ordered from the lowest ice adhesion strength to the highest due to the lack of standards in ice adhesion measurements.
[b] The values have been obtained from [67].

mechanism alone is insufficient to reduce the adhesion strength of ice to PDMS significantly below 100 kPa [45]. There is a limit to how soft bulk silicone materials can be modified to achieve even lower ice adhesion strengths without compromising the mechanical durability of the surface. While optimizing the thickness of the coating can enhance its anti-icing performance, significant improvements require the combination of the hydrophobicity and softness of polysiloxanes with other innovative anti-icing strategies that can further reduce ice adhesion strength. The

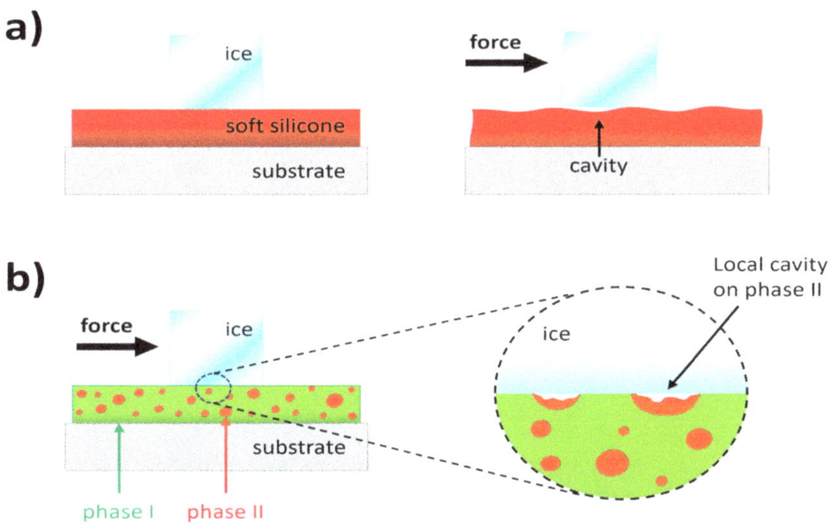

Figure 2.4. Soft silicone materials can be used to lower ice adhesion strength via DI. (a) An illustration of cavity formation in the ice–PDMS interface. (b) Illustration of a two-component polysiloxane coating with areas of high (phase I) and low stiffness (phase II). Crack formation in the softer phase results in remarkably easy ice removal.

lowest recorded ice adhesion strengths on polysiloxane materials were found on silicon elastomer coatings with distinct domains of a softer silicon-based organogel. This system, depicted in figure 2.4(b), utilized the notable difference in stiffness between the two phases to promote crack formation at the ice–coating interface. The resulting ice adhesion strengths achieved remarkable values, dropping below 1 kPa at their lowest.

2.3.2 Hydrated substrates

Another strategy for achieving icephobicity, which contrasts with the approach of increasing hydrophobicity, is to establish a stable layer of liquid water on the surface. In 2014, Dou *et al* developed durable anti-icing coatings that utilized an aqueous lubricating layer, significantly reducing the ice adhesion strength to the surface [46]. These coatings were formed from self-assembling amphiphilic block copolymers with a hydrophobic polyurethane (PU) backbone and pendant groups modified with hydrophilic dimethylolpropionic acid (DMPA). In this structure, the hydrophobic segments remained embedded within the coating, while the hydrophilic groups formed a hydrated corona on the surface. This hydrophilic corona had the capacity to absorb water, creating an aqueous layer that acted as a lubricant, smoothing out surface defects and minimizing mechanical interlocking between ice and the surface. Moreover, the charges on the pendant groups led to an increased ion concentration in the hydration layer. These ions contributed to a reduction in water activity and a lowering of the freezing temperature of the water [47]. In the study by Dou *et al*, the average ice adhesion strength to the coatings was measured at 27 kPa, a sufficiently low value to allow wind speeds of 12 $m\,s^{-1}$ to dislodge ice from the surface. The coating demonstrated resilience at temperatures as low as −53 °C, maintaining stable ice adhesion strength even after 30 icing/de-icing cycles [46].

Anti-icing surfaces employing similar lubricating water layers can remain effective even at extremely low sub-zero temperatures, depending on the type and quantity of ions incorporated into the hydrated layer. As illustrated in figure 2.5, hydrated anti-icing surfaces have been developed using ionizable or hydrophilic

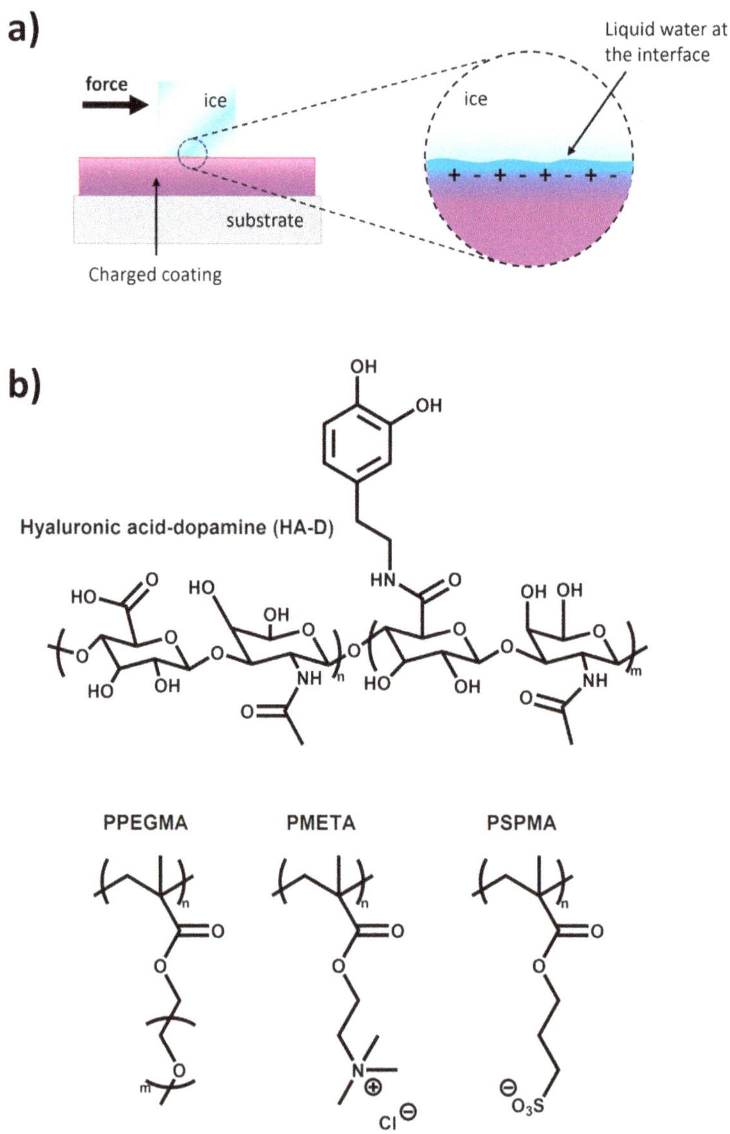

Figure 2.5. (a) An illustration of the interfacial lubricating water layer between the coating and ice. (b) Different functional chemistries for anti-icing via hydrated substrates.

polymeric chains, including hyaluronic acid (HA) [48], poly(acrylic acid) (PAA) [49], poly(ethylene oxide) (PEO) [50], poly(2-(methacryloyloxy)ethyl trimethylammonium) (PMETA) [51], poly(3-sulfopropyl methacrylate) (PSPMA) [58], poly(N,N-dimethylaminoethyl methacrylate) (PDMAEMA) [52], and poly(sulfobetaine methacrylate) (PSBMA) [53, 54]. Notably, these anti-icing coatings can also derive their lubricating layer from the ice itself; thus, evaporation of the water layer should not adversely affect the longevity of the surface. Once icing occurs, the interfacial water is replenished and remains effective as long as the ice persists atop the coating.

2.3.3 Polymer coated substrates

Since the mid-1900s, when the first studies on icephobic surfaces emerged, polymers have been the predominant material choice for anti-icing coatings. An article published in 1967 examined various commercially available plastics concerning their water contact angles and ice adhesion strengths [55]. Similar to polysiloxanes, the benefits of polymeric materials in anti-icing applications are largely attributed to their viscoelastic and hydrophobic properties.

Several soft polymers, beyond polysiloxanes, can effectively reduce ice adhesion strength via the DI mechanism. Bulk polymers characterized by minimal intermolecular interactions—such as the absence of polar groups, bulky side groups, and stereoregularity in the polymer chain—exhibit low glass transition temperatures (T_g) and low elastic moduli. Table 2.1 presents various common bulk polymers, ordered from the lowest to the highest ice adhesion strength. For instance, low-density polyethylene (LDPE) compares favorably to PDMS, with a surface tension of approximately 30 mN m^{-1}, a Young's modulus of 300 MPa, and an ice adhesion strength of 180 kPa. In the same experimental set-up, PDMS demonstrated a surface tension of 20 mN m^{-1}, a Young's modulus of 2 MPa, and an ice adhesion strength of 185 kPa [56].

Among all hydrophobic smooth coatings, perfluorinated polymers have consistently been the primary focus. Despite their higher T_g values and stronger hydrogen bonding with water compared to silicone materials, perfluorinated polymers generally exhibit slightly higher water contact angles. The water-repellent characteristics of these polymers arise from the small overall dipole moment of carbon-fluorine bonds, leading to low electrostatic interactions with water [57]. For instance, the average surface tension and water contact angle for polytetrafluoroethylene (PTFE) are 20 mN m^{-1} and 110°, respectively [58]. Additionally, perfluorinated polymers demonstrate excellent chemical resistance and thermal stability, owing to the stronger bond between carbon and fluorine compared to that between carbon and hydrogen. To address the mechanical durability issues associated with soft polysiloxane coatings, surfaces combining siloxane groups and perfluorinated polymer chains have been proposed. For example, a surface made of polymethyltrifluoropropylsiloxane (PMTFPS) along with polyacrylate copolymers showed improved anti-icing performance compared to the unmodified polyacrylate [59]. However, it is important to note that the measured ice adhesion strengths of around 300 kPa did not fall below the average values recorded for bare PTFE or PDMS.

2.3.4 Polymer brushes

The foundational studies for the development of polymer brush coatings with anti-icing properties date back several decades. In 1985, Murase and Nanishi [60] and again in 1994, Murase *et al* [61] reported the fabrication of an organopolysiloxane-based material containing Li^+ ions, which demonstrated very low ice adhesion strength. They attributed this finding to the presence of both bound and restrained water molecules, a result of the hydrogen bond disruption properties of Li^+ ions. However, these early investigations did not clarify the extent to which the incorporated ions within the polysiloxane matrix contributed to the material's anti-icing properties.

Years later, in 2014, Chernyy *et al* sought to further investigate this phenomenon and enhance understanding of how different counterions influence ice adhesion strength [62]. They prepared various surface-grafted polyelectrolyte brushes using surface-initiated atom-transfer radical polymerization (SI-ATRP) to allow controlled surface modification. The impact of ion incorporation within these polyelectrolyte brush layers on ice adhesion strength was systematically assessed with respect to well-defined surface chemistry.

From a chemical standpoint, five types of methacrylic monomers were polymerized on microscopic glass surfaces via SI-ATRP, enabling surface modification with functional groups such as oligo(ethylene glycol), sulfonate, quaternary ammonium, and carboxylate (as illustrated in figure 2.6). Coatings containing ionizable groups were further subjected to ion exchange to incorporate 13 mono-, bi-, and trivalent ions (H^+, Li^+, Na^+, K^+, Ag^+, Ca^{2+}, La^{3+}, CTA^+, F^-, Cl^-, BF_4^-, SO_4^{2-}, DS^-). The strongest kosmotropes, which are structure-making ions recognized for their significantly negative impact on water structural entropy (such as Li^+ and Na^+), were shown to reduce ice adhesion by 40% and 25%, respectively, at −18 °C. In contrast, no significant influence on ice adhesion was observed for weaker kosmotropes and chaotropes. At −10 °C, all polyelectrolyte coatings exhibited notable mitigation of ice adhesion, with reductions ranging from 20% to 80%, depending on the type of ion used.

He *et al* investigated heterogeneous ice nucleation on polyelectrolyte brush surfaces deposited on gold substrates [51]. The brushes consisted of densely end-grafted polyelectrolyte chains carrying a significant number of ionic groups, as illustrated in figure 2.7. The specific brushes utilized in the study included cationic poly[2-(methacryloyloxy)ethyl trimethylammonium] (PMETA) and anionic poly(3-sulfopropyl methacrylate) (PSPMA). In these brushes, the Cl^- in PMETA and K^+ in PSPMA could be exchanged for various counterions, such as SO_4^{2-}, F^-, acetate (Ac^-), HPO_4^{2-}, Cl^-, Ca^{2+}, Mg^{2+}, guanidinium (Gdm^+), K^+, and Na^+. When water droplets adhered to these polyelectrolyte brush surfaces, a portion of the counterions were released from the brushes due to osmotic pressure, forming a diffusion layer of counterions at the brush/water interface, as shown in figure 2.7. This created an ideal platform for investigating the ion-specific effects on ice nucleation on ionic surfaces. The study's findings indicated that heterogeneous ice nucleation could be modulated by exchanging the counterions of the polymer brushes. The distinct efficiencies of

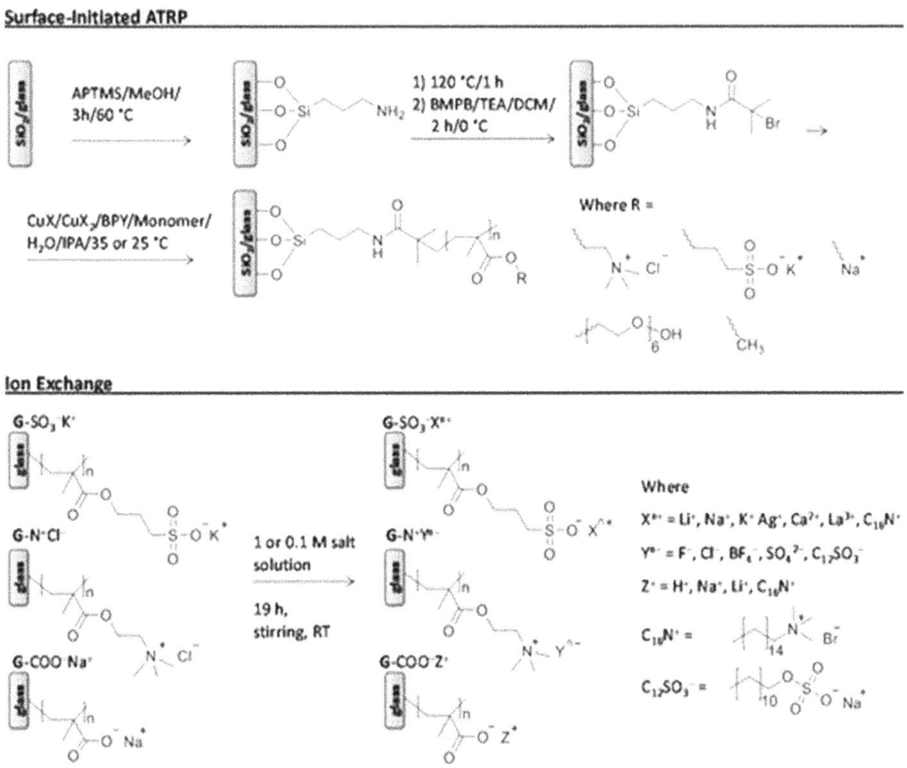

Figure 2.6. The graphical summary of undertaken experimental steps in the preparation of anti-icing polymer brushes using SI-ATRP, as reported by Chernyy *et al* [62]: A two-step solution-phase deposition was conducted involving (3-aminopropyl)trimethoxysilane (APTMS), followed by the reaction of the amine-terminated monolayer with 2-bromo-2-methylpropanoyl bromide (BMPB). Surface-initiated atom transfer radical polymerization (SI-ATRP) of a chosen monomer (KSPM [2], METAC [3], or NaMA [1]) was carried out using a catalyst system composed of CuX and CuX2 (X = Cl– or Br–), with a ligand (BPY) in a 1:1 H_2O/IPA mixture. Finally, the polyelectrolyte brushes were synthesized. (Adapted with permission from [62]. Copyright (2014) the American Physical Society.)

these ions in influencing heterogeneous ice nucleation followed the Hofmeister series. This phenomenon was further interpreted through molecular dynamics (MD) simulations, which demonstrated that the diffused counterions at the brush/water interface effectively regulated the dynamics and structure of the interfacial water, thereby influencing the heterogeneous ice nucleation (HIN) process.

Following the study focused on the ion-specific effects in heterogeneous ice nucleation on polyelectrolyte brush surfaces, the same group of researchers conducted an investigation on polyelectrolyte brush surfaces whose hydration was controllable through counterion exchange [52]. The results of this investigation revealed that ice propagation could also be tuned on the polymer brush surfaces by altering their hydration state. On highly hydrated polymer brush surfaces with hydrophilic counterions, ice propagation was facilitated due to the directional movement of trapped freezable water molecules towards the ice growth front. In

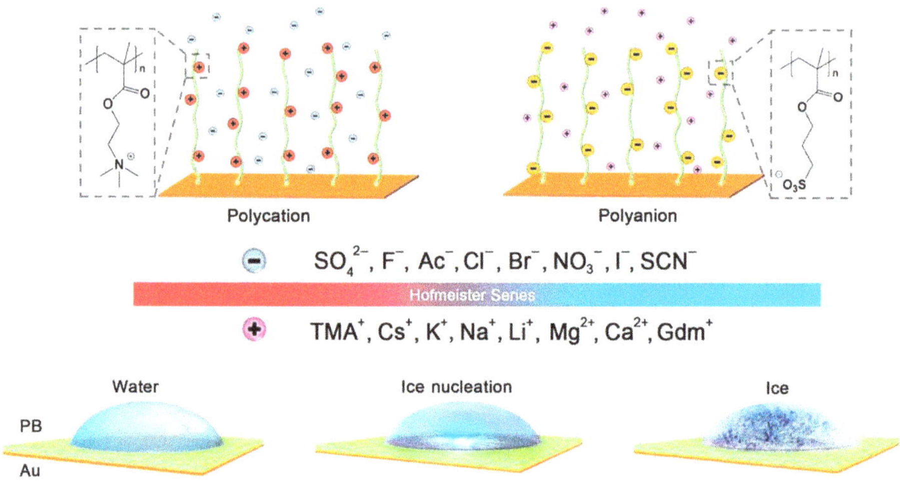

Figure 2.7. Illustration of heterogeneous ice nucleation on cationic and anionic polymer brush surfaces with different counterions based on the work by He *et al* [62]. Cationic poly[2-(methacryloyloxy)-ethyl trimethylammonium] (PMETA) and anionic poly(3-sulfopropyl methacrylate) (PSPMA) brushes have been used to investigate the effect of diffused counterions on tuning heterogeneous ice nucleation. The counterions on the PMETA and PSPMA brush surfaces were successfully exchanged by immersing the brush surfaces into a solution containing the desired counterions. (Reproduced with permission from [51]. Copyright (2016) AAAS.)

stark contrast, on dehydrated polymer brush surfaces with hydrophobic counterions, ice propagation was inhibited by the creation of a 'water depletion region' between droplets. The experimental results demonstrated that the difference in ice propagation rates could reach up to five orders of magnitude.

Liang *et al* prepared an anti-icing coating based on superhydrophilic polyzwitterion brushes [54]. They employed SI-ATRP to synthesize poly(sulfobetaine methacrylate) (PSBMA) brushes on silicon wafers, which are very flat and smooth substrates with easily controllable chemistry. The synthesis process of the SBMA polymer (PSBMA) brushes is illustrated in figure 2.8. In summary, a hydroxylated silicon surface was reacted with 2-bromo-2-methyl-N-[3-(trimethoxysilyl)propyl] propanamide (BrTMOS) as the initiator, forming a stable layer on the substrate that subsequently initiated the grafting polymerization of SBMA to create polyzwitterion brushes. The thickness of the PSBMA brush layers increased linearly with polymerization time, and a transition occurred from a nonassociated hydrophilic regime to a self-associated hydrophobic regime at a layer thickness of approximately 100 nm.

Thermal and chemical analyses revealed that PSBMA contained more nonfreezable bound water than typical polyelectrolytes, such as poly(sodium styrenesulfonate) and poly(2-(methacryloyloxy)ethyl trimethylammonium chloride), resulting in lower ice adhesion strength compared to the latter two. At −20 °C, the PSBMA brush coating exhibited a low ice adhesion strength of 60 kPa, indicating a significant reduction in ice adhesion by up to approximately 75% compared to uncoated silicon wafers. The optimum PSBMA layer thickness for achieving low ice

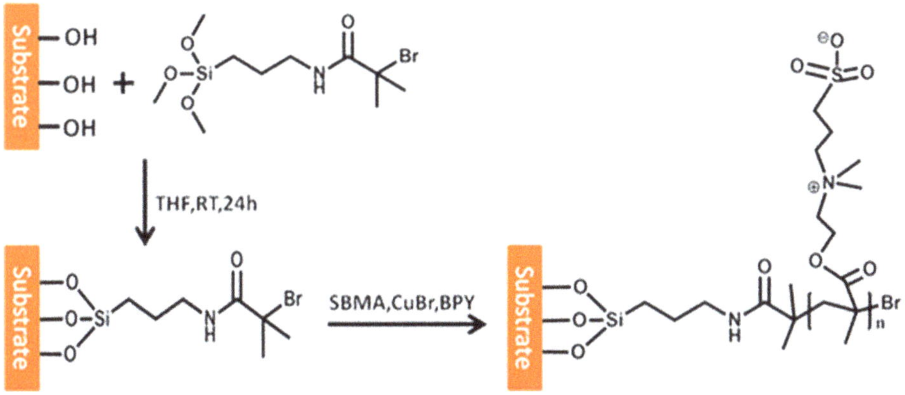

Figure 2.8. Synthesis of PSBMA polymer brushes through SI-ATRP polymerization on smooth silicon substrate based on the work done by Liang *et al* [54] (THF: tetrahydrofuran, RT: room temperature, BPY: 2,2′-bipyridine). (Adapted with permission from [54]. Copyright (2018) the American Physical Society.)

adhesion was found to be around 100 nm. These outcomes suggest that polyzwitterions are excellent candidates for anti-icing coating applications.

From the reviewed studies, it seems obvious that ions are involved in ice formations in various situations (e.g. in the anti-icing polymer brushes that are designed based on them). However, it is still not clear if there is an ion-specific effect of the heterogeneous ice nucleation. Accordingly, the field is extensively open for researchers to profoundly investigate new achievements in anti-icing polymer brushes based on ion-specific effects.

2.3.5 Slippery liquid infused porous surfaces

Generally, there are two types of natural-based anti-wetting surface designs that have inspired researchers. The first type is the traditional lotus leaf surfaces, which exhibit superhydrophobic characteristics due to the creation of air-infused micro- and nanostructured surfaces that minimize the contact area between the surface and liquid. While the underlying concept of this plant has proven effective in fabricating superhydrophobic surfaces, these structures have inherent drawbacks that adversely affect their efficacy in icephobic applications. Specifically, issues such as water condensation, frost formation, and increased ice adhesion arise from the loss of air voids within the porous structure under high humidity conditions. Condensed water domains can act as anchoring points for ice, facilitating its attachment to the surface upon freezing and leading to increased ice adhesion. Additionally, superhydrophobic surface designs are not easily scalable and face challenges under high-pressure impingement and high-temperature applications [63, 64].

The second nature-inspired concept for achieving anti-wetting surfaces addresses these drawbacks by utilizing lubricant-infused ‘slippery’ surfaces, which are inspired by the Nepenthes (tropical pitcher plants) [65, 66]. This unique characteristic of tropical pitcher plants was first investigated in 2004 [67], but the first adaptation of this feature in developing slippery liquid infused porous surfaces (SLIPSs) was carried out in 2011 by Wong *et al* [66]. In their work, synthetic liquid-repellent surfaces were reported that consist of a film of lubricating liquid locked in place by a micro- and nanoporous substrate. The design principles behind their work include the following: (i) the liquid surface is intrinsically smooth and defect-free at the molecular level, (ii) it offers immediate self-repair by wicking into damaged areas of the underlying substrate, (iii) it is largely incompressible, and (iv) it can be tailored to repel immiscible liquids with virtually any surface tension. Their studies revealed that the developed SLIPS formed a smooth, stable interface that nearly eliminated the pinning of the liquid contact line for both high- and low-surface-tension liquids. Additionally, they minimized pressure-induced impalement into the porous structures, self-healed and maintained functionality after mechanical damage, and could even be made optically transparent. SLIPSs allow droplets to contact a layer of lubricant liquid rather than coming into intimate contact with the solid surface. This design results in several beneficial properties: a small sliding angle, low contact angle hysteresis, self-healing capability through capillary wicking upon damage, anti-icing ability, and strength to withstand external pressure [68]. The homogeneous and ultra-smooth nature of SLIPSs makes them particularly attractive for icephobicity (primarily passive anti-icing), as they reduce nucleation sites for ice and water on the surface. Consequently, even if water droplets condense on an SLIPS, they can readily roll off due to the remarkably low contact angle hysteresis [69].

Over the years, significant efforts have been made to optimize the originally developed SLIPSs for icephobicity, resulting in considerable advancements in the field. Heydarian *et al* have categorized SLIPSs into two main groups [70], lubricated micro/nanotextured surfaces and oil-infused polymeric systems. The first category is further divided into two subcategories: ordered textured surfaces and disordered textured surfaces. Various methods have been developed for fabricating SLIPSs, including layer-by-layer (LbL) assembly [71, 72], liquid flame spray (LFS) [73, 74], templating [75, 76], laser lithography [77, 78], and spray and spin methods [79, 80], as well as the creation of magnetic slippery surfaces [81, 82]. Recent studies continue to contribute to the SLIPS field, either focusing on their individual operation or in combination with other anti-/de-icing features. For instance, Liu *et al* combined 'oil film isolating' with electric heating coatings to create a reliable and durable anti-icing coating [83]. To enhance the immobilization of the oil film, they proposed a convenient self-assembly method for constructing closely aligned micropores on the electric heating coating. This approach resulted in a slippery property for the porous electric heating coating after infusing silicone oil into the semi-closed porous structures. A brief description of this method and its steps are illustrated in figure 2.9. More recently, we (Singh *et al*) have shown the use of reticular materials, metal–organic frameworks (MOF) in enhancing the stability of SLIPSs which exhibited excellent anti-icing performance [84].

Cheng *et al* developed a method to prepare photo-thermal slippery surfaces (PSSs) that demonstrate rapid self-repair, exceptional anti-icing and de-icing performance, and excellent stability [85]. These PSSs are created by combining a cost-effective photo-thermal conversion material (black paint), a solid lubricant (paraffin wax), and a porous polyamide (PA) substrate. For instance, the PSSs can

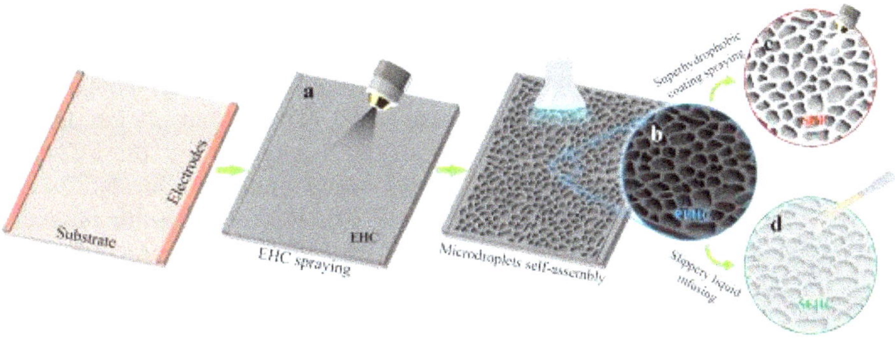

Figure 2.9. Fabrication of the porous electric heating coatings with different functions: (a) smooth electric heating coating (EHC), (b) EHC with close-aligned micropores (PEHC), (c) hydrophobic silica sprayed PEHC (SHP), and (d) slippery silicone oil-infused PEHC (SEHC). (Adapted with permission from [83]. Copyright (2019) Elsevier.)

rapidly self-repair under 1 sun illumination when the surface temperature exceeds 12 °C ± 1.0 °C or under near-infrared (NIR, 808 nm, 300 mW cm^{-12}) when the temperature is higher than –20 °C ± 1.0 °C. Furthermore, at temperatures above –12 °C ± 1.0 °C, ice formation is prevented on the PSSs under 1 sun. Even at lower temperatures, down to –20 °C ± 1.0 °C, ice formation is delayed, taking 212.5 ± 2.5 s to occur.

Despite the advantages of SLIPSs in anti-icing applications, their durability can be compromised, leading to failure in certain operational conditions. SLIPSs achieve liquid repellency primarily through the immiscibility of the lubricants used. A low surface energy lubricant is infused into the surface structure, creating an interface with the immiscible working fluids that results in very low contact angle hysteresis. For effective performance, three criteria must be satisfied: (a) the solid substrate must have a higher affinity for the lubricant than for the droplet; (b) the lubricant and the working fluid must be immiscible; and (c) the lubricant must wick into, spread, and adhere stably within the structure. The primary source of failure for SLIPSs is their inability to retain the lubricant, leading to drainage of the lubricant layer and causing contact line pinning of the working fluid on exposed solid structures [86, 87]. Addressing durability issues and other drawbacks of SLIPSs often requires careful selection of materials, and several studies have been conducted in this regard [84]. The durability of icephobic materials presents a paradox: while they are engineered to prevent ice adhesion and formation, their effectiveness frequently diminishes under prolonged or extreme weather conditions, precisely when such properties are most critical [88–90].

To address the challenges associated with ice adhesion, a new class of 'smart' icephobic coatings has emerged, integrating functionalities such as self-lubrication and phase change materials (PCMs) [91]. PCMs present an innovative solution by undergoing phase transitions that either absorb or release latent heat, thereby counteracting ice formation [92]. While PCMs have been utilized in various applications—including solar panels [93], building materials [94], electronics [95], and road surfaces [96]—their potential in anti-icing coatings remains relatively underexplored [97]. Encapsulated PCMs (EPCMs) offer an effective means of stabilizing these materials by embedding them within protective shells that prevent leakage and facilitate controlled heat release [98]. This encapsulation, typically achieved through micro-encapsulation, creates a barrier around the PCM core, making it easier to integrate into a matrix, which enhances durability during multiple icing cycles (as illustrated in figure 2.10). Recent innovations, such as PCM-impregnated porous metallic structures (PIPMSs), further enhance the icephobic capabilities of SLIPSs by incorporating solid lubricants such as polyethylene glycol (PEG) or coconut oil into the porous matrix [99]. This combination addresses the challenge of lubricant loss, achieving ultra-low ice adhesion (<5 kPa) and withstanding up to 50 icing/de-icing cycles with minimal depletion of the PCM. By releasing latent heat and maintaining a lubricated interface, these materials not only delay ice formation but also support long-term durability in challenging environments. In summary, the integration of EPCMs with various other icephobic coatings —including self-lubricating, photo-thermal, polymeric matrices, and porous metallic substrates—bridges traditional PCM applications with innovative anti-icing needs, offering a promising solution for resilient icephobic surfaces.

2.3.6 Photo-thermal materials

Electro-thermal de-icing has emerged as one of the most widely investigated active de-icing strategies, garnering attention from numerous researchers. This method employs Joule heating, which arises from electrical resistivity, to mitigate ice adhesion and accretion on surfaces [100]. In this approach, a conductive coating layer is deposited onto the surface, and a voltage difference is applied. The resulting flow of electrical current generates heat through the frictional resistance encountered

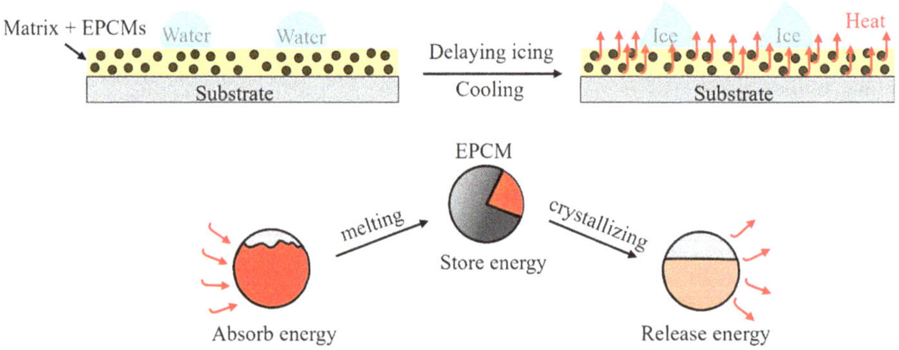

Figure 2.10. Schematic representation of the phase change materials incorporated into a matrix. The (black) shell encapsulated the core (red) functional material, in this case the PCM, enhancing its durability and thermal properties. The material matrix (yellow), in this case a slippery liquid infused porous surface, works as a binder to hold the EPCMs in place while enhances its anti-icing properties. The EPCM working principle relies on absorbing, storing, and releasing energy as heat. EPCMs absorb energy (melt of PCM) from various factors such as outdoor temperature, active heating system etc. When temperatures drop, the stored energy is released (crystallization of PCM) to delay ice formation.

by electrons as they move through the material. This internal heating effectively melts the ice, facilitating de-icing. Despite its advantages, a significant drawback of electro-thermal de-icing is the high energy consumption associated with heat production. To address this challenge, researchers have explored alternative methods that offer greater efficiency, such as surfaces and coatings with photo-thermal effects that convert solar illumination into heat [101].

A pioneering study by Dash *et al* [102] introduced photo-thermal 'traps' designed to mitigate icing. These traps, which are laminates applied to a base substrate, enhance de-icing by converting solar illumination into heat at the interface between the ice and the substrate. The underlying mechanism involves three complementary layers (as depicted in figure 2.11): a selective absorber for solar radiation, a thermal spreader for lateral heat dispersal, and an insulation layer to minimize transverse heat loss.

Upon illumination, the thermal confinement created at the heat spreader results in a rapid increase in surface temperature, leading to the formation of a thin lubricating melt layer that facilitates the removal of ice. According to the findings of this study, solar illumination can induce a temperature increase of up to 33 °C, making this de-icing method highly energy-efficient

Mitridis *et al* [100] proposed metasurfaces with embedded plasmonically enhanced light absorption heating by means of ultrathin hybrid metal-dielectric coatings, acting as a passive and facile strategy for simultaneous de-icing and anti-icing effects solely based on solar illumination as the energy source. In this method, surface transparency is a must and the balance between transparency and absorption is attained by rationally nanoengineered coatings consisting of gold nanoparticle inclusions in a dielectric (titanium dioxide), in which the broadband absorbed solar energy is concentrated into a small volume. As a consequence, a more than 10 °C temperature rise happens in relation to the ambient air at the air–solid interface, which is the place of most ice formation. This leads to retarding water freezing and

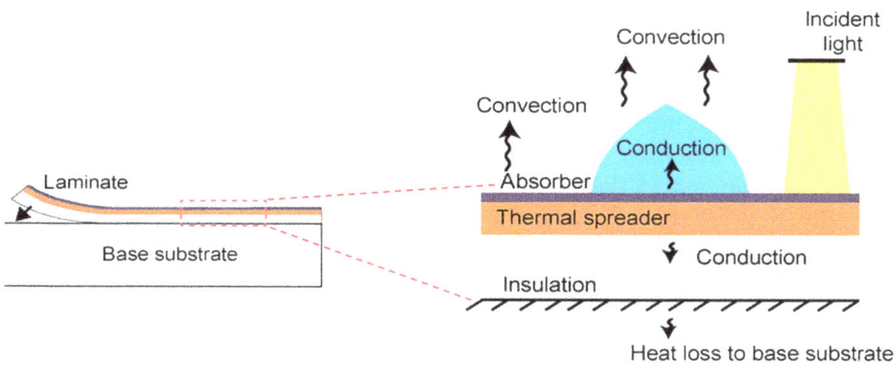

Figure 2.11. Schematic representation of the photo-thermal trap designed by Dash *et al* applied on the base substrate as a laminate and the attributed heat transfer mechanisms. The laminate consists of (top to bottom) highly absorbing cermet, thermal spreader, and insulating layer. (Reproduced with permission from [102]. Copyright (2018) AAAS.)

mitigating ice adhesion (when it occurs) and hence, obtaining anti/de-icing concurrently.

Wu *et al* [103] have employed candle soot to fabricate superhydrophobic photo-thermal icephobic surfaces with high photo-thermal efficiency. The designed coating solution for attaining the photo-thermal icephobic surface consisted of three components as candle soot, silica shell, and grafted PDMS brushes. Candle soot, as an inexpensive material, has been able to provide hierarchical nano/micro-structures capable of trapping sunlight in order to improve light absorption resulting in high photo-thermal efficiency. Also, silica shell was used to strengthen the hierarchical candle soot, and grafted PDMS (as a low-surface-energy material) brushes endowed the surface with superhydrophobicity. When assessing the photo-thermal performance, it has been observed that upon illumination under 1 sun, the surface temperature could increase by 53 °C. This per se could indicate no ice formation at environmental temperature to as low as −50 °C, and the fabricated coating system was able to promptly melt the accumulated frost and ice in 300 s. The photo-thermal feature was even more promoted by superhydrophobicity. In this regard, the superhydrophobic candle soot surface could eliminate the melted water instantly, leaving a dry surface behind, which prevented the accumulation of melted water and thus decreased the heat loss. Second, the surface benefited from self-cleaning, therefore, dust and other contaminants could be wiped away by rain or melted water and the chance of blocking sunlight exposure was minimized. According to the authors' report, the designed photo-thermal icephobic surface, as shown in figure 2.12, exhibited outstanding potential and broad impacts thanks to its cheap base materials, simplicity, eco-friendliness, and high energy efficiency.

Wu *et al* [101] developed hierarchically structured PDMS/reduced graphene oxide (rGO) films that demonstrate excellent anti-icing and de-icing performance. Their study focused on creating crosslinked PDMS/rGO films that can effectively adsorb and convert solar energy for *in situ* de-icing, facilitated by the incorporation of dispersed rGO sheets within the PDMS networks. An diagram of the processes undertaken is provided in figure 2.13(a). The researchers employed two templating methods: an emulsion template to produce porous PDMS/rGO (EPG) films and a sugar/salt template for PDMS/rGO (SPG) films. They further utilized a one-pot dual-template technique to fabricate hierarchically structured PDMS/rGO (HPG) films. The resulting multi-hierarchical-structured surfaces exhibited efficient broad-band solar light absorption and high energy transfer efficiency of 90.4%. This efficiency is attributed to the presence of pores with diameters in the hundreds of microns, as well as micro-scale wrinkles on the internal walls, which create additional traps and longer trajectories for incident light (as shown in figure 2.13(b)).

The findings revealed that films containing 0.3 wt% rGO maintained surface equilibrium temperatures above freezing, even at −40 °C, allowing for easy ice removal under 1 sun illumination. These films demonstrated high de-icing efficiencies across various simulated realistic conditions, highlighting their effectiveness in diverse geographical settings, where available solar illumination can vary significantly based on location, climate, and weather conditions.

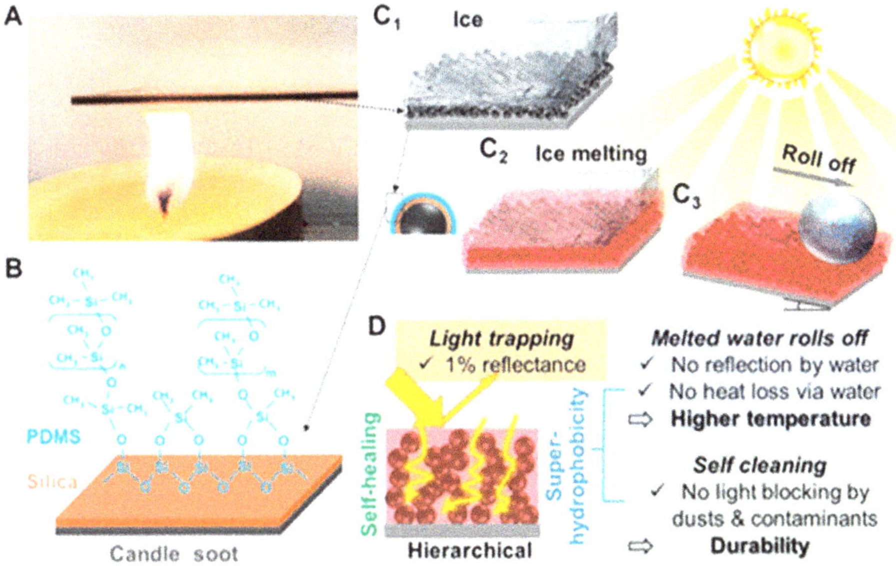

Figure 2.12. Schematic representation of the icephobic surface and fabrication process according to Wu *et al* [103]. (A) Candle soot is deposited onto a glass slide positioned above the candle flame. (B) Structures of the soot particles after being coated with a silica shell, followed by the application of PDMS brushes. (C) Mechanism of the proposed icephobic surface: Under sunlight exposure, the accumulated ice on the surface (C1) melts due to heat generated by the photo-thermal effect (C2), and the melted water then rolls off the inclined surface (C3). (D) The hierarchical structures enhance the surface with light-trapping properties and superhydrophobicity. (Adapted with permission from [103].)

The authors concluded that the advantages of this icephobic system—its broad applicability on a variety of substrates, excellent durability, and low cost—position it as a promising candidate for practical long-term applications in ice management.

In the study by Xie *et al* [104], researchers combined photo-thermal conversion with superhydrophobic materials to achieve exceptional anti-icing and photo-thermal de-icing performance. They developed a carbon-nanowire array on carbon cloth using an electrochemical deposition method, followed by a silanization treatment to create carbon-based materials with micro–nano hierarchical structures. These structures exhibited high-performance photo-thermal superhydrophobic anti-icing and de-icing capabilities, as illustrated in figure 2.14. The results demonstrated that the dense carbon-nanowire array on the surface of the micron carbon fiber provided excellent superhydrophobicity while enhancing light absorption due to the trapping effect induced by the hierarchical structure. The water contact angle measured on the prepared surface reached 155°, and under a solar power density of 1 sun, the temperature increased rapidly to 90 °C. Additionally, the air trapped within the micro–nano hierarchical structures effectively inhibited heat transfer during the icing process, resulting in a significant increase in the icing delay time from 241.22 to 3517.33 s under low-temperature conditions. These findings confirmed the remarkable photo-thermal conversion properties of the developed

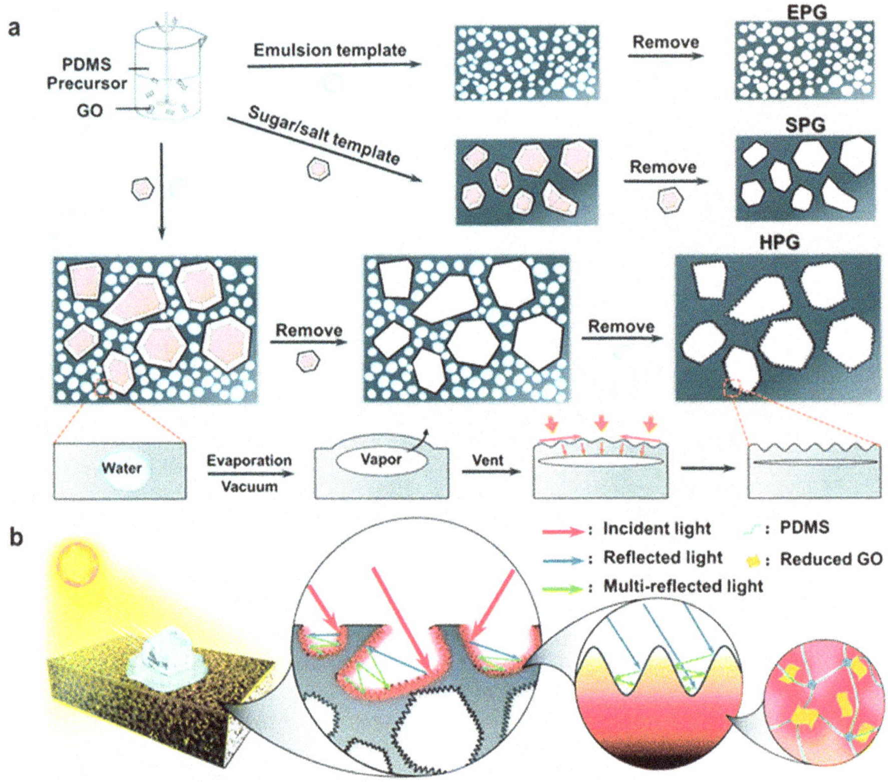

Figure 2.13. (a) Fabrication of porous emulsion-templated PDMS/rGO (EPG), sugar/salt-templated PDMS/rGO (SPG), and hierarchical-structured PDMS/rGO (HPG) films. (b) Schematic illustration of the solar anti-icing/de-icing performance of the HPG film. The dual-scale hierarchical macrostructures promote multiple internal reflections, thereby improving solar-to-heat conversion efficiency. (Adapted with permission from [101]. Copyright (2020) Royal Society of Chemistry.)

surfaces, showcasing their potential for photo-thermal de-icing. The authors propose this method as a promising approach for the next generation of de-icing and anti-icing materials, capitalizing on the benefits of low cost, durability, high-efficiency photo-thermal conversion, and excellent superhydrophobic performance.

Overall, de/anti-icing strategies based on photo-thermal materials and their effects are rather newly investigated and there are a lot of knowledge gaps and open-to-investigate details for future studies in order to verify the existing outcomes and optimizations.

2.4 Combining surface chemistry with additional mechanisms

2.4.1 Nanoscale carbon coatings

Various strategies have been explored utilizing carbon nanotubes (CNTs) and graphene as coatings, which exhibit superhydrophobic properties beneficial for

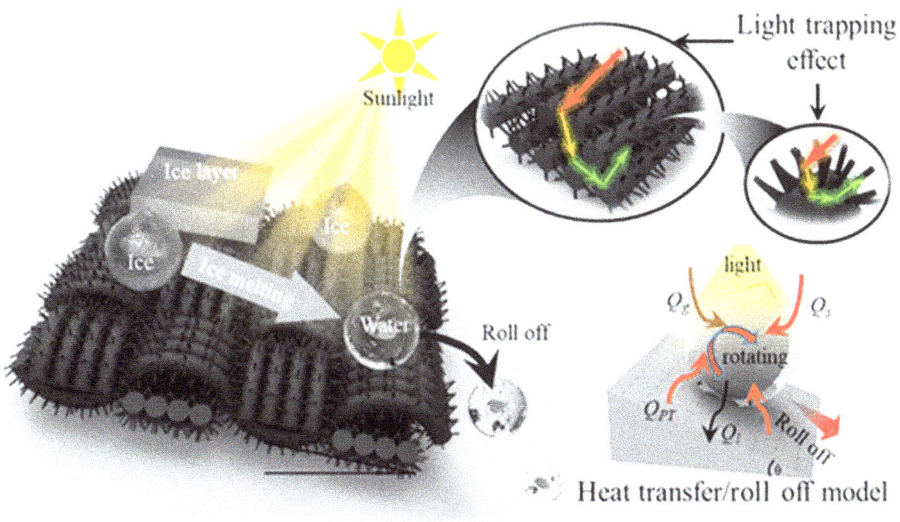

Figure 2.14. Carbon-based materials with micro–nano hierarchical structures with high-performance photothermal superhydrophobic anti-icing/de-icing features. (Adapted with permission from [104]. Copyright (2021) American Chemical Society.)

icephobic applications. These coatings not only provide mechanical durability but also enhance electrical properties [35]. The production of carbon nano-films is relatively simple and cost-effective. In a study by Yao *et al*, icephobic amorphous carbon nano-films were fabricated using ethanol-flame synthesis. These coatings featured multiple micro-textures, achieving a water contact angle of 155° and a sliding-off angle of 5° (figures 2.15(a) and (b)). The icephobic characteristics of the films were evaluated through various experiments, assessing parameters such as freezing time, ice adhesion strength, and dynamic sliding-off angles of droplets [105]. Similarly, Miller *et al* presented an approach involving fluorinated carbon films deposited via a deep reactive ion etching system. To investigate the effect of surface roughness, certain substrates were sandblasted to create textural features prior to film deposition. Their findings indicated a correlation between water contact angles and ice adhesion strength for samples with comparable roughness [108]. Additionally, a composite of graphene nanoribbon (GNR) stacks and epoxy was developed by filling epoxy with GNR stacks. This composite resulted in a conductive polymer suitable for use in de-icing systems, particularly in the aircraft industry [109]. Tour *et al* also employed GNRs in a de-icing heating layer composite, aimed at protecting radio-frequency equipment (figure 2.15(d)) [107]. Later, the same research group introduced an innovative strategy using perfluorododecylated graphene nanoribbons (FDO-GNR). They designed films that function for anti-icing purposes, which can switch to a slippery de-icing mode when lubricants are applied. These films are spray-coated, enabling large-scale manufacturing (figure 2.15(c)) [106].

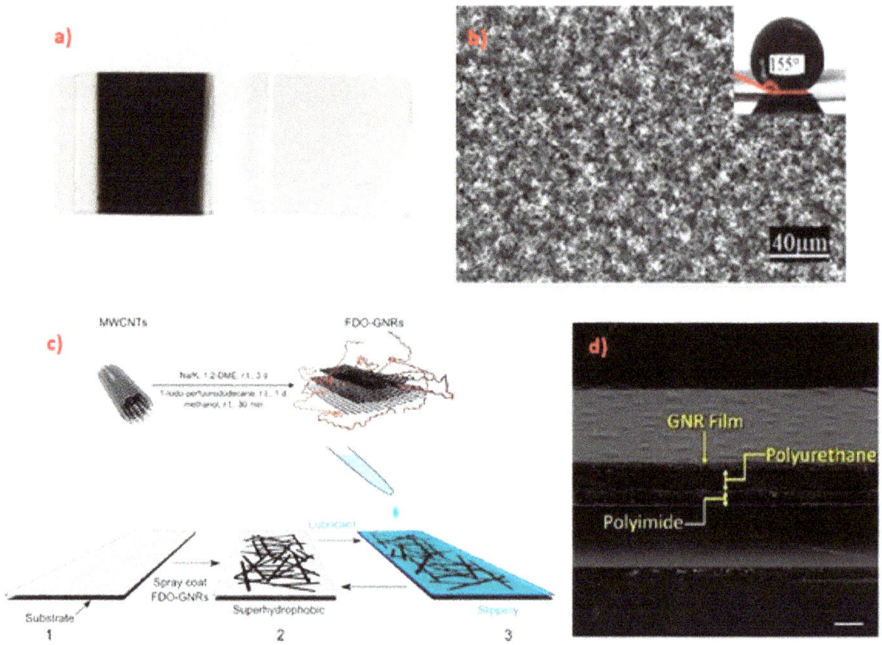

Figure 2.15. (a) Glass substrate with and without carbon nano-film. (b) SEM image of the carbon nano-film. (c) Representation of FDO-GNR synthesis and fabrication of FDO-GNR films. (d) SEM image of the GNR film. The system consists of a polyimide on the bottom, the clear-coat Dulpli-color automotive paint in the middle, and the GNR film as the upper layer. ((a) and (b) Adapted with permission from [105]. Copyright (2019) Elsevier. (c) Adapted with permission from [109]. Copyright (2016) American Chemical Society. (d) Adapted with permission from [108]. Copyright (2014) American Chemical Society.)

2.4.2 Anti-freeze proteins

Observing nature provides valuable insights for developing systems that maintain their properties in cold environments. Numerous living organisms, including plants and animals, exhibit remarkable resilience to cold temperatures, adapting to temperature fluctuations through complex mechanisms involving multiple variables. While these natural adaptations are intricate and challenging to replicate, they serve as a foundation for designing simpler systems to mitigate and delay ice formation.

One significant area of study is the existence of ice-binding proteins (IBPs), which bind to ice and modify its structure. Among these are antifreeze proteins (AFPs), which interact with ice crystals to alter their shape and control ice formation. These properties are crucial for the survival of many organisms in frigid habitats. Although the precise mechanisms by which AFPs operate are still under investigation, the most widely accepted theory is the adsorption-inhibition mechanism. This mechanism posits that AFPs inhibit ice growth by nano-confining water molecules, limiting their movement and preventing them from organizing into ice crystals. This process accounts for two important effects in delaying freezing: thermal hysteresis (TH), which lowers the freezing temperature below the melting point, and ice recrystallization inhibition (IRI), which helps regulate the size of ice crystals.

The physical properties and states of condensation can vary significantly with confinement dimensions, temperature, pressure, and interactions with surfaces. Research suggests that the coexistence of spatially segregated hydrophobic and hydrophilic domains is crucial for delaying water freezing [110]. Successful attempts have been made to create anti-icing surfaces on solid substrates such as aluminum [111] and glass [112] by incorporating AFPs, resulting in decreased supercooling before water freezes on the surface or reduced frost growth. However, challenges remain, including the difficulties associated with the extraction of AFPs, which can drive up material costs, and the limited long-term stability of these proteins, often only lasting a few days on glass [113].

While replicating such complex biological systems in practical applications poses significant challenges, leveraging the fundamental principles behind these mechanisms offers a promising avenue for innovation. Hydrogels, for instance, present interesting opportunities. The properties of hydrogels can be finely tuned through chemical modifications, allowing control over the interfacial regions, water content, and mobility. This tunability makes hydrogels a useful platform for designing bio-inspired strategies to control the organization of water molecules in a system.

In particular, organohydrogels—composed of classical hydrogels infused with an oil phase—offer a promising approach to integrate materials that exhibit anti-icing properties by adjusting the characteristics of interfacial water [114]. In these systems, solvents or ionic compounds can be incorporated between polymer networks, providing a wide range of modifications to inhibit the formation of crystalline water structures [115]. This approach utilizes the concepts developed by AFPs while avoiding the challenges associated with using the proteins themselves. Section 2.4.5 will delve into the chemistries and applications of hydrogels and ionogels in anti-icing strategies in more detail.

2.4.3 Wettable polymeric coatings

Many studies in the past have documented a correlation between high water contact angles and lower ice adhesion strengths [57, 116–118]. In 2010, Meuler *et al* connected these two variables via equation (2.6) which predicts the value of ice adhesion strength τ_{ice} from the receding water contact angle θ_{rec} of a substrate [117]. Hence, research in the field was for a long time focused on maximizing water-repellent properties in order to reach the lowest possible ice adhesion strength to a surface. As the limitations of SHS have become more evident, the field has started to accept that a high water contact angle does not guarantee icephobicity of a surface [119–122].

In some cases, smooth hydrophilic surfaces, such as silicon wafers, have shown lower ice adhesion strengths and higher freezing delay times compared to their hydrophobic and/or superhydrophobic counterparts in the same study [119, 120]. More notable is, however, the appearance of the first hydrated surfaces in 2014 [46, 48]. These wettable coatings with a continuous layer of water on top have proven to be able to reduce ice adhesion strength down to the scale of 10 kPa while simultaneously staying functional in record-breaking low temperatures. It has also been observed that polymers with flexible mobile chains on the surface of the material induce interfacial slippage that can reduce ice adhesion strength to below 1 kPa [10]. In the work of Golovin *et al*, several elastomers were embedded with mobile polymer chains or oils to study the effect of crosslink density and interfacial slippage on the ice adhesion and durability of the polymer substrates. Although most of the samples were hydrophobic fluorinated or silicone-based elastomers, remarkably, a series of hydrophilic oil embedded PU rubbers showed the lowest ice adhesion strength with values below 10 kPa:

$$\tau_{\text{ice}} = \beta\gamma(1 + \cos(\theta_{\text{rec}})). \tag{2.6}$$

In equation (2.6), β is an experimental constant and γ is the surface tension of water.

2.4.4 Non-wettable polymeric coatings

Major classes of non-wettable polymers reported for anti-icing applications are outlined in this section.

2.4.4.1 Fluorinated materials

Fluorinated polymers account for the most commonly and widely studied classes of anti-icing materials. The high content of $-CF_3$ groups in their structures, results in remarkably low surface energy, much lower than liquid water (<72 mN m^{-1}), and consequently fluorinated materials constitute candidates for anti-icing applications. The capability of poly(tetrafluoroethylene) to act as icephobic material has been systematically studied. The hydroxyl groups present on the oxide surfaces (i.e. Al_2O_3, SiO_2, TiO_2) or metals, are capable of reacting with functional groups of molecules, making them good substrates to obtain fluorinated materials. At the same time, the group orientation has attracted much attention. Yang *et al* reported that ice adhesion on a bare Al surface was measured at $\sim$1500 kPa. Both smooth and rough fluorinated surfaces were found to have very little ice adhesion, namely $\sim$100 and 200 kPa if compared to bare Al. The ice accretion is minimal on very high morphological non-homogeneities, such as micrometric porous superhydrophobic polyvinylidene fluoride (PVDF), due to the low adhesion and limited contact area between the water drops and coating.

Although morphology plays a critical role, good anti-icing properties have also been obtained by orienting fluorinated functionalizing groups. By covalently bonding the functionalized fluorinated molecules to the surface of the substrate, it is possible to form monolayers of self-assembling molecules (figure 2.16). In that way, by orienting fluorinated or other hydrophobic groups with low free energy groups in the desired direction, particularly towards the water interface, the water repellency effect is maximized and ice adhesion strength is decreased [123].

Recently, Chen *et al* fabricated an anti-icing coating capable of tolerating extreme environmental conditions, such as ultra-low or high temperatures, strong acids or alkalis and long-term exposure to sunlight. This coating is developed by

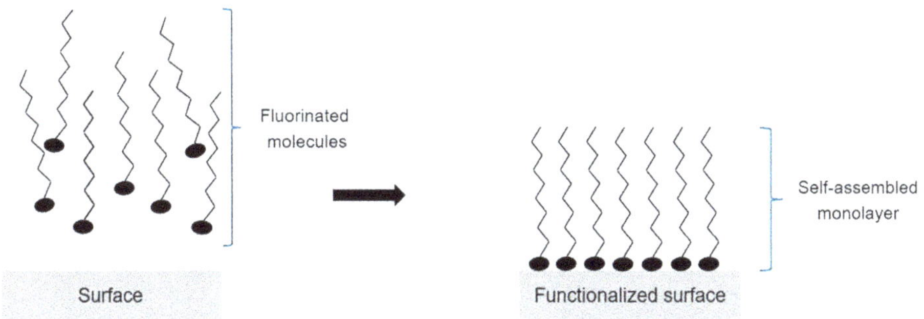

Figure 2.16. Self-assembly of functionalized molecules forming a monolayer on a solid surface.

incorporation of fluorinated amphiphilic copolymers (PFA-PVP-PDMS) and photo-thermal nanocarbon fibers (NFs), which transfuse the de-icing properties to the coating, into a PDMS matrix. It is very important that both the morphology and the de-icing performance are maintained after exposure to extreme environmental conditions, attributable to the tough coating materials and the intrinsic anti-icing properties of materials rather than artificial surface structure.

2.4.4.2 Non-fluorinated materials

Fluorine-free materials are increasingly recognized for their environmental benefits, leading to a surge of interest in recent years. Consequently, many researchers have focused on developing low-cost, eco-friendly preparation techniques for producing fluorine-free hydrophobic coatings, which hold significant potential for large-area applications, particularly in outdoor settings such as anti-icing surfaces.

Xie *et al* reported a facile spray-coating approach for the fabrication of non-fluorinated and durable anti-icing superhydrophobic coating by the combination of an aqueous suspension of hexadecyl polysiloxane, prepared by hydrolytic condensation of hexadecyltrimethoxysilane (HDTMS, 98%) and tetraethoxysilane (TEOS, 99.9%) onto the DE microparticles in water. These microparticles were modified by natural diatomaceous earth (DE@HD-POS) and a waterborne PU aqueous solution. The PU/DE@HD-POS coatings showed excellent superhydrophobicity with a high contact angle and a low sliding angle. The coatings demonstrated robust mechanical stability while exhibiting outstanding anti-icing performance, characterized by notably longer freezing times and reduced ice adhesion strength on diverse substrates substrates [124].

Xie *et al*, in another study, described non-fluorinated and durable photo-thermal superhydrophobic coatings based on natural attapulgite nanorods for efficient anti-icing and de-icing. A coating consisting of oxidative polymerized polypyrrole and hydrolytically condesated hexadecyl polysiloxane was formed on the attapulgite nanorods. The photo-thermal superhydrophobic coatings were deposited by spray-coating. The coatings demonstrated excellent superhydrophobicity due to the highly hierarchical micro-/nanostructure and the low surface energy. Moreover, the coatings showed good mechanical, chemical and thermal stability [125].

Wu *et al* presented transparent, fluorine-free, and antireflective hydrophobic coating created via the hydrolysis of TEOS combined with the hydrophobic modification of methyl MQ silicone resin (Me-MQ). Me-MQ is noted for its excellent mechanical properties, high-temperature resistance, outstanding durability, and remarkable waterproof characteristics, largely due to its long-chain spherical molecular structure. This coating also demonstrated exceptional self-cleaning capabilities, as contaminants were easily removed with water. Furthermore, it exhibited excellent durability, mechanical stability, and strong anti-ultraviolet performance [126].

2.4.4.3 Hydrophobic anti-wetting polymeric surfaces

Parts of this section have been reproduced with permission from [136]. Copyright (2022) Elsevier.

For instance, we know that lotus leaf represents remarkable anti-wetting self-cleaning properties due to the presence of hydrophobic epicuticular wax and isotropic dual (micro–nano) scale roughness papillae. Sun *et al* in their research used PDMS for its hydrophobic properties and used a nano-casting process to attain a structure similar to lotus leaf, transforming hydrophobic PDMS into a super-hydrophobic polymeric surface. The water contact angle of the PDMS before casting was 105° and after fabrication of the lotus leaf replication surface was 160° [127].

2.4.4.4 Hydrophilic anti-wetting polymeric surfaces

Parts of this section have been reproduced with permission from [136]. Copyright (2022) Elsevier.

Like hydrophobic surfaces, hydrophilic surfaces also exist naturally. Snail shell, fish scales, pitcher plants, shark skin, etc, are some common examples of such surfaces. Low drag to the surfaces and maintenance of characteristic self-cleaning properties are the result of natural hydrophilic surfaces.

Syafiq *et al* in their research prepared a smart coating using polypropylene glycol (PPG) in order to raise the surface energy of a glass substrate. Titanium dioxide was incorporated in the polymer, so as to improve surface roughness. The wettability analysis of this coating showed that the surface provided a water contact angle as low as 5° initially which suddenly reduces to 0° after 10 s of exposure. The major advantage of this type of coating is that it does not form any water streak marks over the surface of glass and provides high transparency and self-cleaning properties [128].

2.4.4.5 Flexible anti-wetting polymeric surfaces

Parts of this section have been reproduced with permission from [136]. Copyright (2022) Elsevier.

It is crucial for a coating to maintain its structural integrity and its functional properties, even upon subjection to a load or harsh environmental conditions. Polymeric materials are well known for their flexibility and adhesive properties. Polymeric materials are appropriate candidates for such coatings as their structural integrity remains intact and protects the surfaces from harsh environmental conditions. Adding flexibility characteristics to a coating material extends its horizons of application. Usually, the flexibility of a polymeric coating is determined by the amount of binder present in the mixture and is an inverse function of this amount. Flexibility does not mean that the hardness of the coating is sacrificed, a coating material can possess both hardness and flexibility.

Choi *et al* in their research prepared a transparent flexible hard coating using organic–inorganic molecular nanocomposite siloxane linked by epoxy polymers. This particular coating provides a surface with a plastic-like modulus along with glass-like scratch resistance, as well as remarkable optical transparency [129].

2.4.5 Hydrogel and ionogel based surfaces

Once ice has been formed on a substrate, it is necessary to reduce its adhesion to the surface so that it can be removed passively, say under the influence of gravity or by drag force. In this way, the detrimental effects of ice accretion on the substrates can be avoided. The various forces of attraction between ice and the substrate are the coulomb and the van der Waals forces [130]. Ice adhesion can be reduced by weakening the atomic interactions between the surface and water molecules. This can be done by increasing the hydrophobicity of the substrate by covering the substrate by low-energy molecules. Ice adhesion can also be reduced by introduction of an insulating layer between the ice and the substrate, thereby preventing direct contact between the two. An example of this is liquid infused surfaces (LISs) and SLIPSs [66, 131]. Moving further, the ice–substrate interface always has some voids which can act as crack initiators and create stress concentration which aids in the propagation of crack at the ice–substrate interface.

In addition to this, gels also offer great potential in anti-icing applications. There is a significant amount of liquid in the gels which forms an insulating layer between the ice and the substrate and hence reduces the ice adhesion. Second, the gels reduce the apparent surface elastic modulus and cause a stiffness mismatch between the ice and the substrate. This initiates the development of interfacial cracks and helps to lower the ice adhesion [132]. The physical properties of the gels are determined by the property of liquid contained in the gel. For instance, hydrogels display hydrophilic properties as they contain plenty of water. Ionogels on the other hand exhibit high electrical conductivity as they are composed of conductive ionic liquid. In the following sections, hydrogels and ionogels are discussed in more detail.

2.4.5.1 Hydrogels

Hydrogels consist of a large quantity of water. Since water freezes at around 0 °C at normal atmospheric pressure, the common hydrogels are not able to show a high resistance to freezing at low substrate temperatures. However, by introducing suitable additives and altering the polymer networks, the hydrogels can retain their softness and other gel characteristics even at sub-zero temperatures [133]. Such modified hydrogels can also be used for anti-icing applications since they act as a lubricant between the ice and the substrate and reduce the ice adhesion. It is known that salts help to reduce the freezing point of water and are used widely to melt the ice present on roads. Inspired by this, Li *et al* [134] developed electrolyte hydrogel (EH) surfaces by infusing salted water solution into a polyvinyl alcohol (PVA) hydrogel matrix. The resulting EH surfaces prevented ice and frost formation for an extremely long time and diminished the ice adhesion strength to an ultra-low value (Pascal level) at very low temperatures (–48.4 °C). The authors demonstrated by molecular dynamics simulations and experiments that such an extreme performance of EH surfaces is credited to the diffusion of ions to the interface between the EH and ice, as shown in the figure 2.17. The ice formed on the EH surfaces can be removed by gravity due to the ultra-low ice adhesion strength. Moreover, the salts used in the EH can be replenished by various sources such as seawater which make these EH surfaces a promising candidate in coastal infrastructure and shiphulls [134].

Inspired by antifreeze proteins, He *et al* prepared a PDMS grafted polyelectrolyte hydrogel which can be used in anti-icing applications [114]. The authors reported that by tuning the arrangement of the hydrophobic PDMS and the specificity of the ions, the properties of the interfacial water can be controlled. This controllability of the interfacial water properties provided multiple benefits. First it helped in inhibiting the ice nucleation as the ice nucleation temperature was reduced below −30 °C. Second it lowered the rate of ice propagation and finally the ice adhesion strength was also reduced (below 20 kPa).

2.4.5.2 Ionogels

Ionogels are composed of polymer networks and ionic liquids. Ionogels have been utilized widely for self-cleaning surfaces, ionic conductors and as electrolytes for batteries. Zhuo *et al* prepared an anti-icing ionogel consisting of crosslinked gelatin and 1-butyl-3-methylimidazolium bromide (BMImBr) [135]. The resultant ionogel surface inhibited ice nucleation and also controlled the ice growth rate. The authors showed that on such a surface, ice grows in an unconventional way from the water droplet–air interface towards the substrate, as shown in figure 2.18. This results in a spherical cap ice tip rather than a normal pointy cap ice. The authors further illustrated through molecular dynamics simulations as well as experiments that due to the inward ice growth and the presence of ionic liquid, an interfacial liquid layer is generated which helps to lower ice adhesion and prevent frost formation.

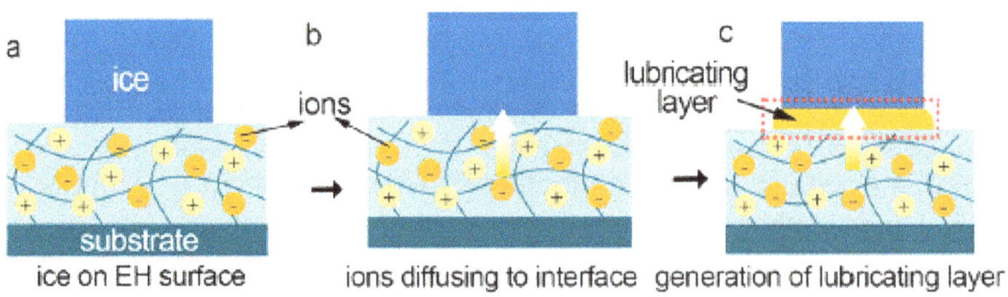

Figure 2.17. Self-generation of the lubricating layer on the EH surface. (Adapted from [134]. CC BY 4.0.)

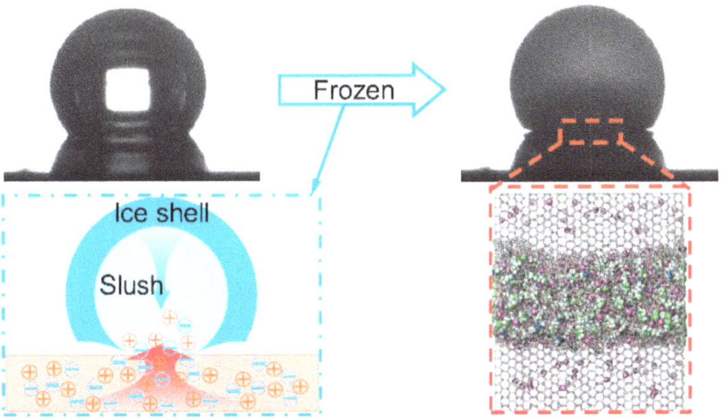

Figure 2.18. Schematic showing unconventional inward ice growth on an ionogel surface leading to a spherical ice cap and a concentrated ionic liquid aqueous interface. (Adapted with permission from [135]. Copyright (2020) American Chemical Society.)

The anti-icing potential of ionogels has been realized very recently and hence their application is currently limited. Also, some of the ionic liquids may leak into the environment and hence may be hazardous. Hence, a large effort is being made into the study of ionogels and their applications in icing considering their encouraging results in the limited studies carried out thus far.

2.5 Conclusions and perspectives

This review highlights the state-of-the-art surface chemistries relevant for anti-icing applications, detailing their advantages and limitations. Broadly, surface chemistries can be classified into fluorinated and fluorine-free categories, both of which can manifest in liquid-repellent forms that effectively reduce ice adhesion. These chemistries can be applied to surfaces in diverse forms, ranging from thin self-assembled monolayers to thicker polymers or nanocomposites.

Among the fluorinated polymers, various coatings based on materials such as polytetrafluoroethylene (PTFE) and polyvinylidene fluoride (PVDF) have been utilized. Some coatings incorporate perfluorinated end-groups ($–CF_3$), which contribute to achieving the lowest interfacial surface energy. Such coatings are recognized for their excellent chemical resistance and thermal stability. However, due to the environmental hazards associated with fluorinated compounds, there has been a significant push toward developing fluorine-free hydrophobic coatings that are more eco-friendly.

Fluorine-free polymers, particularly silicone materials such as soft polysiloxane, are commonly used in icephobic applications due to their low surface tension and energy, stemming from the inorganic Si–O bonds in their chemical structure. These silicone-based coatings can achieve ice adhesion strengths as low as 10 kPa, attributed to their softness, which induces a deformation incompatibility between the ice and the coated substrate.

Beyond polymer coatings and short-chain molecules, polyelectrolyte brushes have been prepared using surface grafting techniques. These brushes have demonstrated tunable heterogeneous ice nucleation properties by exchanging counterions, and the ice propagation rate can be adjusted by altering the hydration state of the polymer brush surfaces.

Another approach to reducing ice adhesion involves introducing a liquid layer between the ice and substrate. This liquid layer can be created using various hydrophilic, hydrophobic, or ionizable polymeric chains, with surfaces showing ice adhesion strengths below 30 kPa. SLIPSs exemplify this strategy, where a textured solid is impregnated with a lubricating liquid. SLIPSs minimize direct contact between the substrate and impacting droplets, leading to low contact angle hysteresis and allowing droplets to roll off easily, which reduces ice nucleation sites. However, SLIPSs face durability challenges, as the lubricant can be drained over time, resulting in contact line pinning of droplets.

Combining soft materials with liquid layers, gels have also emerged as potential anti-icing solutions. These gels, which can be hydrogels containing water or ionogels containing ionic liquids, provide insulation between the ice and substrate, thus reducing adhesion forces. While the application of gel-based coatings is currently limited, their promising results in preliminary studies warrant further exploration.

Recently, photo-thermal methods have been introduced for de-icing applications, utilizing Joule heating to eliminate ice on substrates. Despite its apparent simplicity, this method faces significant challenges due to high energy consumption for heat production. There is a pressing need for improved design and manufacturing of surfaces and coatings that can efficiently convert solar irradiation into heat and retain it to facilitate ice melting and removal. The integration of plasmonic effects and photo-thermal absorption with surface chemistries has shown promise in enhancing anti-icing characteristics.

Based on this review, several research opportunities for further exploration include:

- Despite tremendous progress with anti-icing features, the progress on combining durability with anti-icing characteristics needs improvement particularly when it comes to harsh applications such as those encountered in the aerospace industries and transport sectors in general.
- There is a clear need to develop halogen-free chemistries that prevent contamination of low surface tension materials, as these tend to jeopardize the anti-icing characteristics.
- There is a need to consider multi-functionality such as maintaining high or low thermal conductivity without comprising anti-icing characteristics. Molecular scale phonon-scattering and electronic transport needs to be investigated to address this.

Last, but perhaps not least, surface manufacturing techniques need to consider sustainability and scalability simultaneously in order to develop industry-ready anti-icing solutions.

References

[1] Gupta M C and Mulroney A 2020 Ice adhesion and anti-icing using microtextured surfaces *Ice Adhesion* (Beverley, MA: Scrivener) pp 389–415

[2] Grizen M and Tiwari M K 2020 Icephobic surfaces *Ice Adhesion* (Beverley, MA: Scrivener) pp 417–66

[3] Maitra T, Tiwari M K, Antonini C, Schoch P, Jung S, Eberle P and Poulikakos D 2014 On the nanoengineering of superhydrophobic and impalement resistant surface textures below the freezing temperature *Nano Lett.* **14** 172–82

[4] Maeda N 2021 Brief overview of ice nucleation *Molecules* **26** 392

[5] Li C, Liu Z, Goonetilleke E C and Huang X 2021 Temperature-dependent kinetic pathways of heterogeneous ice nucleation competing between classical and non-classical nucleation *Nat. Commun.* **12** 4954

[6] Ickes L, Welti A, Hoose C and Lohmann U 2015 Classical nucleation theory of homogeneous freezing of water: thermodynamic and kinetic parameters *Phys. Chem. Chem. Phys.* **17** 5514–37

[7] Niedermeier D, Shaw R A, Hartmann S, Wex H, Clauss T, Voigtländer J and Stratmann F 2011 Heterogeneous ice nucleation: exploring the transition from stochastic to singular freezing behavior *Atmos. Chem. Phys.* **11** 8767–75

[8] Eberle P, Tiwari M K, Maitra T and Poulikakos D 2014 Rational nanostructuring of surfaces for extraordinary icephobicity *Nanoscale* **6** 4874–81

[9] Fletcher H L 1962 *The physics of rainclouds*. By N H Fletcher. Cambridge: Cambridge University Press, 1962. pp x, 386; 99 Figures; 6 Plates. 65s *Q. J. R. Meteorolog. Soc.* **88** 559–9

[10] Golovin K, Kobaku S P R, Lee D H, DiLoreto E T, Mabry J M and Tuteja A 2016 Designing durable icephobic surfaces *Sci. Adv.* **2** e1501496

[11] Singh V, Zhang J, Chen J, Salzmann C G and Tiwari M K 2023 Precision covalent organic frameworks for surface nucleation control *Adv. Mater.* **35** 2302466

[12] Boinovich L, Emelyanenko A M, Korolev V V and Pashinin A S 2014 Effect of wettability on sessile drop freezing: when superhydrophobicity stimulates an extreme freezing delay *Langmuir* **30** 1659–68

[13] He Z, Zhuo Y, Zhang Z and He J 2021 Design of icephobic surfaces by lowering ice adhesion strength: a mini review *Coatings* **11** 1343

[14] Liu L, Wang S, Zeng X, Pi P and Wen X 2021 Dropwise condensation by nanoengineered surfaces: design, mechanism, and enhancing strategies *Adv. Mater. Interfaces* **8** 2101603

[15] Marmur A 2016 Non-wetting fundamentals *Non-wettable Surfaces: Theory, Preparation, and Applications* ed R H A Ras and A Marmur (London: The Royal Society of Chemistry)

[16] Li S, Huang J, Chen Z, Chen G and Lai Y 2017 A review on special wettability textiles: theoretical models, fabrication technologies and multifunctional applications *J. Mater. Chem.* A **5** 31–55

[17] Yong J, Chen F, Yang Q, Huo J and Hou X 2017 Superoleophobic surfaces *Chem. Soc. Rev.* **46** 4168–217

[18] Chen F, Wang Y, Tian Y, Zhang D, Song J, Crick C R, Carmalt C J, Parkin I P and Lu Y 2022 Robust and durable liquid-repellent surfaces *Chem. Soc. Rev.* **51** 8476–583

[19] He M, Wang J, Li H and Song Y 2011 Super-hydrophobic surfaces to condensed micro-droplets at temperatures below the freezing point retard ice/frost formation *Soft Matter* **7** 3993–4000

[20] Tourkine P, Le Merrer M and Quéré D 2009 Delayed freezing on water repellent materials *Langmuir* **25** 7214–6

[21] Jung S, Dorrestijn M, Raps D, Das A, Megaridis C M and Poulikakos D 2011 Are superhydrophobic surfaces best for icephobicity? *Langmuir* **27** 3059–66

[22] Lupi L and Molinero V 2014 Does hydrophilicity of carbon particles improve their ice nucleation ability? *J. Phys. Chem.* A **118** 7330–7

[23] Meuler A J, McKinley G H and Cohen R E 2010 Exploiting topographical texture to impart icephobicity *ACS Nano* **4** 7048–52

[24] Kulinich S A and Farzaneh M 2009 How wetting hysteresis influences ice adhesion strength on superhydrophobic surfaces *Langmuir* **25** 8854–6

[25] Nosonovsky M and Hejazi V 2012 Why superhydrophobic surfaces are not always icephobic *ACS Nano* **6** 8488–91

[26] Azimi Yancheshme A, Momen G and Jafari Aminabadi R 2020 Mechanisms of ice formation and propagation on superhydrophobic surfaces: a review *Adv. Colloid Interface Sci.* **279** 102155

[27] Gao L and McCarthy T J 2008 Teflon is hydrophilic. Comments on definitions of hydrophobic, shear versus tensile hydrophobicity, and wettability characterization *Langmuir* **24** 9183–8

[28] Antonini C, Villa F, Bernagozzi I, Amirfazli A and Marengo M 2013 Drop rebound after impact: the role of the receding contact angle *Langmuir* **29** 16045–50

[29] Zheng L, Li Z, Bourdo S, Khedir K R, Asar M P, Ryerson C C and Biris A S 2011 Exceptional superhydrophobicity and low velocity impact icephobicity of acetone-functionalized carbon nanotube films *Langmuir* **27** 9936–43

[30] Wang Y, Liu J, Li M, Wang Q and Chen Q 2016 The icephobicity comparison of polysiloxane modified hydrophobic and superhydrophobic surfaces under condensing environments *Appl. Surf. Sci.* **385** 472–80

[31] Kulinich S A and Farzaneh M 2009 Ice adhesion on super-hydrophobic surfaces *Appl. Surf. Sci.* **255** 8153–7

[32] Kreder M J, Alvarenga J, Kim P and Aizenberg J 2016 Design of anti-icing surfaces: smooth, textured or slippery? *Nat. Rev. Mater.* **1** 15003

[33] Antonini C, Innocenti M, Horn T, Marengo M and Amirfazli A 2011 Understanding the effect of superhydrophobic coatings on energy reduction in anti-icing systems *Cold Reg. Sci. Technol.* **67** 58–67

[34] Li W, Zhan Y and Yu S 2021 Applications of superhydrophobic coatings in anti-icing: theory, mechanisms, impact factors, challenges and perspectives *Prog. Org. Coat.* **152** 106117

[35] Huang X, Tepylo N, Pommier-Budinger V, Budinger M, Bonaccurso E, Villedieu P and Bennani L 2019 A survey of icephobic coatings and their potential use in a hybrid coating/active ice protection system for aerospace applications *Prog. Aerosp. Sci.* **105** 74–97

[36] Wu X, Zhao X, Ho J W C and Chen Z 2019 Design and durability study of environmental-friendly room-temperature processable icephobic coatings *Chem. Eng. J.* **355** 901–9

[37] 2007 *Physics and Chemistry of Ice* ed M Kuhs (London: Royal Society of Chemistry)

[38] Farhadi S, Farzaneh M and Kulinich S A 2011 Anti-icing performance of superhydrophobic surfaces *Appl. Surf. Sci.* **257** 6264–9

[39] Golovin K, Dhyani A, Thouless M D and Tuteja A 2019 Low-interfacial toughness materials for effective large-scale deicing *Science* **364** 371–5

[40] Reedy E D and Stavig M E 2020 Interfacial toughness: dependence on surface roughness and test temperature *Int. J. Fract.* **222** 1–12

[41] Chen N, Kim D H, Kovacik P, Sojoudi H, Wang M and Gleason K K 2016 Polymer thin films and surface modification by chemical vapor deposition: recent progress *Annu. Rev. Chem. Biomol. Eng.* **7** 373–93

[42] Zhuo Y, Xiao S, Amirfazli A, He J and Zhang Z 2021 Polysiloxane as icephobic materials—the past, present and the future *Chem. Eng. J.* **405** 127088

[43] He Z, Xiao S, Gao H, He J and Zhang Z 2017 Multiscale crack initiator promoted super-low ice adhesion surfaces *Soft Matter* **13** 6562–8

[44] Irajizad P, Al-Bayati A, Eslami B, Shafquat T, Nazari M, Jafari P, Kashyap V, Masoudi A, Araya D and Ghasemi H 2019 Stress-localized durable icephobic surfaces *Mater. Horiz.* **6** 758–66

[45] Wang C, Fuller T, Zhang W and Wynne K J 2014 Thickness dependence of ice removal stress for a polydimethylsiloxane nanocomposite: Sylgard 184 *Langmuir* **30** 12819–26

[46] Dou R, Chen J, Zhang Y, Wang X, Cui D, Song Y, Jiang L and Wang J 2014 Anti-icing coating with an aqueous lubricating layer *ACS Appl. Mater. Interfaces* **6** 6998–7003

[47] Koop T, Luo B, Tsias A and Peter T 2000 Water activity as the determinant for homogeneous ice nucleation in aqueous solutions *Nature* **406** 611–4

[48] Chen J, Luo Z, Fan Q, Lv J and Wang J 2014 Anti-ice coating inspired by ice skating *Small* **10** 4693–9

[49] Chen J, Dou R, Cui D, Zhang Q, Zhang Y, Xu F, Zhou X, Wang J, Song Y and Jiang L 2013 Robust prototypical anti-icing coatings with a self-lubricating liquid water layer between ice and substrate *ACS Appl. Mater. Interfaces* **5** 4026–30

[50] Heydari G, Tyrode E, Visnevskij C, Makuska R and Claesson P M 2016 Temperature-dependent deicing properties of electrostatically anchored branched brush layers of poly (ethylene oxide) *Langmuir* **32** 4194–202

[51] He Z, Xie W J, Liu Z, Liu G, Wang Z, Gao Y Q and Wang J 2016 Tuning ice nucleation with counterions on polyelectrolyte brush surfaces *Sci. Adv.* **2** e1600345

[52] Liu Z, He Z, Lv J, Jin Y, Wu S, Liu G, Zhou F and Wang J 2017 Ion-specific ice propagation behavior on polyelectrolyte brush surfaces *RSC Adv.* **7** 840–4

[53] Li C, Li X, Tao C, Ren L, Zhao Y, Bai S and Yuan X 2017 Amphiphilic antifogging/anti-icing coatings containing POSS-PDMAEMA-b-PSBMA *ACS Appl. Mater. Interfaces* **9** 22959–69

[54] Liang B, Zhang G, Zhong Z, Huang Y and Su Z 2019 Superhydrophilic anti-icing coatings based on polyzwitterion brushes *Langmuir* **35** 1294–301

[55] Landy M and Freiberger A 1967 Studies of ice adhesion: I. Adhesion of ice to plastics *J. Colloid Interface Sci.* **25** 231–44

[56] Christopher J and Wohl D H B 2019 *Contamination Mitigating Polymeric Coatings for Extreme Environments* **vol 284** (Cham: Springer Nature)

[57] Menini R and Farzaneh M 2011 Advanced icephobic coatings *J. Adhes. Sci. Technol.* **25** 971–92

[58] Mark J E 2007 *Physical Properties of Polymers Handbook* (Cham: Springer Nature)

[59] Li X, Zhao Y, Li H and Yuan X 2014 Preparation and icephobic properties of polymethyltrifluoropropylsiloxane–polyacrylate block copolymers *Appl. Surf. Sci.* **316** 222–31

[60] Murase H and Nanishi K 1985 On the relationship of thermodynamic and physical properties of polymers with ice adhesion *Ann. Glaciol.* **6** 146–9
[61] Murase H, Nanishi K, Kogure H, Fujibayashi T, Tamura K and Haruta N 1994 Interactions between heterogeneous surfaces of polymers and water *J. Appl. Polym. Sci.* **54** 2051–62
[62] Chernyy S, Järn M, Shimizu K, Swerin A, Pedersen S U, Daasbjerg K, Makkonen L, Claesson P and Iruthayaraj J 2014 Superhydrophilic polyelectrolyte brush layers with imparted anti-icing properties: effect of counter ions *ACS Appl. Mater. Interfaces* **6** 6487–96
[63] Jia Y, Yang Y, Cai X and Zhang H 2024 Recent developments in slippery liquid-infused porous surface coatings for biomedical applications *ACS Biomater. Sci. Eng.* **10** 3655–72
[64] Wang Y, Xue J, Wang Q, Chen Q and Ding J 2013 Verification of icephobic/anti-icing properties of a superhydrophobic surface *ACS Appl. Mater. Interfaces* **5** 3370–81
[65] Latthe S S, Sutar R S, Bhosale A K, Nagappan S, Ha C-S, Sadasivuni K K, Liu S and Xing R 2019 Recent developments in air-trapped superhydrophobic and liquid-infused slippery surfaces for anti-icing application *Prog. Org. Coat.* **137** 105373
[66] Wong T-S, Kang S H, Tang S K Y, Smythe E J, Hatton B D, Grinthal A and Aizenberg J 2011 Bioinspired self-repairing slippery surfaces with pressure-stable omniphobicity *Nature* **477** 443–7
[67] Bohn H F and Federle W 2004 Insect aquaplaning: nepenthes pitcher plants capture prey with the peristome, a fully wettable water-lubricated anisotropic surface *Proc. Natl Acad. Sci.* **101** 14138–43
[68] Huang C and Guo Z 2019 Fabrications and applications of slippery liquid-infused porous surfaces inspired from nature: a review *J. Bionic Eng.* **16** 769–93
[69] Li J, Ueda E, Paulssen D and Levkin P A 2019 Slippery lubricant-infused surfaces: properties and emerging applications *Adv. Funct. Mater.* **29** 1802317
[70] Heydarian S, Jafari R and Momen G 2021 Recent progress in the anti-icing performance of slippery liquid-infused surfaces *Prog. Org. Coat.* **151** 106096
[71] Sunny S, Vogel N, Howell C, Vu T L and Aizenberg J 2014 Lubricant-infused nanoparticulate coatings assembled by layer-by-layer deposition *Adv. Funct. Mater.* **24** 6658–67
[72] Zhu G H, Cho S-H, Zhang H, Zhao M and Zacharia N S 2018 Slippery liquid-infused porous surfaces (SLIPS) using layer-by-layer polyelectrolyte assembly in organic solvent *Langmuir* **34** 4722–31
[73] Juuti P *et al* 2017 Achieving a slippery, liquid-infused porous surface with anti-icing properties by direct deposition of flame synthesized aerosol nanoparticles on a thermally fragile substrate *Appl. Phys. Lett.* **110** 161603
[74] Subramanyam S B, Rykaczewski K and Varanasi K K 2013 Ice adhesion on lubricant-impregnated textured surfaces *Langmuir* **29** 13414–8
[75] Liu X, Gu H, Wang M, Du X, Gao B, Elbaz A, Sun L, Liao J, Xiao P and Gu Z 2018 3D printing of bioinspired liquid superrepellent structures *Adv. Mater.* **30** 1800103
[76] Villegas M, Cetinic Z, Shakeri A and Didar T F 2018 Fabricating smooth PDMS microfluidic channels from low-resolution 3D printed molds using an omniphobic lubricant-infused coating *Anal. Chim. Acta* **1000** 248–55
[77] Yong J, Huo J, Yang Q, Chen F, Fang Y, Wu X, Liu L, Lu X, Zhang J and Hou X 2018 Femtosecond laser direct writing of porous network microstructures for fabricating super-slippery surfaces with excellent liquid repellence and anti-cell proliferation *Adv. Mater. Interfaces* **5** 1701479

[78] Yong J, Chen F, Yang Q, Fang Y, Huo J, Zhang J and Hou X 2017 Nepenthes inspired design of self-repairing omniphobic slippery liquid infused porous surface (SLIPS) by femtosecond laser direct writing *Adv. Mater. Interfaces* **4** 1700552

[79] Liu M, Hou Y, Li J, Tie L and Guo Z 2018 Transparent slippery liquid-infused nanoparticulate coatings *Chem. Eng. J.* **337** 462–70

[80] Wang N, Xiong D, Lu Y, Pan S, Wang K, Deng Y and Shi Y 2016 Design and fabrication of the lyophobic slippery surface and its application in anti-icing *J. Phys. Chem.* C **120** 11054–9

[81] Irajizad P, Hasnain M, Farokhnia N, Sajadi S M and Ghasemi H 2016 Magnetic slippery extreme icephobic surfaces *Nat. Commun.* **7** 13395

[82] Irajizad P, Ray S, Farokhnia N, Hasnain M, Baldelli S and Ghasemi H 2017 Remote droplet manipulation on self-healing thermally activated magnetic slippery surfaces *Adv. Mater. Interfaces* **4** 1700009

[83] Liu X, Chen H, Zhao Z, Yan Y and Zhang D 2019 Slippery liquid-infused porous electric heating coating for anti-icing and de-icing applications *Surf. Coat. Technol.* **374** 889–96

[84] Singh V, Zhang J, Mandal P, Hou D, Papakonstantinou I and Tiwari M K 2024 Designing impact resistance and robustness into slippery lubricant infused porous surfaces *Adv. Mater.* **36** 2409818

[85] Cheng S, Guo P, Wang X, Che P, Han X, Jin R, Heng L and Jiang L 2022 Photothermal slippery surface showing rapid self-repairing and exceptional anti-icing/deicing property *Chem. Eng. J.* **431** 133411

[86] Laney S K, Michalska M, Li T, Ramirez F V, Portnoi M, Oh J, Thayne I G, Parkin I P, Tiwari M K and Papakonstantinou I 2021 Delayed lubricant depletion of slippery liquid infused porous surfaces using precision nanostructures *Langmuir* **37** 10071–8

[87] Villegas M, Zhang Y, Abu Jarad N, Soleymani L and Didar T F 2019 Liquid-infused surfaces: a review of theory, design, and applications *ACS Nano* **13** 8517–36

[88] Heydarian S, Maghsoudi K, Jafari R, Gauthier H and Momen G 2022 Fabrication of liquid-infused textured surfaces (LITS): the effect of surface textures on anti-icing properties and durability *Mater. Today Commun.* **32** 103935

[89] Mousavi S M, Sotoudeh F, Chun B, Lee B J, Karimi N and Faroughi S A 2024 The potential for anti-icing wing and aircraft applications of mixed-wettability surfaces—a comprehensive review *Cold Reg. Sci. Technol.* **217** 104042

[90] Liu C, Li Y, Lu C, Liu Y, Feng S and Liu Y 2020 Robust slippery liquid-infused porous network surfaces for enhanced anti-icing/deicing performance *ACS Appl. Mater. Interfaces* **12** 25471–7

[91] Shamshiri M, Jafari R and Momen G 2022 An intelligent icephobic coating based on encapsulated phase change materials (PCM) *Colloids Surf.* A **655** 130157

[92] Yuan Y, Xiang H, Liu G and Liao R 2021 Fabrication of phase change microcapsules and their applications to anti-icing coating *Surf. Interfaces* **27** 101516

[93] Irfan Lone M and Jilte R 2021 A review on phase change materials for different applications *Mater. Today Proc.* **46** 10980–6

[94] Fokaides P A, Kylili A and Kalogirou S A 2015 Phase change materials (PCMs) integrated into transparent building elements: a review *Mater. Renew. Sustain. Energy* **4** 6

[95] Kandasamy R, Wang X-Q and Mujumdar A S 2007 Application of phase change materials in thermal management of electronics *Appl. Therm. Eng.* **27** 2822–32

[96] Bentz D P and Turpin R 2007 Potential applications of phase change materials in concrete technology *Cem. Concr. Compos.* **29** 527–32
[97] Azimi Yancheshme A, Allahdini A, Maghsoudi K, Jafari R and Momen G 2020 Potential anti-icing applications of encapsulated phase change material–embedded coatings; a review *J. Energy Storage* **31** 101638
[98] Do T, Ko Y G, Chun Y and Choi U S 2015 Encapsulation of phase change material with water-absorbable shell for thermal energy storage *ACS Sustain. Chem. Eng.* **3** 2874–81
[99] Yang D, Bao R, Clare A T, Choi K-S and Hou X 2024 Phase change surfaces with porous metallic structures for long-term anti/de-icing application *J. Colloid Interface Sci.* **660** 136–46
[100] Mitridis E *et al* 2018 Metasurfaces leveraging solar energy for icephobicity *ACS Nano* **12** 7009–17
[101] Wu C *et al* 2020 Highly efficient solar anti-icing/deicing via a hierarchical structured surface *Mater. Horiz.* **7** 2097–104
[102] Dash S, de Ruiter J and Varanasi K K 2018 Photothermal trap utilizing solar illumination for ice mitigation *Sci. Adv.* **4**
[103] Wu S *et al* 2020 Superhydrophobic photothermal icephobic surfaces based on candle soot *Proc. Natl Acad. Sci.* **117** 11240–6
[104] Xie Z *et al* 2021 Carbon-based photothermal superhydrophobic materials with hierarchical structure enhances the anti-icing and photothermal deicing properties *ACS Appl. Mater. Interfaces* **13** 48308–21
[105] Xu Y *et al* 2019 Icephobic behaviors of superhydrophobic amorphous carbon nano-films synthesized from a flame process *J. Colloid Interface Sci.* **552** 613–21
[106] Wang T *et al* 2016 Passive anti-icing and active deicing films *ACS Appl. Mater. Interfaces* **8** 14169–73
[107] Volman V *et al* 2013 Radio-frequency-transparent, electrically conductive graphene nanoribbon thin films as deicing heating layers *ACS Appl. Mater. Interfaces* **6** 298–304
[108] Zou M *et al* 2011 Effects of surface roughness and energy on ice adhesion strength *Appl. Surf. Sci.* **257** 3786–92
[109] Raji A-R *et al* 2016 Composites of graphene nanoribbon stacks and epoxy for joule heating and deicing of surfaces *ACS Appl. Mater. Interfaces* **8** 3551–6
[110] Salvati Manni L *et al* 2019 Soft biomimetic nanoconfinement promotes amorphous water over ice *Nat. Nanotechnol.* **14** 609–15
[111] Gwak Y *et al* 2015 Creating anti-icing surfaces via the direct immobilization of antifreeze proteins on aluminum *Sci. Rep.* **5** 12019
[112] Kasahara K, Waku T, Wilson P W, Tonooka T and Hagiwara Y 2020 The inhibition of icing and frosting on glass surfaces by the coating of polyethylene glycol and polypeptide mimicking antifreeze protein *Biomolecules* **10** 259
[113] Xu Y, Rong Q, Zhao T and Liu M 2020 Anti-Freezing multiphase gel materials: bioinspired design strategies and applications *Giant* **2** 100014
[114] He Z *et al* 2020 Bioinspired multifunctional anti-icing hydrogel *Matter* **2** 723–34
[115] Gao H *et al* 2017 Adaptive and freeze-tolerant heteronetwork organohydrogels with enhanced mechanical stability over a wide temperature range *Nat. Commun.* **8** 15911
[116] Koivuluoto H, Hartikainen E and Niemelä-Anttonen H 2020 Thermally sprayed coatings: novel surface engineering strategy towards icephobic solutions *Materials* **13** 1434

[117] Meuler A J *et al* 2010 Relationships between water wettability and ice adhesion *ACS Appl. Mater. Interfaces* **2** 3100–10
[118] Hejazi V, Sobolev K and Nosonovsky M 2013 From superhydrophobicity to icephobicity: forces and interaction analysis *Sci. Rep.* **3** 2194
[119] Chen J *et al* 2012 Superhydrophobic surfaces cannot reduce ice adhesion *Appl. Phys. Lett.* **101** 111603
[120] Jung S *et al* 2011 Are superhydrophobic surfaces best for icephobicity? *Langmuir* **27** 3059–66
[121] Kulinich S A, Farhadi S, Nose K and Du X W 2011 Superhydrophobic surfaces: are they really ice-repellent? *Langmuir* **27** 25–9
[122] Wu X, Silberschmidt V v, Hu Z-T and Chen Z 2019 When superhydrophobic coatings are icephobic: role of surface topology *Surf. Coat. Technol.* **358** 207–14
[123] Tagliaro I, Cerpelloni A, Nikiforidis V-M, Pillai R and Antonini C 2022 On the development of icephobic surfaces: bridging experiments and simulations *The Surface Wettability Effect on Phase Change* (Cham: Springer International) pp 235–72
[124] Xie H *et al* 2021 Waterborne, non-fluorinated and durable anti-icing superhydrophobic coatings based on diatomaceous earth *New J. Chem.* **45** 10409–17
[125] Xie H *et al* 2022 Non-fluorinated and durable photothermal superhydrophobic coatings based on attapulgite nanorods for efficient anti-icing and deicing *Chem. Eng. J.* **428** 132585
[126] Wu Y *et al* 2022 Nonfluorinated, transparent, and antireflective hydrophobic coating with self-cleaning function *Colloids Surf.* A **634** 127919
[127] Sun M *et al* 2005 Artificial lotus leaf by nanocasting *Langmuir* **21** 8978–81
[128] Syafiq A, Vengadaesvaran B, Pandey A K and Abd Rahim N 2018 Superhydrophilic smart coating for self-cleaning application on glass substrate *J. Nanomater.* **2018**
[129] Choi G M *et al* 2017 Flexible hard coating: glass-like wear resistant, yet plastic-like compliant, transparent protective coating for foldable displays *Adv. Mater.* **29** 1700205
[130] Xiao S, Skallerud B H, Wang F, Zhang Z and He J 2019 Enabling sequential rupture for lowering atomistic ice adhesion *Nanoscale* **11** 16262–9
[131] Kim P *et al* 2012 Liquid-infused nanostructured surfaces with extreme anti-ice and anti-frost performance *ACS Nano* **6** 6569–77
[132] Kevin G *et al* 2021 Designing durable icephobic surfaces *Sci. Adv.* **2** e1501496–e16
[133] Jian Y *et al* 2021 Biomimetic anti-freezing polymeric hydrogels: keeping soft-wet materials active in cold environments *Mater. Horiz.* **8** 351–69
[134] Li T *et al* 2020 Self-deicing electrolyte hydrogel surfaces with Pa-level ice adhesion and durable antifreezing/antifrost performance *ACS Appl. Mater. Interfaces* **12** 35572–8
[135] Zhuo Y, Xiao S, Håkonsen V, He J and Zhang Z 2020 Anti-icing ionogel surfaces: inhibiting ice nucleation, growth, and adhesion *ACS Mater. Lett.* **2** 616–23
[136] Paras and Kumar A 2022 Anti-wetting polymeric coatings *Encyclopedia of Materials: Plastics and Polymers* vol 2 (Amsterdam: Elsevier) pp 786–95

Chapter 3

Conceptual framework for ice adhesion investigation on surfaces

Luca Stendardo, Irene Tagliaro, Valérie Pommier-Budinger, Marc Budinger and Carlo Antonini

The development of icephobic surface coatings relies on accurate and reproducible characterization of their properties, including ice adhesion testing. However, a reliable method to quantify ice adhesion on surfaces is still lacking. In general, ice adhesion is quantified by the average shear strength, which is the ratio of the force required to remove the ice to the ice–substrate contact area. However, this measure does not take into account important aspects such as stress concentrations and fracture propagation mode. This ultimately results in discrepancies up to an order of magnitude in the measurements of ice adhesion from different laboratories, even on traditional surfaces such as aluminum or steel.

This chapter explains the different fracture mechanisms that can occur at the ice–substrate interface and provides an analysis of the horizontal push test, one of the most commonly used ice adhesion test systems. For each fracture mechanism, a specific evaluation method is developed combining experimental and numerical approaches, allowing a more accurate characterization of ice adhesion.

Ultimately, this work contributes to the standardization of ice adhesion testing, aiming for a robust, universally applicable evaluation of surface icephobicity.

3.1 Introduction

While efficient anti-icing coatings might significantly delay the formation of ice on exposed surfaces, in extreme weather conditions ice nucleation and freezing are likely to occur. De-icing systems are therefore essential to guarantee complete protection from the severe consequences of icing. A reduced ice adhesion on the affected surfaces is essential for an efficient de-icing process.

Ice adhesion is generally expressed in the literature in terms of the shear adhesion strength: $\tau_{\text{adh}} = F/A$, where F is the ice removal force and A is the area of the

doi:10.1088/978-0-7503-6009-8ch3

interface between ice and substrate [1]. Icephobic materials or coatings typically exhibit a shear ice adhesion strength below a threshold, often arbitrarily set at 100 kPa [2]. Traditional structural materials such as aluminum or steel show ice adhesion strengths of up to 1600 kPa [3], while natural forces, such as wind or gravity, can remove ice when its adhesion to the substrate is below 10 kPa [2, 4]. An icephobic material can therefore be considered effective if the ice adhesion strength is reduced by one to two orders of magnitude.

Over recent years, various studies have revealed the limitations of such a simple approach to measuring ice adhesion. Tests performed on identical set-ups and on the same substrates have demonstrated a systematic scatter of ±20% in the measurements of ice adhesion [5]. These discrepancies can reach up to an order of magnitude when comparing results from different test methods, as shown in various literature reviews [6, 7]. The difficulty of comparing ice adhesion results from different test typologies was demonstrated by a recent round-robin study involving fifteen test facilities with harmonized parameters [8]. Although the study included identical, centrally produced surface samples, the authors concluded that traditional methods are unlikely to provide a unified test standard for ice adhesion measurements.

The literature commonly describes the ice adhesion with the shear adhesion strength, $\tau_{\mathrm{adh}} = F/A$, regardless of the utilized test method, the size of the ice block, the ice type (impact ice or freezer ice), or the fracture mechanism [6, 7]. Measuring F/A (Pa) implicitly assumes that the fracture mechanism is controlled by critical stress, which is not always correct. While it is tempting to standardize ice adhesion measurements similarly to contact angle measurements for wetting properties, the F/A value can be misleading if the fracture mechanism is not understood.

The relationship between surface contact angle and ice adhesion (measured as F/A) is also not completely clear. Over the years there have been various attempts to correlate ice adhesion to wetting properties, such as the contact angle hysteresis or the receding contact angle [9–13]. While some promising trends could be identified (especially on smooth surfaces), superhydrophobic, rough, or soft surfaces do not exhibit a direct link between wetting and ice adhesion properties [14].

This chapter aims to discuss important aspects of ice adhesion testing. Particularly, the fracture mechanics at the ice–substrate interface is discussed and the derived model is applied to laboratory-scale static push tests, such as the horizontal shear test. By integrating experimental and numerical methods, this chapter proposes a conceptual framework for a thorough assessment of surface icephobicity. Additionally, this work seeks to highlight common errors and misconceptions in ice adhesion testing, emphasizing the importance of more reliable and reproducible tests for designing and fabricating coatings and materials that resist icing.

3.1.1 Relevant aspects in ice adhesion testing

This section analyses the physics of ice adhesion and discusses the most important physical parameters in ice adhesion testing.

3.1.1.1 Ice type

Rønneberg *et al* [5] investigated the effect of the ice type on the measured ice adhesion value. Three identical aluminum samples were covered with precipitation ice, in-cloud ice, and bulk water ice, respectively, and tested under constant environmental conditions on a centrifugal ice adhesion test system. The measured adhesion strengths were 0.78 ± 0.10 MPa for precipitation ice, 0.53 ± 0.12 MPa for in-cloud ice, and 0.28 ± 0.08 MPa for bulk water ice, correlating inversely with the ice density. Despite identical test conditions, relatively high standard deviations were observed, ranging from 15% to 30% of the mean value. Similarly high scatter was observed in other studies [7, 15].

3.1.1.2 Temperature

A general trend observed in the literature is that ice adhesion increases with decreasing substrate temperature [7, 15–17]. This phenomenon is typically attributed to the presence of the so-called 'quasi-liquid layer', a liquid film of water on the surface of the ice, which can be observed even at temperatures below the freezing point [18]. The nanoscale thickness of the liquid layer at the interface decreases with decreasing temperature [1], resulting therefore in an increase in ice adhesion. The presence of the liquid layer on the ice surface has been observed down to temperatures of −25 °C in stationary conditions [19, 20].

At lower temperatures, cohesive failure within the bulk ice, rather than adhesive failure at the interface, becomes more prevalent due to the reduction of the thickness of the quasi-liquid water layer [21]. As the temperature decreases, ice becomes more brittle, and its bulk mechanical properties can affect the ice adhesion results. With decreasing substrate temperature the increase in ice adhesion due to the disappearing of the liquid layer is contrasted by temperature-induced cracking at the ice–substrate interface [1]. However, the increase in ice adhesion due to the vanishing liquid water layer is more significant than the cracking induced by low temperatures [1]. Makkonen's study also explores how the presence of the quasi-liquid layer at temperatures near the freezing point renders ice ductile, a phenomenon known as ice creeping [22]. In this ductile state, thermal interface stresses are diffused into the bulk of the ice without causing fracturing [1].

In ice adhesion testing, interfacial properties can be greatly affected by the substrate temperature. To minimize unwanted variations in the measured ice adhesion value, precise temperature control of the substrate during the ice adhesion test is essential.

3.1.1.3 Stress distributions

Stress concentrations in ice adhesion experiments complicate data comparison across various measurement methods and test systems. While it is acknowledged that the stress state at the interface is nonuniform in most ice adhesion tests, this factor is typically not accounted for in the majority of studies [7]. Therefore, the discrepancies—up to one order of magnitude—reported in the literature can be primarily attributed to the presence of stress concentrations [23].

The study of Work and Lian [7] highlighted the limited number of studies that analyse the interface stresses with finite element analysis (FEA). Lou *et al* [24] were among the first to analyse the stress concentrations at the ice–substrate interface. The ice adhesion strength, expressed as F/A, is corrected by a stress concentration factor θ, derived by static-stress numerical FEA.

A few studies have already attempted to exploit stress concentrations to reduce the ice adhesion strength to substrates, highlighting the importance of the topic. Hollow structure within the bulk of the substrate material can facilitate the detachment of the ice [25–27]. Likewise, the combination of soft and rigid materials can promote the formation of cracks, reducing ice adhesion [28–31].

Recently developed discontinuity-enhanced icephobic surfaces exploit the elasticity discontinuities at the edge between soft and rigid materials to reduce ice adhesion [28]. On soft-rigid composite surfaces, stress is concentrated along the discontinuity line instead of solely on the soft or rigid areas. This concentration of stress results in micro-cracking along the discontinuities, which subsequently leads to macro-cracks, initially on the rigid areas and then on the soft areas. Moreover, increasing the density of these elastic discontinuities further promotes the initiation of cracks, thereby reducing ice adhesion.

3.1.1.4 Ice–substrate interface area

The literature on ice adhesion testing often lacks clarity about the size of the ice samples used. This raises a pertinent question: does the size of the ice sample influence the measured adhesion strength?

From the perspective of linear elastic fracture mechanics, an ice layer of finite thickness covering a plate can be modeled as a 'bimaterial bonded joint' [32]. According to this theory, ice detachment can be governed by two contrasting fracture criteria: the stress or toughness condition [32–36]. The stress condition dominates small interfaces, where the full interface fails instantaneously without crack propagation [33, 37, 38] (see figure 3.1).

In larger interfaces, ice detachment is characterized by a gradual crack propagation and is controlled by toughness [34] (figure 3.2). As a consequence, the measurement of the adhesion strength as F/A is only meaningful in stress-dominated conditions, i.e. when the interface fails instantaneously. Otherwise, in a

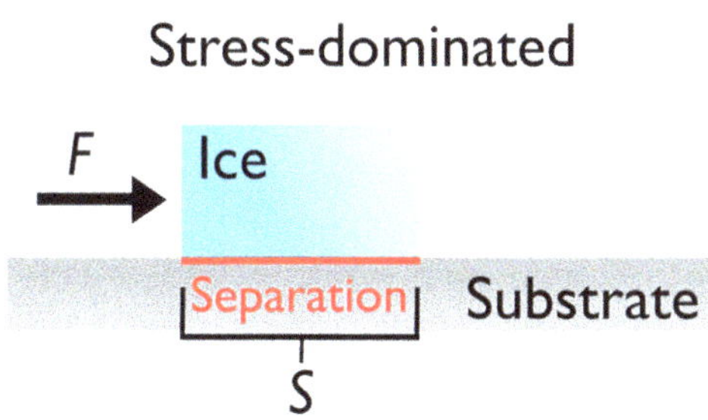

Figure 3.1. The stress-dominated interface fracture. Small interfaces fail by instantaneous detachment, without crack propagation. (Adapted from [39]. CC BY 4.0.)

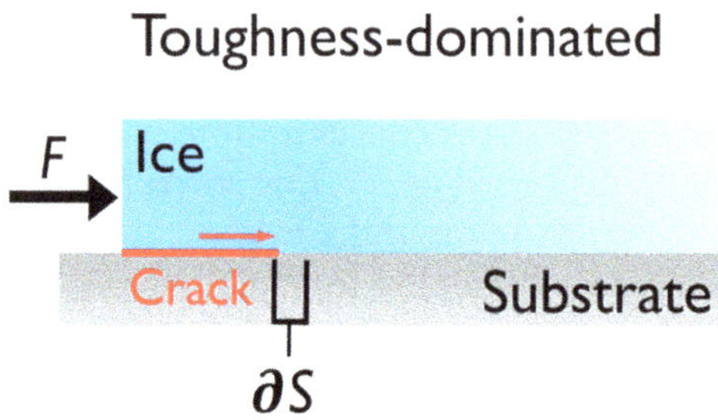

Figure 3.2. Large interfaces fail by toughness-dominated fracture. A gradual crack propagation is visible in this case. (Adopted from [39]. CC BY 4.0.)

toughness-dominated regime, fracture propagation is controlled by strain energy. Most of the studies, however, do not consider this difference in fracture mechanisms.

The effectiveness of low interfacial toughness (LIT) materials has been demonstrated by Golovin *et al* [33]. The study showed how the force required to remove extensive areas of ice can be drastically reduced by employing specific coatings that reduce the resistance to toughness-dominated fracture propagation. In their tests ice samples of varying length from 0.5 cm to 1 m were employed. The authors were able to show that after a critical length of the ice sample, the force required to detach the ice no longer increased. In other words, a critical length was found after which the ice removal force becomes independent of the interface area [33]. For ice dimensions below the critical length, the fracture is controlled by stress. In contrast, for dimensions above the critical length the fracture is toughness-dominated and the force F becomes independent of the interface area A.

3.1.1.5 Fracture mechanism

In an instantaneous, stress-dominated fracture, the critical stress threshold must be surpassed across the entire interface:

$$\tau \geqslant \tau_c, \forall\, x \in S, \tag{3.1}$$

or, in the case of a tensile loading,

$$\sigma \geqslant \sigma_c, \forall\, x \in S, \tag{3.2}$$

where x indicates a location of the interface S [32] (see figure 3.1). This fracture behavior is typically observed for short and thick ice blocks, where the stress is uniform across the interface [37].

The critical stress value that leads to detachment corresponds, therefore, to the minimum stress τ_{min} (see equations (3.1) and (3.2)). In common ice adhesion tests, however, the average adhesion strength $\tau_{ave} = F_{max}/A$ is calculated, where A indicates the entire ice–substrate interface. This measure cannot be considered a pure surface property, but it rather depends on the geometry of the ice adhesion test system, such as the size of the ice block and the force application point. A characterization of the pure ice adhesion strength τ_{min} can be achieved by correcting the average stress τ_{ave} with a stress concentration factor. This is discussed in detail in section 3.3.

In contrast, with high stress concentrations at the interface and with low ice thicknesses, the fracture is controlled by toughness. The propagation of the crack at the interface is controlled by strain energy and the toughness is a measure of the resistance by the interface to crack propagation. In other words, the toughness is the maximum potential strain energy that can be accumulated before the interfacial joint fails. A higher toughness implies that more strain energy is necessary to propagate the interfacial crack.

A toughness-dominated crack opening typically starts from a separation of finite area ΔS (see figure 3.2). In this case, ΔS is typically a fraction of S, meaning that the opening is only partial [32, 34]. A partial crack opening generally occurs when the 'coupled criterion' is met [32]:

$$\tau \geqslant \tau_c, \forall\, x \in \Delta S \tag{3.3}$$

$$\Delta W_m - \Delta W_p \geqslant G_c \Delta S, \tag{3.4}$$

where G_c stands for the interfacial toughness, ΔW_m is the external work performed by the force and supplied to the ice block during the fracture, and $-\Delta W_p = W_p(0) - W_p(S)$ is the variation in potential strain energy between the state before and after the initial opening.

Once this initial opening has occurred, the crack propagation is controlled only by interfacial toughness [32, 37]. The crack propagation mechanism can be explained by the energy criterion proposed by Griffith [34, 40]. According to this theory, an infinitesimal propagation of the crack ∂S (see figure 3.2) results in a change in potential energy ∂W_p. The differential energy release rate can therefore be defined as

$$\frac{\partial W_m}{\partial S} - \frac{\partial W_p}{\partial S} = G \geqslant G_c. \tag{3.5}$$

This condition needs to be satisfied for every incremental crack area ∂S for the crack to continue its propagation.

An important takeaway from the fracture mechanics theory is that the different fracture mechanisms must be properly identified to allow an accurate characterization of the ice adhesion. In a stress-dominated fracture mechanism, the stress controls the detachment and, therefore, ice adhesion can be expressed as a strength (in units of Pa). On the other hand, in a toughness-controlled fracture mechanism, energy (in units of J) governs the separation of the ice and the typically measured ice adhesion strength becomes meaningless.

An important aspect to consider when comparing ice adhesion results from substrates is that the interface is not always subjected to the same loading condition. The fractured surfaces can move differently relative to each other depending on the type of loading to which they are subjected. There are three principal crack opening modes [42] (see figure 3.3): (I) normal, (II) shearing or sliding, and (III) tearing. The

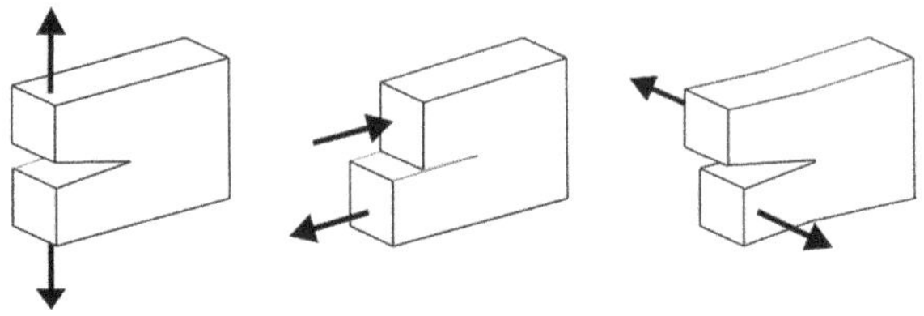

Figure 3.3. Three different crack opening modes: (I) normal, (II) in-plane shearing or sliding, and (III) out-of-plane shearing or tearing. (Adapted from [41]. CC BY 4.0.)

three modes are not mutually exclusive, thus experimentally measured values for toughness (G_c) are generally a combination of different modes (G_I, G_{II}, and G_{III}). Numerical analysis can help to quantify the relative contributions and identify the dominating opening mode.

3.2 Experimental and numerical methods

3.2.1 Experimental set-up

The horizontal shear test used for the analysis was designed in-house and has been described elsewhere [39, 41, 43]. It consists of an environmental chamber, a cooling system for samples, and an adhesion force measurement system consisting of an actuator and a force gauge (see figure 3.4). The environmental chamber is filled with nitrogen gas to prevent frost formation on the cooled components and on the surface sample. A constant environment is created inside the chamber, with ambient temperature T_{amb}= 20 °C and relative humidity RH< 3%.

The sample cooling system is composed of a liquid-cooled heat sink, thermoelectric cells (or Pertier cells), and an aluminum plate that acts as a thermal equalizer and sample holder. The aluminum plate is in contact with the cold side of the thermoelectric cells and can be cooled to a surface temperature of approximately −30 °C. The sample temperature is controlled by a PID controller (TE-Technology TC-36-25-RS232), while the heat sink is cooled by a recirculating chiller (Thermo Fisher ThermoChill II). The thermoelectric cells are arranged side by side, to provide an effective cooling area of about 40 × 80 mm^2 on which the samples are placed.

The ice is formed on the substrates by pouring a few drops of deionized water into cylindrical nylon (PA.6) molds. The inner diameter of the molds ranges from 8 to 14 mm (see figure 3.5). After bringing the sample to freezing temperatures, a conditioning time of approximately 15–20 min is typically required to ensure complete freezing of the water contained in the mold.

To remove the ice, a force is applied to the outside of the mold containing the ice, effectively creating stress at the interface between ice and substrate. The force gauge (Mark-10 model M5-20) is pushed against the ice column by a linear actuator (Newport LTA-HL with Conex-CC controller). The load cell applies pressure

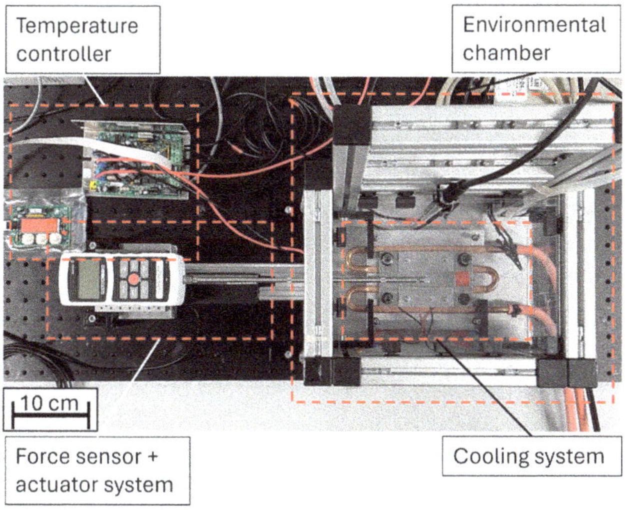

Figure 3.4. The horizontal shear test. The environmental chamber is filled with nitrogen gas to prevent condensation on the cooled components. A force gauge, together with a linear actuator, are employed to measure the ice adhesion force. The sample temperature is controlled by a PID controller. (Adapted from [53].)

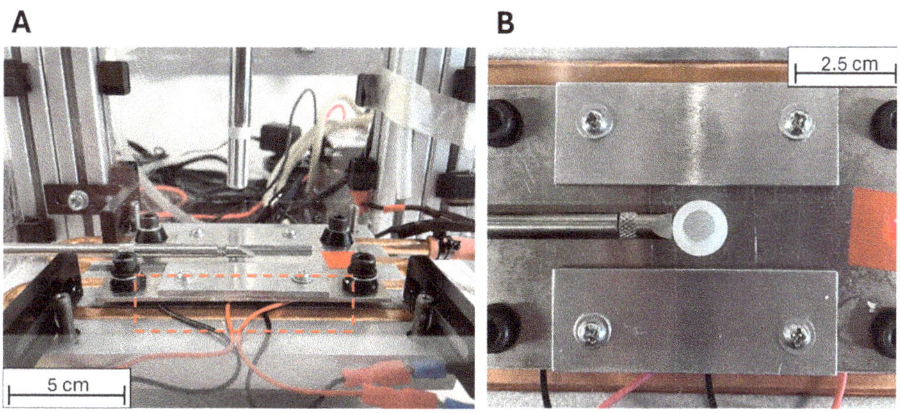

Figure 3.5. (A) The sample cooling system. The redo box indicates the location where the two thermoelectric cells are positioned, below the sample holder. (B) Nylon mold containing ice placed on top of the substrate to be investigated. (Adapted from [53].)

through a metallic rod that terminates in a wedge-shaped tip of 1 mm size (see figure 3.5). The sensor has a peak force of 100 N with a resolution of 0.01 N.

The linear actuator is operated at a velocity of 10 μm s^{-1}. Preliminary experiments have shown that in this regime, dynamic effects can be excluded and the quasi-static deformation of the ice under load can be recorded [39, 41]. A high-speed camera (PHOTRON NOVA FASTCAM S6) was used to classify the fracture mechanisms. Adequate magnification of the image is ensured by a Venus Laowa 100 mm f/2.8 2× Ultra Macro lens and a JJC Focus Extension Tube (20 mm).

3.2.2 Numerical model

The numerical FEA model simulates the ice adhesion experiment in its three main components: the substrate, the ice column, and the nylon mold (see figure 3.6(A)) [39, 41].

The model copies the dimensions and main properties of the experimental ice adhesion test. The force is modeled on the outside of the mold as a uniform pressure on an area of 1×1 mm^2 (see figure 3.6(B)), as this area corresponds to the contact area of the pushing rod head on the outside of the nylon mold in the experimental set-up. The contact between the ice and the nylon mold and between the ice and the substrate was set as 'bonded', whereas a 'frictionless' contact condition was imposed between the mold and the substrate. The selected material properties are summarized in table 3.1.

The numerical model has been employed both for the calculation of the stress states and the toughness at the interface. For evaluating the stress states, the model is used without simulating the actual detachment of the ice. In this case, the force is

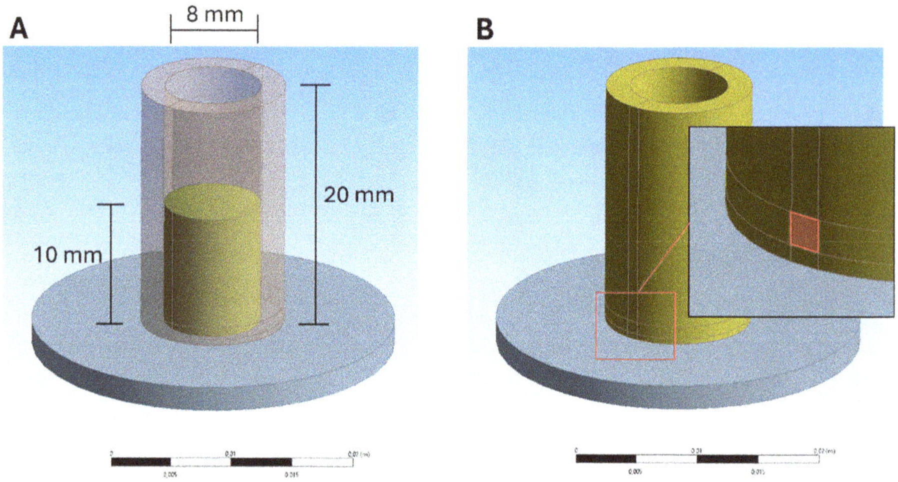

Figure 3.6. Schematic of the numerical model. The dimensions shown in (A) can be adapted to the corresponding experimental case. (B) The force is modeled as a homogeneous pressure over an area of 1×1 mm^2. (Adapted from [41]. CC BY 4.0.)

Table 3.1. Material parameters used in the numerical simulations. The properties of ice have been taken from [44] and correspond to 'freezer ice'.

Material	ρ (kg m^{-3})	E (GPa)	ν
Ice	897	9	0.31
Nylon	1140	1.5	0.39
Aluminum	2700	68	0.36

applied on the outside of the mold in the model, which then derives the static-stress distribution at the ice–substrate interface. The stresses can be evaluated at every node of the interface, allowing easier data analysis.

On the other hand, for the toughness measurements a crack is modeled at the interface and its length is varied throughout the simulation process from fully attached to fully detached. To achieve this, the 'ice' body was divided into two sub-bodies. One part is assumed to be separated from the substrate, while the other part is still attached (see figure 3.7).

The toughness is then calculated by applying a constant force on the outside of the mold and simultaneously moving the crack line along the interface. From the point of view of an energy balance, there are three contributions to the energetic state of the system: the potential energy due to the elastic deformation of the mold and the ice, the mechanical energy dissipated by the propagation of the crack, and the work performed by external forces [45, 46]. In preliminary experiments it was shown that for a similar configuration [39, 41], the crack propagation speed on rigid substrates is in the range of 3–5 $m\,s^{-1}$, while the force probe is operated at 10 $\mu m\ s^{-1}$. Therefore, the experimental system can be assumed to be in a quasi-static condition, and the external work performed by the force probe can be neglected. The terms ΔW_m and ∂W_m in equations (3.4) and (3.5) can therefore be removed. This implies that the potential strain energy is entirely converted into the propagation of the interface crack.

To calculate the toughness value, the experimentally required force to propagate the crack along the entire interface is measured. It is then assumed that for this complete propagation to take place, the differential strain energy release rate must

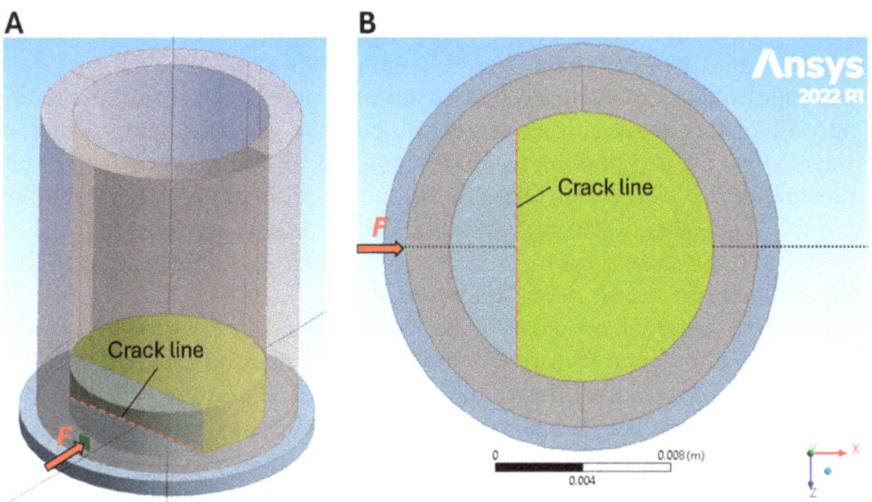

Figure 3.7. (A) The crack propagation is simulated by dividing the ice block into two sub-bodies. The body that is set as separated from the substrate appears in light green, while the bonded body appears in yellow. (B) Throughout the simulation, the crack line is advanced in the direction of the force application line. (Adapted from [41]. CC BY 4.0.)

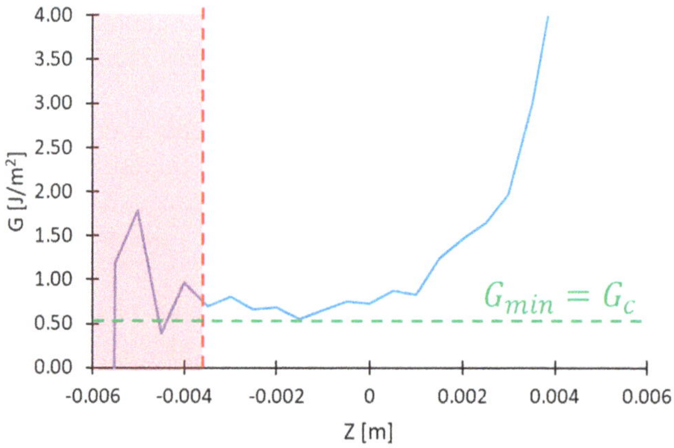

Figure 3.8. Interfacial toughness is determined by calculating the minimum energy release rate for a given external force applied to the mold. The Z-axis is the centerline coordinate of the ice–substrate interface, with the force applied at $Z = -0.006$ m. Results within the red area of the graph are excluded due to mesh dependency. (Adapted from [53].)

exceed the critical value (toughness) along the entire interface ($G > G_c$). The experimentally determined force is then used as an input for the numerical model, where the propagation of the crack with this constant force applied is simulated. For every incremental step of the crack, the strain energy difference is calculated. The interfacial toughness is then defined as the minimum of the calculated strain energy differences along the interface, $G_{min} = G_c$ (see figure 3.8). Additional details on the used methods can be found in [39, 41].

3.3 Results and discussion

Ice adhesion on a commercially available aluminum alloy, Al-6060, is measured multiple times under constant environmental and surface temperature conditions. The purpose of this analysis is to investigate how the geometric parameters of the test system affect the ice adhesion results (see figure 3.9), and to reveal the limitations of expressing the ice adhesion in terms of the commonly used average shear strength, $\tau_{ave} = F/A$. The tests were performed under constant ambient ($T_{amb} = 20$ °C, RH < 3%) and substrate ($T_{sub} = -10$ °C) conditions. A freezing time of 20 min was allowed before the tests were performed.

The characteristic dimensions of the horizontal shear test are shown in figure 3.9. The aluminum alloy sample was tested for its ice adhesion strength at different pushing heights h (1–4 mm) and with molds of different diameters D (8 and 14 mm). At least five repetitions for each combination of h and D were performed. The force F required to detach the ice from the substrate is measured and then divided by the contact area A. The average shear strength $\tau_{ave} = F/A$ for different combinations of h and D is shown in figure 3.10.

A change in the pushing height h evidently results in a different average shear strength, even under constant environmental conditions and on identical substrates. Therefore, τ_{ave} is not only a property of the substrate, but it also depends on the type

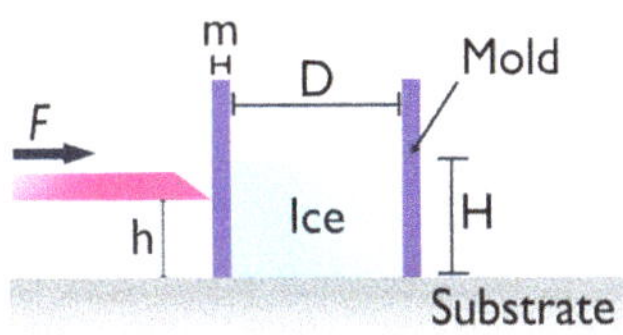

Figure 3.9. Horizontal push test with characteristic dimensions. The distance between force probe and substrate is indicated with h, the ice column height with H, the diameter of the mold with D, and the thickness of the mold with m. (Adapted from [39]. CC BY 4.0.)

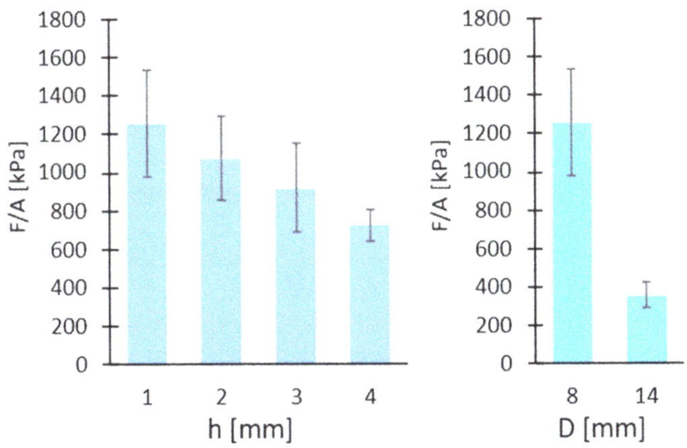

Figure 3.10. Comparison of the ice adhesion values obtained on aluminum alloy for different pushing heights h and mold diameters D. The adhesion strength is calculated by dividing the experimentally obtained removal force F by the interface area A. (Adapted from [41]. CC BY 4.0.)

of experiment performed and on the geometry of the test system. The comparison of the ice adhesion value obtained with ice columns of different diameters shows that, even if the expression $\tau_{\mathrm{ave}} = F/A$ already takes into account the interface area, the average shear strength is not independent of the interface area. In fact, the comparison of the adhesion values obtained with ice blocks of 8 and 14 mm diameter shows average strength values of 1257 ± 279 kPa and 358 ± 69 kPa, respectively, a four-fold difference.

This simple test has shown how relatively small changes in the test system configuration can result in differences in the measured adhesion value of up to 400%. A more in-depth analysis is therefore required to derive a more reliable ice adhesion measurement that takes into account the geometry of the test system and the occurrence of different fracture mechanisms.

3.3.1 Stress analysis

A general description of a physical function can be given in the following form [47]:

$$y = f(d_1, d_2, \ldots, p_1, p_2, \ldots), \tag{3.6}$$

where y stands for the physical quantity to be modeled, and d and p are the characteristic dimensions and the boundary conditions, respectively. The variables d and p can be considered independent, while y is the dependent variable.

The Buckingham Π-theorem [48, 49] allows one to write equation (3.6) as

$$\pi_0 = F(\pi_1, \pi_2, \dots, \pi_q), \tag{3.7}$$

where π_i indicate dimensionless variables. The number q can be calculated as $q = n - m$, with n standing for the number of variables and m for the number of physical units. The dependent variable y can be then written as

$$\pi_0 = y \prod d_i^{a_i} p_i^{b_i}. \tag{3.8}$$

The exponents a_i and b_i are such that π_0 is dimensionless. It follows that if the dimensionless parameters $\pi_i \in \{1, \dots, q\}$ of the function F are constant, then also π_0 is constant.

In the case of a stress-dominated interface fracture, the characteristic equation can be written as follows:

$$\frac{\tau}{F} = f(D, h, H), \tag{3.9}$$

where τ/F is considered a single variable, given the direct proportionality of τ and F. The geometry of the test system is given by D, h, and H, and considering that $q = n - m = 2$, equation (3.9) can be written as

$$\frac{\tau_{\min} D^2}{F} = f\left(\frac{h}{D}, \frac{H}{D}\right). \tag{3.10}$$

Multiplying the left-hand side by a dimensionless constant leads to

$$\frac{\tau_{\min} D^2 \pi}{4F} = f\left(\frac{h}{D}, \frac{H}{D}\right) \tag{3.11}$$

and ending with

$$\frac{\tau_{\text{ave}}}{\tau_{\min}} = f\left(\frac{h}{D}, \frac{H}{D}\right). \tag{3.12}$$

The left-hand side of equation (3.12) can be considered a factor that expresses the concentration of the stresses at the interface; the higher this factor, the more stress concentrations are present. Moreover, it puts the average shear stress ($\tau_{\text{ave}} = F/A$), in relation to the minimum interface shear stress ($\tau_{\min}$). This ratio of average to minimum shear stress is called shear stress intensity factor (SSIF) [39, 41]. According to the theory presented in section 3.1.1.5, the minimum shear stress needs to overcome the critical stress value to cause a stress-dominated fracture. Therefore, the minimum shear stress is the value of interest when characterizing ice adhesion in a stress-dominated fracture regime. Equation (3.12) also shows how this SSIF only depends on two geometrical parameters of the test system, h/D and H/D, which can be easily measured and adjusted when setting up the ice adhesion test.

A numerical FEA model (see section 3.2) is employed to compute the SSIF values for different combinations of h/D and H/D. To obtain the SSIF for each pair of geometric parameters, structural simulations are carried out: an arbitrary force is applied in the model on the outside of the mold, and the ratio of average to minimum shear stress, $\tau_{\text{ave}}/\tau_{\text{min}}$, is calculated at the interface mesh nodes (see figure 3.11).

The SSIF is schematically represented in figure 3.12(A). For improved clarity, the plot shows the distribution of the stresses along the center axis of the ice–substrate interface; the calculation of the SSIF is done, however, based on the entire interface. A 2D-map which shows the SSIF for different combinations of h/D and H/D is shown in figure 3.12(B). A cubic interpolation method was used to draw the contour lines based on single simulation points [39].

While doing ice adhesion experiments, the SSIF is determined based on the pushing height, h/D, and the ice column height, H/D. Once the average shear strength is known (determined as $\tau_{\text{ave}} = F/A$), the minimum shear stress is calculated as

$$\tau_{\text{min}} = \frac{\tau_{\text{ave}}}{\text{SSIF}}. \tag{3.13}$$

The contour plots described in this section are valid for cylindrical ice shapes, but the theory can also be extended to other ice sample shapes (see section 3.4). The dimensional analysis and the derived expressions are valid for high density ice (with a Young's modulus of E_{ice}> 4.5 GPa), and for rigid surfaces with a Young's modulus of E_{sub}> 35 GPa, such as metals, resins, and composite materials. Additional details on this analysis can be found in [39].

An experimental validation of the numerical results on Al-6060 aluminum alloy is shown in table 3.2. The experiments were conducted for different pushing heights

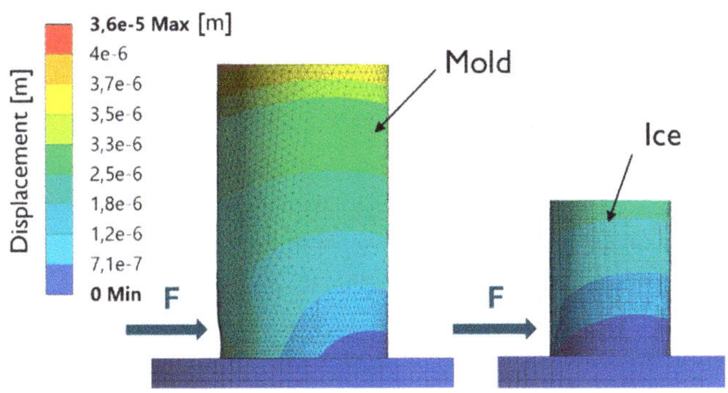

Figure 3.11. Representation of the structural simulation to obtain the SSIFs. The color scheme shows the deformation of the mold and of the ice column when a force is applied on the outside. In this figure, the displacements have been magnified 30 times. (Adapted from [41]. CC BY 4.0.)

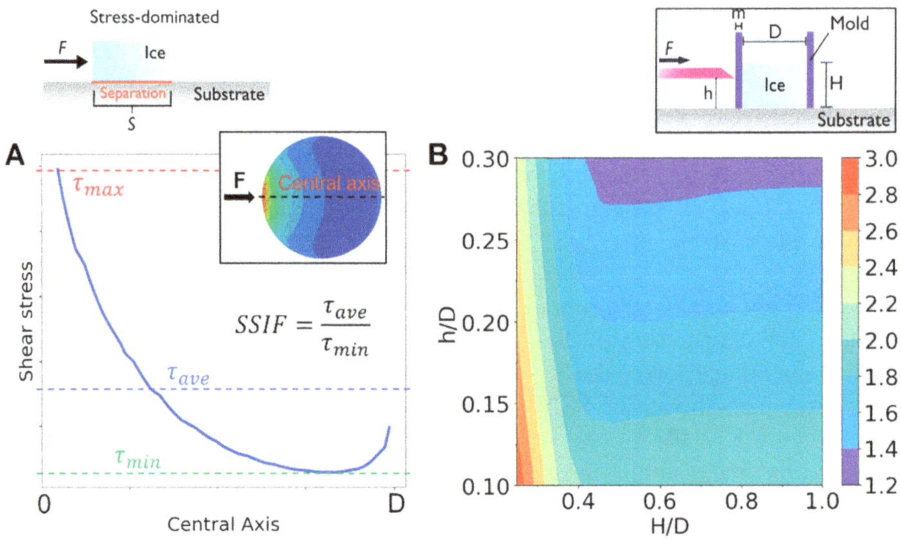

Figure 3.12. (A) Definition of the SSIF. An example of the shear stress distribution along the centerline of the ice–substrate interface is shown. (B) Contour map of the SSIF as a function of the test system parameters h/D and H/D. Inset: geometrical parameters of the horizontal shear test. (Adapted from [39]. CC BY 4.0.)

Table 3.2. Validation of the SSIF contour map. In each configuration, the minimum shear stress, τ_{min}, is determined by the numerical model. For τ_{ave} and τ_{min}, the standard deviation over the mean value ($\sigma(x)/\overline{x}$) was calculated. (Adapted from [39]. CC BY 4.0.)

	Ice diameter (mm)	h/D	H/D	Applied force (N)	SSIF = $\tau_{\text{ave}}/\tau_{\text{min}}$	τ_{ave} (kPa)	τ_{min} (kPa)
	8	0.125	0.525	55.02	1.9	1095	576
	8	0.25	0.625	48.56	1.5	966	644
	8	0.375	0.625	38.82	1.2	772	644
	10	0.1	0.4	91.36	1.9	1163	581
	12	0.083	0.46	107.34	1.8	949	527
$\sigma(x)/\overline{x}$						13.6%	7.5%

and ice column shapes, and each measurement was carried out in a fracture regime dominated by stress. Therefore, a constant minimum shear stress (τ_{min}) should be predicted by the model throughout the experiments. As was discussed before, this value depends only on the substrate and it is independent of the test system geometry.

The standard deviation of the minimum shear stress was reduced to ±7.5%, a considerable improvement compared to the standard deviation of the applied force and of the average shear stress, respectively ±38.7% and ±13.6%. With the help of the numerical model, the dependence of the ice adhesion value on the test system configuration could be reduced.

3.3.2 Toughness analysis

A similar approach to the stress analysis is employed here to derive a quantitative toughness measurement method. In the case of the horizontal shear test, the toughness can be expressed in the following way:

$$\frac{G}{F^2} = g(D, h, H, m, E_{\text{ice}}, E_{\text{mold}}). \tag{3.14}$$

Given the direct proportionality of G and F^2, G/F^2 is considered a single variable. The geometry of the test system is described by the parameters D, h, H, and m, while the Young's moduli of the ice and of the mold are given by E_{ice} and E_{mold}. The Buckingham Π-theorem allows to rewrite equation (3.14) by considering $q = n - m = 4$ as

$$\frac{G_c D^3}{F^2} E_{\text{ice}} = g\left(\frac{h}{D}, \frac{H}{D}, \frac{m}{D}, \frac{E_{\text{ice}}}{E_{\text{mold}}}\right). \tag{3.15}$$

Here, the interfacial toughness is expressed by G_c and F stands for the force required to propagate the interface crack along the entire interface. Additional details on the derivation of equation (3.15) can be found in [39].

The left-hand side of equation (3.15) can be defined as the 'toughness parameter' (TP), analogous to the SSIF in the stress-dominated fracture case. In fact, this TP depends on the test system geometry and can be used to determine the interfacial toughness experimentally with the horizontal shear test. However, to draw contour lines similar to the stress-dominated case, it is required to reduce the number of variables of equation (3.15). To achieve this, three distinct values of m/D are considered in this study, and a separate contour plot is presented for each value. Additionally, as only nylon is considered as a mold material, $E_{\text{ice}}/E_{\text{mold}}$ can be considered a constant of the test system and eliminated from the equation. Equation (3.15) then becomes

$$\text{TP} = \frac{G_c D^3}{F^2} E_{\text{ice}} = g\left(\frac{h}{D}, \frac{H}{D}\right). \tag{3.16}$$

Reducing the number of variables in the TP function limits its applicability. The calculated TP is valid solely for cylindrical ice blocks and nylon molds with specific dimensions. Furthermore, the model's validity is constrained by the substrate's elasticity, requiring a Young's modulus of at least 35 GPa. However, the underlying conceptual framework can be adapted to different ice block geometries and mold properties without compromising its fundamental validity [39].

For each combination of h/D, H/D, and m/D, the TPs according to equation (3.16) were calculated by the 3D numerical model. The results are represented in the contour plot in figure 3.13.

The TP plots represent a simple method to determine the interfacial toughness of an ice–substrate interface by measuring the required force F to propagate an

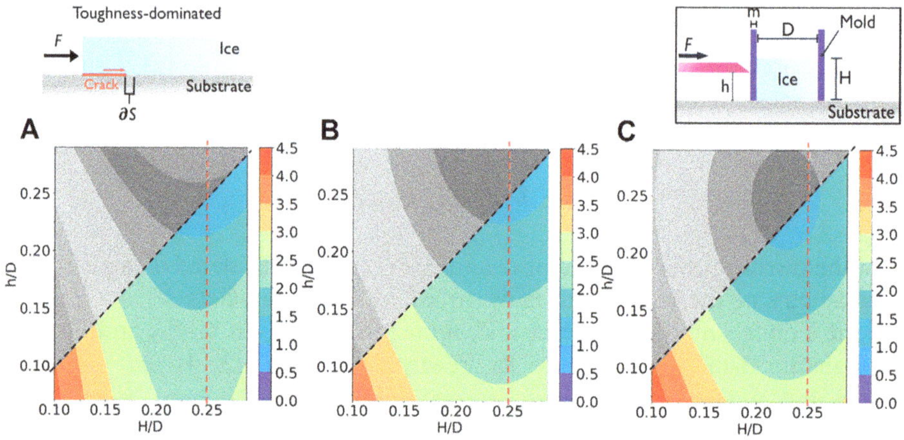

Figure 3.13. TPs were numerically calculated for three aspect ratios (m/D): 0.2, 0.167, and 0.143. These ratios correspond to a 2 mm thick mold with diameters D of 10, 12, and 14 mm, respectively. Regions where the normalized pushing height (h/D) exceeds the normalized ice column height (H/D) are shaded gray. The red dotted lines denote the boundary where H/D is equal to 0.25. (Adapted from [39]. CC BY 4.0.)

interface crack. The correct TP is selected based on the geometric parameters of the horizontal push test (h/D, H/D, and m/D), and the corresponding interfacial toughness G_c can be calculated by knowing the ice block's elastic modulus (E_{ice}).

On this particular test system, preliminary tests on Al-6060 aluminum alloy revealed that for ice column heights exceeding 0.25 times the diameter, the fracture mechanism is consistently stress-dominated [39]. To ensure accurate toughness measurements, the ice column height should be maintained below this threshold. This aligns with th eexisting literature, which indicates that a low ice thickness-to-length ratio promotes toughness-dominated detachment [33, 37, 50].

By adjusting the ice sample thickness, it is possible to control the fracture mechanism. This allows for the measurement of both adhesion strength and interfacial toughness using a single test system. Neglecting this variable in experimental design can lead to an uncontrolled fracture mechanism and significant measurement errors in ice adhesion studies.

The interfacial toughness model presented in this study is experimentally validated by a series of ice adhesion tests performed on aluminum (Al-6060), with the previously shown horizontal push test. All the adhesion tests were carried out by measuring the ice removal force F in a toughness-dominated regime. The results are presented in table 3.3.

The results show an independence of the ice removal force from the interface diameter, which is indicative of a toughness-dominated fracture regime. For each case, the interfacial toughness is calculated as described above, by converting the ice removal force F in an interfacial toughness value. The so-obtained G_c value for Al-6060 aluminum alloy is 0.46 ± 0.12 J m^{-2}, which is within the interfacial toughness values typically reported for aluminum substrates [36, 51, 52].

Table 3.3. Interfacial toughness values (G_c) were computed from experimental ice adhesion measurements on Al-6060 aluminum alloy. Numerically calculated TPs were employed to convert the crack propagation force into interfacial toughness values. The standard deviation relative to the mean toughness value ($\sigma(x)/\overline{x}$) is included in the table. (Adapted from [39]. CC BY 4.0.)

	Diameter (mm)	h/D	H/D	Applied force (N)	TP	Toughness (G_c) (J m^{-2})
	10	0.1	0.2	41.12	2.4	0.45
	10	0.1	0.25	53.98	2.1	0.68
	12	0.08	0.22	51.94	2.5	0.43
	14	0.07	0.19	59.74	2.8	0.40
	14	0.07	0.21	54.70	2.7	0.32
$\sigma(x)/\overline{x}$						26.4%

3.3.3 Conceptual framework

The model described in this chapter was developed for a horizontal shear test that uses ice columns of cylindrical shape, contained in nylon molds. The numerically derived contour plots therefore have a restricted validity to the chosen geometry. However, the approach shown in this chapter can be seen as a conceptual framework that can be generalized and applied to any sort of mechanical push test.

This framework can be applied to a different test system geometry by recalculating numerically the functions f and g (see equations (3.12) and (3.16)). New SSIFs must be defined, and the strain energy release rate must be analysed for the new test system geometry (as a function of h/D and H/D). The approach is summarized in figure 3.14.

In both cases, the experimentally determined ice removal force F is used in conjunction with geometric parameters to calculate the corresponding SSIF or TP. Subsequently, the minimum interfacial shear stress ($\tau_{\min}$) or interfacial toughness (G_c), is computed. The relationship between force F and $\tau_{\min}$ or G_c is defined by the functions f and g, which are derived from FEA.

3.4 Conclusions

The traditional approach of using average adhesion strength (F/A) to assess ice adhesion, even if useful for comparative analysis among different materials, does not fully capture the complexities of ice adhesion phenomena, particularly when considering the effects of stress concentrations and crack propagation. This work introduces a novel framework for the horizontal shear test. It distinguishes between two primary fracture mechanisms: stress-dominated and toughness-dominated. For each mechanism, a numerical correction factor is employed to minimize the influence of test system geometry.

The shear stress intensity factor (SSIF) allows for the calculation of the true adhesion strength (τ_c), while the toughness parameter (TP) relates the force required to propagate an interface crack to the total interfacial toughness (G_c). This

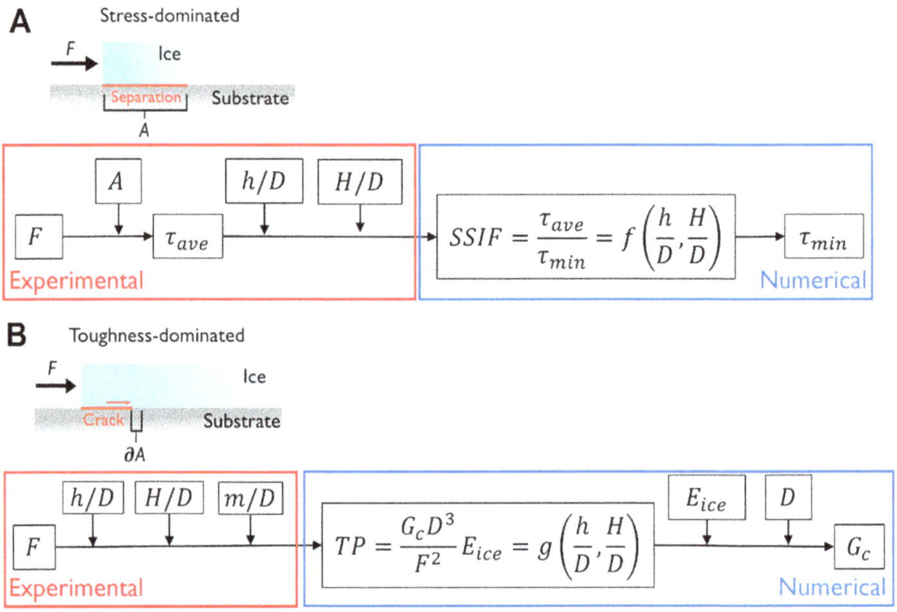

Figure 3.14. Conceptual framework for ice adhesion measurements. The approach combines experimental and numerical results and can be adapted to other ice sample geometries. (Adapted from [39]. CC BY 4.0.)

comprehensive approach provides a more accurate assessment of ice adhesion compared to traditional methods.

While the current methodology is specifically developed for cylindrical ice shapes and a fixed mold material, it provides a robust foundation for understanding ice adhesion. This novel framework can potentially be adapted to a wider range of experimental conditions, contributing to the standardization of ice adhesion measurements.

Acknowledgments

This project has received funding from the European Union's Horizon 2020 research and innovation program under the Marie Skłodowska-Curie grant agreement No. 956703 (SURFICE Smart surface design for efficient ice protection and control).

The authors are grateful to Giulia Gastaldo, Pablo F Ibáñez Ibáñez, and Catalina Ospina for their help and fruitful discussions that led to this book chapter.

References

[1] Makkonen L 2012 Ice adhesion—theory, measurements and countermeasures *J. Adhes. Sci. Technol.* **26** 413

[2] Hejazi V, Sobolev K and Nosonovsky M 2013 From superhydrophobicity to icephobicity: forces and interaction analysis *Sci. Rep.* **3** 2194

[3] Golovin K, Kobaku S P R, Lee D H, DiLoreto E T, Mabry J M and Tuteja A 2016 Designing durable icephobic surfaces *Sci. Adv.* **2** e1501496

[4] Zhuo Y, Xiao S, Amirfazli A, He J and Zhang Z 2021 Polysiloxane as icephobic materials—the past, present and the future *Chem. Eng. J.* **405** 127088
[5] Rønneberg S, Laforte C, Volat C, He J and Zhang Z 2019 The effect of ice type on ice adhesion *AIP Adv.* **9** 055304
[6] Rønneberg S, He J and Zhang Z 2020 The need for standards in low ice adhesion surface research: a critical review *J. Adhes. Sci. Technol.* **34** 319
[7] Work A and Lian Y 2018 A critical review of the measurement of ice adhesion to solid substrates *Prog. Aerosp. Sci.* **98** 1
[8] Rehfeld N *et al* 2024 Round-Robin study for ice adhesion tests *Aerospace* **11** 106
[9] Meuler A J, Smith J D, Varanasi K K, Mabry J M, McKinley G H and Cohen R E 2010 Relationships between water wettability and ice adhesion *ACS Appl. Mater. Interfaces* **2** 3100
[10] Meuler A J, McKinley G H and Cohen R E 2010 Exploiting topographical texture to impart icephobicity *ACS Nano* **4** 7048
[11] Susoff M, Siegmann K, Pfaffenroth C and Hirayama M 2013 Evaluation of icephobic coatings—screening of different coatings and influence of roughness *Appl. Surf. Sci.* **282** 870
[12] He Z, Vågenes E T, Delabahan C, He J and Zhang Z 2017 Room temperature characteristics of polymer-based low ice adhesion surfaces *Sci. Rep.* **7** 42181
[13] Cui W, Jiang Y, Mielonen K and Pakkanen T A 2019 The verification of icephobic performance on biomimetic superhydrophobic surfaces and the effect of wettability and surface energy *Appl. Surf. Sci.* **466** 503
[14] Sarkari N M *et al* 2024 Experimental debate on the overlooked fundamental concepts in surface wetting and topography vs ice adhesion strength relationships *J. Colloid Interface Sci.* **682** 825–48
[15] Scavuzzo R J and Chu M L 1987 Structural properties of impact ices accreted on aircraft structures *NASA Contractor Report* 179580 National Aeronautics and Space Administration
[16] Soltis J, Palacios J, Eden T and Wolfe D 2015 Evaluation of ice-adhesion strength on erosion-resistant materials *AIAA J.* **53** 1825
[17] Druez J, Nguyen D D and Lavoie Y 1986 Mechanical properties of atmospheric ice *Cold Reg. Sci. Technol.* **13** 67
[18] Rosenberg R 2005 Why is ice slippery? *Phys. Today* **58** 50
[19] Amirfazli A and Antonini C 2017 Fundamentals of anti-icing surfaces *Non-Wettable Surfaces: Theory, Preparation, and Applications* (London: Royal Society of Chemistry)
[20] Chen J, Dou R, Cui D, Zhang Q, Zhang Y, Xu F, Zhou X, Wang J, Song Y and Jiang L 2013 Robust prototypical anti-icing coatings with a self-lubricating liquid water layer between ice and substrate *ACS Appl. Mater. Interfaces* **5** 4026
[21] Guerin F, Laforte C, Farinas M-I and Perron J 2016 Analytical model based on experimental data of centrifuge ice adhesion tests with different substrates *Cold Reg. Sci. Technol.* **121** 93
[22] Weertman J 1983 Creep deformation of ice *Annu. Rev. Earth Planet. Sci.* **11** 215
[23] Schulz M and Sinapius M 2015 Evaluation of different ice adhesion tests for mechanical deicing systems *SAE Technical Paper* 2015–01–2135
[24] Lou D, Hammond D and Pervier M-L 2014 Investigation of the adhesive properties of the ice–aluminum interface *J. Aircr.* **51** 1051

[25] He Z, Zhuo Y, Wang F, He J and Zhang Z 2020 Design and preparation of icephobic PDMS-based coatings by introducing an aqueous lubricating layer and macro-crack initiators at the ice–substrate interface *Prog. Org. Coat.* **147** 105737
[26] He Z, Xiao S, Gao H, He J and Zhang Z 2017 Multiscale crack initiator promoted super-low ice adhesion surfaces *Soft Matter* **13** 6562
[27] He Z, Zhuo Y, Wang F, He J and Zhang Z 2019 Understanding the role of hollow sub-surface structures in reducing ice adhesion strength *Soft Matter* **15** 2905
[28] Ibáñez P F I, Stendardo L, Ospina C, Chaudhary R, Tagliaro I and Antonini C 2024 Discontinuity-enhanced icephobic surfaces for low ice adhesion *J. Colloid Interface Sci.* **679** 403–10
[29] Irajizad P, Al-Bayati A, Eslami B, Shafquat T, Nazari M, Jafari P, Kashyap V, Masoudi A, Araya D and Ghasemi H 2019 Stress-localized durable icephobic surfaces *Mater. Horiz.* **6** 758
[30] Wang J, Wu M, Liu J, Xu F, Hussain T, Scotchford C and Hou X 2021 Metallic skeleton promoted two-phase durable icephobic layers *J. Colloid Interface Sci.* **587** 47
[31] Jiang X, Lin Y, Xuan X, Zhuo Y, Wu J, He J, Du X, Zhang Z and Li T 2023 Stiffening surface lowers ice adhesion strength by stress concentration sites *Colloids Surf.* A **666** 131334
[32] Martin E, Vandellos T, Leguillon D and Carrère N 2016 Initiation of edge debonding: coupled criterion versus cohesive zone model *Int. J. Fract.* **199** 157
[33] Golovin K, Dhyani A, Thouless M D and Tuteja A 2019 Low-interfacial toughness materials for effective large-scale deicing *Science* **364** 371
[34] Leguillon D 2002 Strength or toughness? A criterion for crack onset at a notch *Eur. J. Mech.* A **21** 61
[35] Gastaldo G, Palanque V, Budinger M and Pommier-Budinger V 2022 Stress and energy release rate influence on ice shedding with resonant electro-mechanical de-icing systems *33rd Congress of the International Council of the Aeronautical Sciences (Stockholm, Sweden, 4–9 September)* hal–03823920
[36] Palanque V, Villeneuve E, Budinger M, Pommier-Budinger V and Momen G 2022 Cohesive strength and fracture toughness of atmospheric ice *Cold Reg. Sci. Technol.* **204** 103679
[37] Huré M, Olivier P and Garcia J 2022 Effect of Cassie–Baxter versus Wenzel states on ice adhesion: a fracture toughness approach *Cold Reg. Sci. Technol.* **194** 103440
[38] Yang Q and Cox B 2005 Cohesive models for damage evolution in laminated composites *Int. J. Fract.* **133** 107
[39] Stendardo L, Gastaldo G, Budinger M, Tagliaro I, Pommier-Budinger V and Antonini C 2024 Why the adhesion strength is not enough to assess ice adhesion on surfaces *Appl. Surf. Sci.* **672** 160740
[40] Zehnder A T 2013 Griffith theory of fracture *Encyclopedia of Tribology* (Boston, MA: Springer) p 1570
[41] Stendardo L, Gastaldo G, Budinger M, Pommier-Budinger V, Tagliaro I, Ibáñez-Ibáñez P F and Antonini C 2023 Reframing ice adhesion mechanisms on a solid surface *Appl. Surf. Sci.* **641** 158462
[42] Perez N 2017 Linear-elastic fracture mechanics *Fracture Mechanics* (Cham: Springer International) pp 79–130
[43] Stendardo L, Gastaldo G, Budinger M, Antonini C, Pommier-Budinger V and Ospina Patiño A C 2023 Dynamic and static test methods: quantifying the shear strength at the interface of iced substrates *Technical Paper* 2023-01–1451 SAE International, Vienna

[44] Macklin W C 1962 The density and structure of ice formed by accretion *Q. J. Royal Met. Soc.* **88** 30

[45] Rice J R 1968 A path independent integral and the approximate analysis of strain concentration by notches and cracks *J. Appl. Mech.* **35** 379

[46] The J-integral *Fracture Mechanics* https://fracturemechanics.org/j-integral.html

[47] Sanchez F, Budinger M and Hazyuk I 2017 Dimensional analysis and surrogate models for the thermal modeling of multiphysics systems *Appl. Therm. Eng.* **110** 758

[48] Buckingham E 1914 On physically similar systems; illustrations of the use of dimensional equations *Phys. Rev.* **4** 345

[49] Taylor E S 1974 *Dimensional Analysis for Engineers* (Oxford: Oxford University Press)

[50] Sills R B and Thouless M D 2013 The effect of cohesive-law parameters on mixed-mode fracture *Eng. Fract. Mech.* **109** 353

[51] Yeong Y H, Milionis A, Loth E, Sokhey J and Lambourne A 2015 Atmospheric ice adhesion on water-repellent coatings: wetting and surface topology effects *Langmuir* **31** 13107

[52] Pervier M A and Hammond D W 2019 Measurement of the fracture energy in mode I of atmospheric ice accreted on different materials using a blister test *Eng. Fract. Mech.* **214** 223

[53] Stendardo L 2025 *PhD Thesis* University of Milano—Bicocca

Chapter 4

Ice adhesion dynamics in the tensile mode

Ali Raza Shaikh and Ilia Roisman

This chapter explores the mechanics of tensile ice adhesion under dynamic conditions, emphasizing the interplay of environmental factors and substrate properties using experimental and numerical methods. Ice adhesion is a critical challenge across industries such as aviation, telecommunications, infrastructure, and energy, where ice accretion can reduce operational efficiency and increase the chances of hazards. An extensive experimental investigation was conducted to analyse tensile ice adhesion using aluminum, copper–zinc alloy, and polyamide substrates under varying ambient temperatures (from −10 °C to −20 °C). High-speed imaging (40 000 FPS) revealed that delamination initiates at edges and propagates across the ice–substrate interface, influenced by substrate properties and environmental conditions. Key findings include the dependence of ice adhesion strength on temperature, with lower temperatures enhancing adhesion due to increased intermolecular forces and hydrogen bonding. Substrate surface characteristics, such as roughness and energy, significantly influence adhesion strength, with metals generally exhibiting stronger adhesion compared to polymers. Surface area also influences ice adhesion due to the substrate geometry, including round patterns. Meanwhile, numerical simulations validated experimental observations, identifying stress concentrations at critical points of delamination. These insights help us to understand ice adhesion and design effective anti-icing coatings, although challenges remain regarding durability, environmental impact, and cost. Future research should explore high-speed imaging techniques, varying strain rates, and diverse environmental conditions to deepen understanding. This study provides a comprehensive framework for advancing anti-icing strategies, combining experimental and computational approaches to address real-world ice adhesion challenges in industries such as aviation, wind energy, infrastructures, and telecommunications.

doi:10.1088/978-0-7503-6009-8ch4

4.1 Introduction

Ice accretion is a widely recognized physical phenomenon that occurs under cold ambient conditions due to various mechanisms. These include the impact of supercooled droplets, ice crystals, and snowflakes, as well as the processes of condensation and vapor deposition [1–3]. The nature of ice accretion is significantly influenced by the interaction between ice and solid surfaces [4, 5]. One of the critical aspects of this interaction is the adhesion strength of ice to these substrates, which plays a vital role in determining the effectiveness of different ice removal techniques. Understanding the factors that affect ice adhesion is essential for developing efficient methods to manage and mitigate ice accumulation on various structures and systems [6, 7].

Ice formation and build-up on exposed surfaces are vital considerations that can greatly affect the operational efficiency of numerous systems and structures. This issue is relevant to many applications including aircraft [8, 9], helicopters [10, 11], ships, offshore oil platforms [12], power lines [13], telecommunications equipment [14], and wind turbines [15, 16]. Previous studies have shown that the surface chemistry, including surface roughness, surface energy, and water contact angle, greatly influences ice adhesion strength [17, 18]. Nonetheless, ice adhesion strength also depends heavily on other factors such as ambient temperature, wind speed, relative humidity, and ice type [19, 20]. Therefore, it is very important to have extensive experimental studies to understand ice adhesion and calculate its adhesion strength in various conditions.

Over the past two decades, extensive research has been dedicated to studying ice adhesion through various experimental set-ups. Most of the studies have employed shear or centrifugal testing mechanisms to evaluate the adhesive strength of ice on different surfaces [21–25]. Despite the valuable insights gained from these methods, there remains a significant need for more comprehensive investigations into tensile ice adhesion. Conducting tensile ice adhesion tests is crucial for a deeper understanding of the behavior and characteristics of ice adhesion under tensile forces. This knowledge is essential for developing more effective ice mitigation strategies in applications where tensile forces are predominant. In the past 20 years, various anti-icing strategies have been developed, encompassing icephobic and superhydrophobic coatings [26–30]. These innovations are designed to reduce ice accumulation and improve the efficiency and safety of various systems and structures exposed to freezing conditions. Despite the progress made, several key issues remain unresolved. The durability, environmental impact, and associated costs of these coatings remain significant issues that hinder their widespread adoption. Consequently, it is important to conduct comprehensive evaluations of the performance of different anti-icing coatings. Such assessments should utilize diverse ice adhesion test rigs, including those designed to measure tensile, shear, and centrifugal adhesion forces. This thorough testing is essential to ensure the coating's durability, performance, and suitability under harsh environments for commercial applications across various industries.

This book chapter presents the experimental work on ice adhesion in tensile mode using different substrates and sizes under various ambient temperature conditions.

This systematic approach provides and in-depth study of the mechanisms governing ice adhesion in tensile mode, providing valuable insights into the complex interactions among substrate properties, substrate sizes, and environmental conditions. The chapter also discusses the interfacial delamination phenomenon of ice on solid substrates by utilizing a high-speed imaging system, highlighting key mechanisms influencing ice adhesion dynamics. The observations provide foundational insights and set the stage for further investigation into the complexities of ice–substrate interactions.

4.2 Experimental set-up

The experimental set-up for investigating ice adhesion strength in tensile mode consists of four main components. These include the substrate-carrying part, a pulling mechanism with a force sensor, a high-speed video system, and a data acquisition system.

The schematic diagram of the designed experimental set-up is shown in figure 4.1. The pneumatic system is connected to a cylindrical substrate, which is frozen with a cylindrical ice sample at the interface. Air pressure goes in with the inlet valve and gets out with the outlet valve. The ice sample is formed in a transparent glass reservoir using a clear ice box.

The ice is formed at the solid substrate, and one side of the ice is connected to a pneumatic cylinder. The set-up includes a system for the movement of the pneumatic cylinder, which enables the separation of the substrate from the ice block. The pneumatic system is connected to a force sensor, and the plate holder carries a sample to make the ice adhesion test. The force orientation and ice adhesion sample mold are shown in figure 4.2.

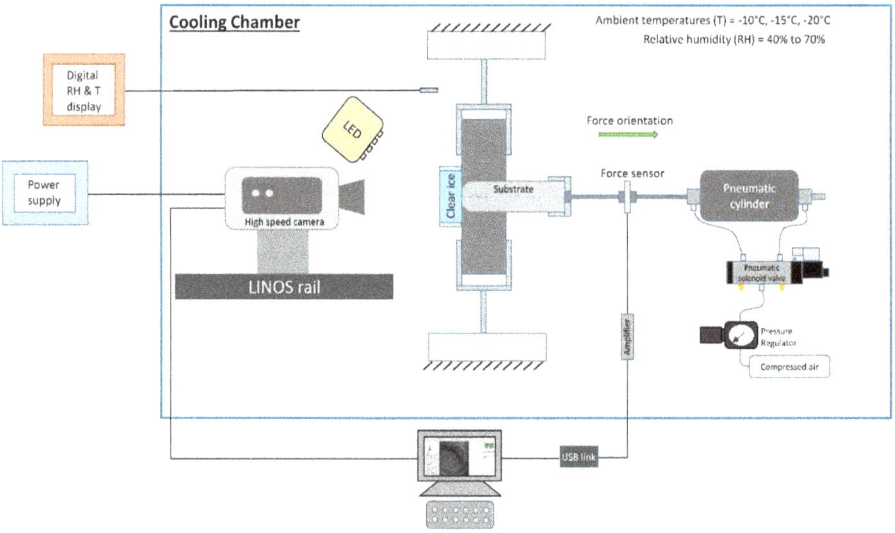

Figure 4.1. Schematic diagram of the experimental set-up. (Adapted from [63].)

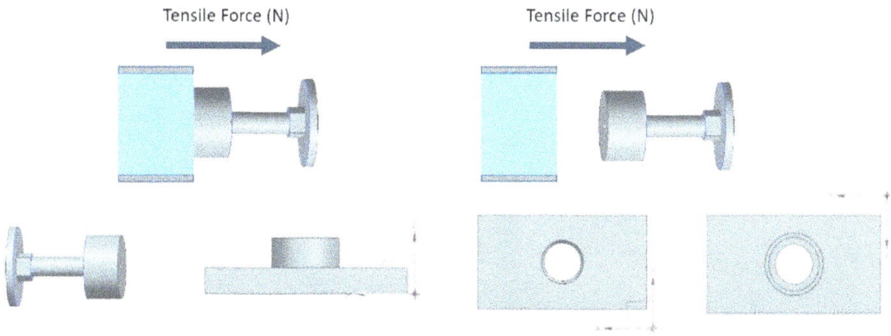

Figure 4.2. Orientation of pulling mechanism and ice adhesion sample molds. (Adapted from [63].)

The cylindrical substrate and water container together undergo a gradual freezing process within a clear ice box, maintained at temperatures ranging from −6 °C to −8 °C, spanning a duration of approximately 36–48 h. This deliberate freezing procedure is implemented to minimize the formation of air bubbles and impurities within the ice and obtain a very clear ice interface to observe through the ice at the interface between the ice and surface.

Transparent ice is achieved through the utilization of a custom-designed clear ice box featuring two distinct sections, namely the top and bottom compartments, along with a mechanism for segregating transparent ice from opaque ice. Test samples are positioned within the upper compartment, initiating a freezing process that progresses downward. The upper section is equipped with small apertures strategically positioned to direct air bubbles toward the lower section, thereby facilitating the formation of transparent ice incorporating the test samples.

To conduct the ice adhesion test, the sample is inserted into the holder mounted in the experimental set-up. It is then allowed to stabilize for 30–40 min to reach the desired ambient temperature. Subsequently, the tensile pull test is performed, and the desired experimental results are obtained.

Ice adhesion measurements are carried out in tensile mode, as shown in the ice adhesion set-up in figure 4.3, employing three distinct substrates: aluminum, copper–zinc alloy, and polyamide, at various ambient temperatures (−10 °C, −15 °C, −20 °C) within the cooling chamber after freezing the samples. Additionally, three different surface sizes are utilized for each substrate at −10 °C. Each substrate undergoes ice adhesion tests at an approximate strain rate of 30 mm s^{-1}, which is relatively quite fast because the high-speed camera system has a short time to record the test. The test sample is pulled through the ice using pressurized air via a pneumatic system operating at a constant pressure of two bars. The relative humidity exhibited variability, fluctuating between 40% and 70% across distinct temperature conditions.

All three samples are prepared through the machining process as shown in figure 4.4. The samples are mounted on a machine and have a constant speed of 920 mm rev^{-1}. The cutting tool feed rate is around 0.016 mm rev^{-1}.

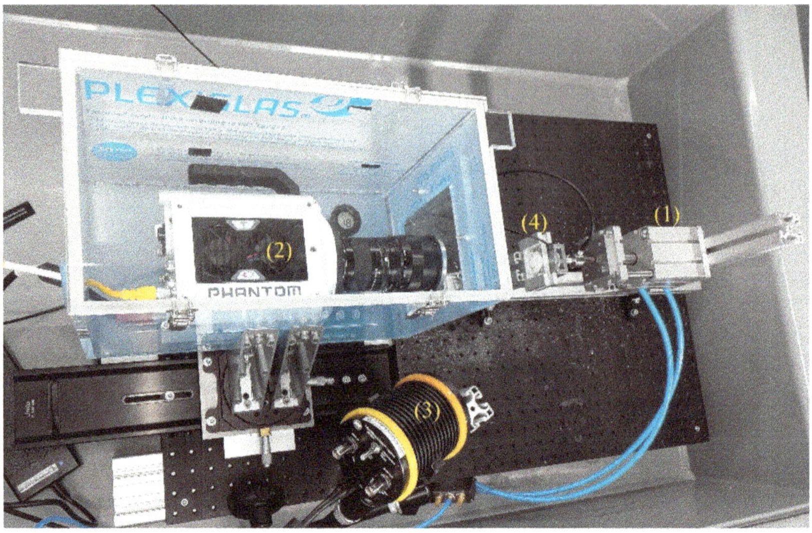

Figure 4.3. Ice adhesion experimental set-up inside the freezer containing (1) a pulling mechanism with a pneumatic air pressurized system, (2) a high-speed camera, (3) an LED, and (4) a sample holder. (Adapted from [63].)

Figure 4.4. The as-prepared samples via the machining process on Al, CuZn, and PA. (Adapted from [63].)

The tactile method was employed for the determination of surface roughness on Al, CuZn, and PA surfaces. The tactile method for determining surface roughness involved physical contact between a measuring instrument and the surface analysed. The contact profilometer was used, which has a diamond-tipped stylus. Stylus was dragged across the surface to measure its topographical variations. As the stylus moves over the peaks and valleys of the surface, vertical displacements are recorded, providing a detailed profile of the surface roughness. The recorded data included R_a (average, or arithmetic average of profile height deviations from the mean line), and R_z (maximum peak to valley height of the profile, within a single sampling length; average R_z value over assessment length) (table 4.1).

The measurements of the water contact angle were conducted at 25 °C using a Kruss DSA 100 goniometer, employing the Ellipse-tangent model on Al, CuZn, and

Table 4.1. Average roughness values of various surfaces. (Adapted from [63].)

Surfaces	R_a (μm)	R_z (μm)
Bare Al	0.41	2.37
Bare CuZn	0.98	6.43
Bare PA	1.53	9.19

Table 4.2. Static water contact angles of various surfaces. (Adapted from [63].)

Surfaces	WCA (°)
Bare Al	78 ± 6
Bare CuZn	84 ± 3
Bare PA	86±4

PA surfaces. A 4-L droplet of distilled water was deposited on the surface of the samples for the determination of the contact angle. The measurements were repeated three times on each sample (table 4.2).

4.3 Results and discussion

4.3.1 The process of ice adhesion

The ice adhesion is a very spontaneous process, thus to visualize the ice detachment process the Phantom T3610 at 40 000 FPS is used. The ice adhesion process is observed at various temperatures such as −10 °C, −15 °C and −20 °C. The study was conducted using aluminum, CuZn alloy, and polyamide substrates. Figure 4.5 shows the interface between the ice and substrates at −10 °C while detaching the substrate from the ice. It can be observed that the visibility of the number of cracks/fractures in the ice at the interface is higher after detachment, especially for the Al and CuZn substrates. Moreover, figure 4.6 shows the interface between the ice and substrates at −15 °C, before and after detachment. It can be observed that the cracks/fractures are not visible in ice after detachment. Similarly, figure 4.7 shows ice adhesion observations at −20 °C in which the cracks/fractures are also not visible in ice after detachment. This indicates that the ice has stronger bonds at lower temperatures that can resist initiating cracks/fractures in ice after detachment.

Figure 4.5 shows that after detachment from Al and CuZn substrates, both the crack propagation and the fracture occur on the ice surface. In particular, with the Al substrate, ice has completely fractured.

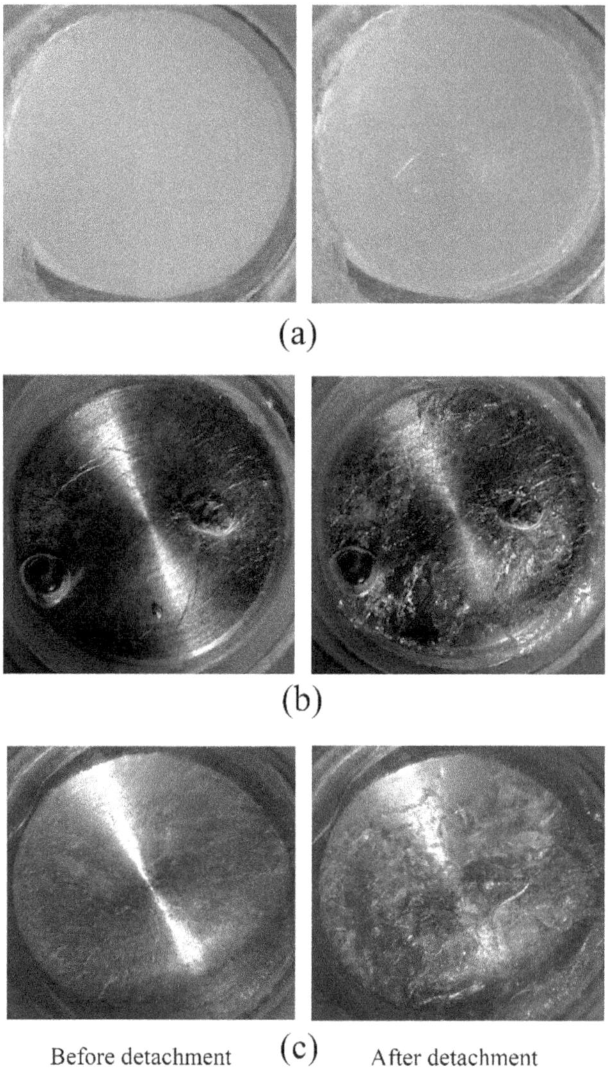

Figure 4.5. Images showing the process of ice adhesion before and after the test with different substrates at −10 °C: (a) PA, (b) CuZn and (c) Al. (Adapted from [63].)

4.3.2 Ice adhesion versus temperature

Ice adhesion has been studied for several decades using different environmental conditions including temperature, relative humidity, and different conditions in icing wind tunnels [5, 31–39]. In nature, clouds contain super-cooled water droplets that can immediately form ice if they are in contact with any solid surface. To remove such ice it has been observed that ice adhesion may depend on various environmental conditions.

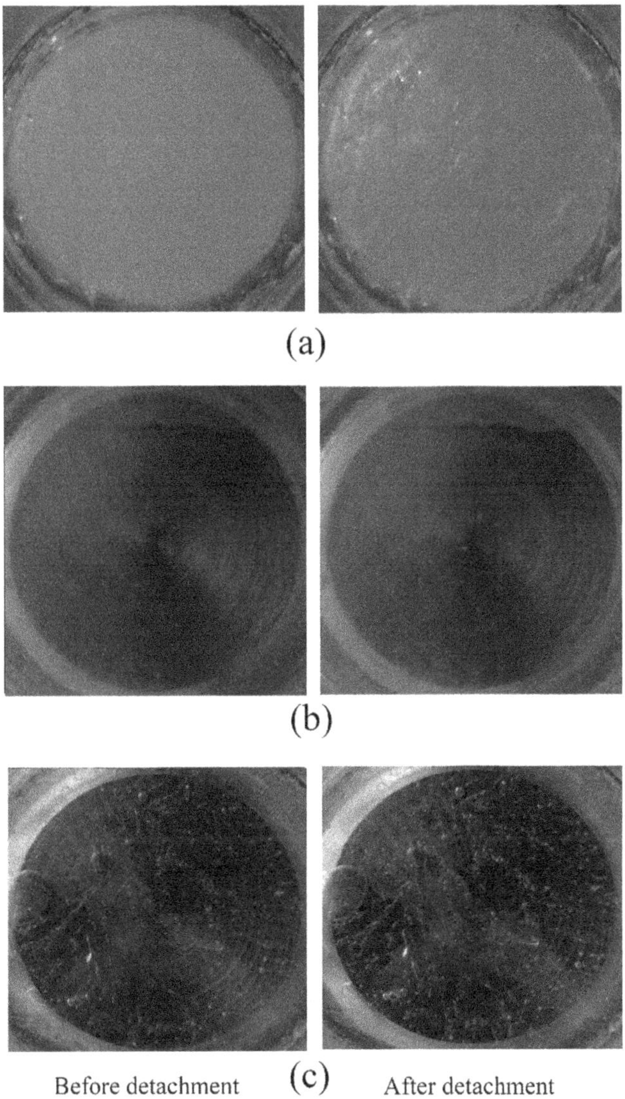

Figure 4.6. Images showing the process of ice adhesion before and after the test with different substrates at −15 °C: (a) PA, (b) CuZn and (c) Al. (Adapted from [63].)

There is a strong dependence of the ice adhesion strength on temperature. Decreasing ambient temperature can have several effects. As the temperature decreases, the adhesive strength between ice and a substrate typically increases. This is because lower temperatures lead to a decrease in the mobility of molecules at the interface, resulting in stronger intermolecular forces, hydrogen bonds, and enhanced adhesion [40–43].

Similarly, at lower temperatures, the freezing of water molecules on the surface can create a stronger bond with the substrate, increasing the adhesion strength. The

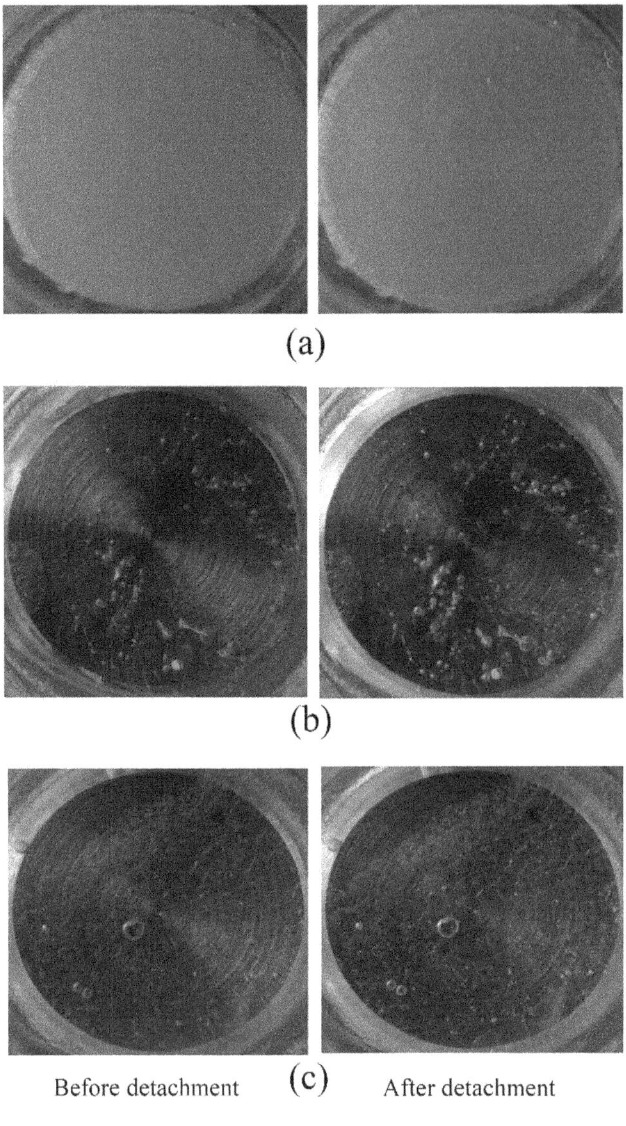

Figure 4.7. Images showing the process of ice adhesion before and after the test with different substrates at −20 °C: (a) PA, (b) CuZn and (c) Al. (Adapted from [63].)

formation of ice crystals at colder temperatures can also lead to a larger contact area between the ice and the substrate, further enhancing adhesion [44–50].

The experimental results of the ice adhesion while detaching at 30 mm s^{-1} approximately show that the average ice adhesion force and strength increases with decreasing ambient temperature, see figure 4.8. The ice adhesion and temperature exhibit a linear relationship in the range of −10 °C to −20 °C.

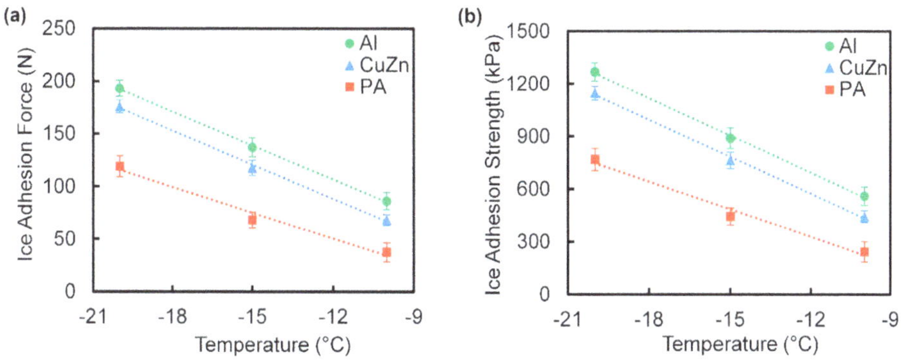

Figure 4.8. (a) Ice adhesion force and (b) ice adhesion strength at different ambient temperatures with 14 mm substrate diameter. Error bars represent the standard deviation. (Adapted from [63].)

4.3.3 Ice adhesion versus surface area

In ideal scenarios, the ice adhesion strength should be the same regardless of the surface size. This is because adhesion strength is an intrinsic property of the interface between the solid surface and ice. However, in practice, deviations from ideal conditions (e.g., surface geometry, texture, ice adhesion method, and surface modification method) can lead to significant variations in the actual measured adhesion strength of ice. As described earlier, the surfaces used were prepared via the machining process and it is clear that the machining process creates round patterns on surfaces. These concentric patterns introduce micro-grooves and circular ridges, which can act as stress concentrators or crack initiators at the ice–surface interface. As such, they may promote easier ice delamination under tensile loading (figure 4.9). Before freezing the contact between water and surfaces tends to show a Wenzel state. After freezing, it potentially shows that the Wenzel state is partially shifted to the Cassie–Baxter state. This transition may be driven by a combination of factors, including thermal contraction of the substrate and volumetric expansion of water during freezing, which together can trap air pockets or form microvoids at the interface, reducing effective contact area and thus adhesion strength. This phenomenon of transition between Wenzel–Cassie or Cassie–Wenzel transition during freezing has been well documented in the literature [51–59]. Increasing area can cause an increasing number of round patterns on each surface. Therefore, the overall average ice adhesion may decrease by increasing the surface area. The specific ice adhesion method (e.g., tensile) and ice sample preparation (e.g., confined mold) may introduce edge effects and internal cracking of ice, influencing overall average ice adhesion strength, as noted in prior studies [60–62]. Additionally, the high delamination speed may introduce dynamic effects, influencing the overall average ice adhesion strength.

The ice adhesion results show that the overall adhesive force increases by increasing substrate area, as shown in figure 4.10(a). Meanwhile, a linear trend is

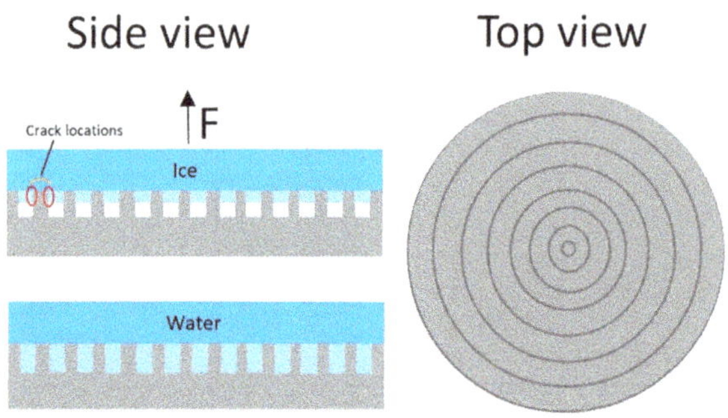

Figure 4.9. Schematic showing a side view and top view of the machining patterned surface. A side view shows before freezing when water can fill the gaps. After freezing due to the thermal contraction of surfaces, water expansion at the interface could produce air pockets/microvoids. For detachment of the surface from ice, the crack locations are shown which can potentially make the delamination process easy. (Adapted from [63].)

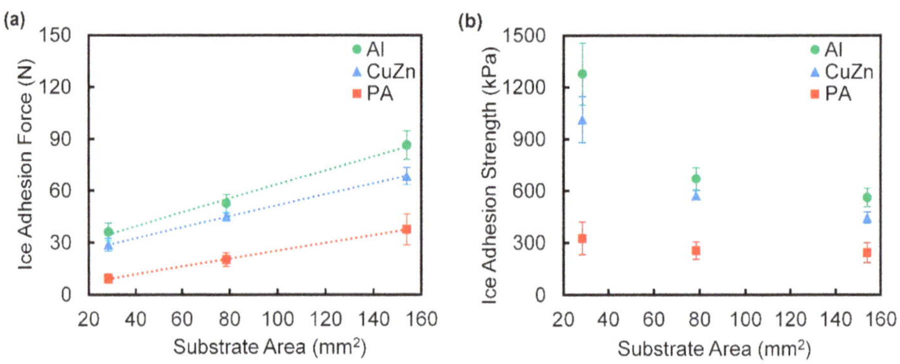

Figure 4.10. (a) Average values of ice adhesion force and (b) average values of ice adhesion strength with 6, 10, and 14 mm substrates diameters. The error bars represent the standard deviation. (Adapted from [63].)

observed between the ice adhesion force and substrate area. In addition, larger substrate surface areas lead to more round patterns, resulting in lower average ice adhesion strength, as shown in figure 4.10(b). Figure 4.10(b) shows that the smallest substrate area has the highest ice adhesion strength and a larger standard deviation, due to less uniformity and fewer round patterns. Additionally, material properties also play a role. Metal substrates (Al and CuZn, for instance) exhibit increased average ice adhesion compared to polymers like PA, as shown in figure 4.10(b), likely due to differences in surface energy, stiffness, and interfacial bonding strength.

4.3.4 Finite element model of ice adhesion dynamics

The numerical model is developed by using the LS-DYNA simulation software. The standard, constant-stress hexahedral elements are used in the model. The ice is a cylinder with a radius $R = 9.5$ mm and a height $h = 9$ mm. The model can be seen in figure 4.11. The blue arrows on the bottom show the application of the pulling force over a circle of radius $r = 7$ mm concentric to the cylinder's axis, whereas the black crosses represent a constraining of the remaining nodes of the bottom surface. A force of 583 N is uniformly applied over the blue nodes (table 4.3).

The simulation results are shown in figure 4.12 with the ice (height of 9 mm, diameter of 19 mm and the substrate attached has a diameter of 14 mm) being pulled by 583 N with the timestamp unit in ms. The simulation ran from 0 to 1 s. The values in the legend correspond to the effective or von Mises stress and are given in GPa (1e −3 GPa is 1 MPa). Additionally, it can be observed from the simulation results that there are differences in stress distributions between timestamps, if the simulation was static or relatively slow detachment, the stress distribution would be the same all the time. However, since the force is applied almost instantaneously in this simulation, it can be seen that the shape changes due to dynamic effects. Meanwhile, in this simulation, the Poisson's ratio of the ice also plays a role, as deformation in other directions influences inertia as well.

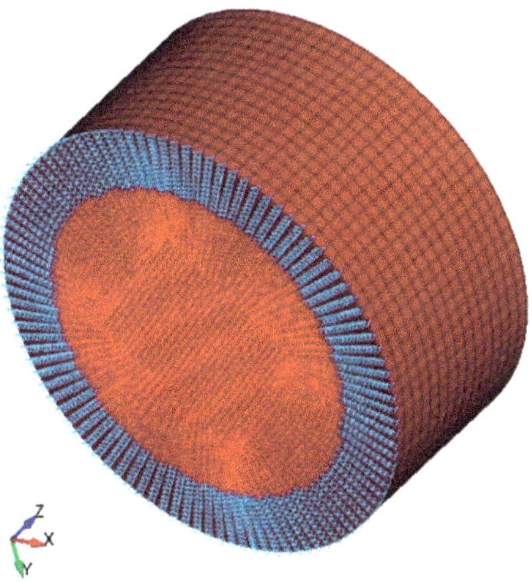

Figure 4.11. LS-DYNA model of the ice cylinder. (Adapted from [63].)

Table 4.3. Density, elastic modulus, and Poisson's ratio of the ice. (Adapted from [63].)

	ρ (kg m^{-3})	E (GPa)	v
Ice	900	9	0.33

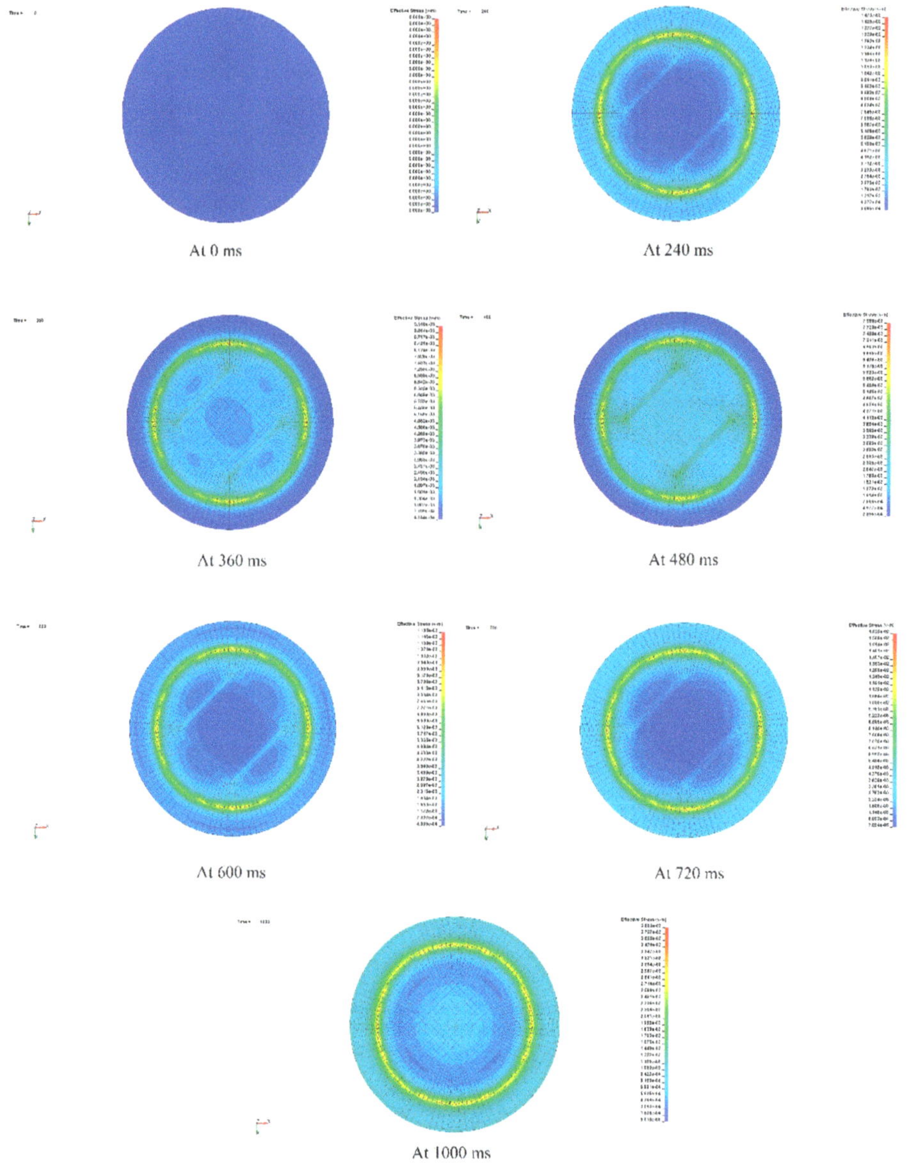

Figure 4.12. Tensile stresses at the ice–substrate interface corresponding to the tensile force applied to a bottom view (i.e. the surface attached to the substrate) the moment the ice detaches. The stresses are given in GPa (that is, 1e−3 GPa is 1 MPa) at various time frames in ms. (Adapted from [63].)

As shown in the simulation results, strong stress concentrations at the edge from the sides of the ice–substrate interface can be observed, which explains that the edge effects are present in such a geometry or mold type while detaching. In experimental work, it is clear that delamination starts from one side and propagates until it reaches the other side. The simulation results can also provide the stress concentrations at different points at the ice–substrate interface which clearly show which points are critical for crack propagation. In this simulation, the effects of patterns are not considered.

4.4 Conclusion

This chapter provides a comprehensive exploration of the phenomenon of ice adhesion, with an emphasis on tensile mode testing under varied environmental and substrate conditions. The findings are pivotal for advancing the understanding of ice adhesion mechanisms and guiding the development of efficient anti-icing strategies. Key conclusions are as follows:

- *Ice adhesion and environmental factors*: The adhesion strength of ice is strongly influenced by environmental conditions, particularly ambient temperature. Lower temperatures lead to increased adhesive strength due to enhanced intermolecular forces, hydrogen bonding, and larger contact areas. Relative humidity also plays a significant role, in affecting ice formation and adhesion properties.
- *Influence of substrate properties*: The nature of the substrate material, its surface roughness, energy, and water contact angle—substantially influence ice adhesion. Metal substrates, such as aluminum and copper–zinc alloy, exhibit higher ice adhesion strength compared to polymers such as polyamide, likely due to their higher surface energy and structural characteristics.
- *Impact of surface area*: Ice adhesion strength exhibits a dependence on surface area. Larger surfaces, while showing higher overall adhesion force, demonstrate a reduction in adhesion strength due to stress distribution effects and the increased presence of crack-initiating patterns. Smaller surfaces exhibit higher adhesion strength but less uniform stress distribution.
- *High-speed imaging insights*: The use of high-speed imaging (up to 40 000 FPS) enabled visualization of ice delamination dynamics. Observations revealed that delamination starts at the edges and propagates across the interface. The fracture mechanics and crack propagation can be seen through images at −10 °C, −15 °C, and −20 °C for Al, CuZn, and PA, which provides a basis for new and interesting avenues to further investigate this phenomenon.
- *Numerical simulation validation*: Finite element modeling provided valuable insights into the stress distribution at the ice–substrate interface during detachment. Simulations revealed significant stress concentrations at the edges, which correspond well with experimental observations of crack propagation during delamination.
- *Practical implications and future directions*: The current studies underscore the importance of tailoring anti-icing coatings and mitigation strategies to specific substrate materials and environmental conditions. The development of durable, cost-effective, and environmentally friendly coatings remains a critical challenge. Future studies should focus on high-speed imaging at even higher frame rates, varying strain rates, and conducting experiments under diverse conditions to refine the understanding of ice adhesion dynamics.

This work establishes a foundation for designing more effective anti-icing systems for applications across various industries, including aviation, wind energy, and telecommunications. Integrating experimental observations with numerical simulations provides a robust framework for addressing ice adhesion challenges in real-world environments.

Notes

The authors declare no competing financial interest.

Acknowledgments

This project has received funding from the European Union's Horizon 2020 research and innovation program under the Marie Skłodowska-Curie Grant Agreement 956703 (SURFICE Smart Surface Design for Efficient Ice Protection and Control). Special thanks are extended to AeroTex UK LLP for providing simulations for ice adhesion.

References

[1] List R 1977 Ice accretions on structures *J. Glaciol.* **19** 451–65

[2] Makkonen L 2012 Ice adhesion—theory, measurements and countermeasures *J. Adhes. Sci. Technol.* **26** 413–45

[3] Struk P M, Broeren A P, Tsao J-C, Vargas M, Wright W B, Currie T, Knezevici D and Fuleki D 2012 Fundamental ice crystal accretion physics studies *Int. Conf. on Aircraft and Engine Icing and Ground Deicing*

[4] Petrenko V F 1997 Study of the surface of ice, ice/solid and ice/liquid interfaces with scanning force microscopy *J. Phys. Chem.* B **101** 6276–81

[5] Raraty L E and Tabor D 1958 The adhesion and strength properties of ice *Proc. R. Soc.* A **245** 184–201

[6] Matsumoto K and Daikoku Y 2009 Fundamental study on adhesion of ice to solid surface: discussion on coupling of nano-scale field with macro-scale field *Int. J. Refrig.* **32** 444–53

[7] Matsumoto K, Akimoto T and Teraoka Y 2010 Study of scraping force of ice growing on cooling solid surface *Int. J. Refrig.* **33** 419–27

[8] Mishchenko L, Hatton B, Bahadur V, Taylor J A, Krupenkin T and Aizenberg J 2010 Design of ice-free nanostructured surfaces based on repulsion of impacting water droplets *ACS Nano* **4** 7699–707

[9] Lynch F T and Khodadoust A 2001 Effects of ice accretions on aircraft aerodynamics *Prog. Aerosp. Sci.* **37** 669–767

[10] Dutta P K, Ryerson C C and Pergantis C 2005 Thermal deicing of polymer composite helicopter blades *Mater. Sp. Appl.* **851** 563

[11] Boluk Y 1996 Adhesion of freezing precipitates to aircraft surfaces *Report* 00780031 Transportation Research Board

[12] Ryerson C C 2009 Assessment of superstructure ice protection as applied to offshore oil operations safety *Techincal Report* ADA495836 Defense Technical Information Center

[13] Volat C, Farzaneh M and Leblond A 2005 De-icing/anti-icing techniques for power lines: current methods and future direction *11th International Workshop on Atmospheric Icing of Structures*

[14] Yamauchi G, Takai K and Saito H 2000 PTEE based water repellent coating for telecommunication antennas *IEICE Trans. Electron.* **83** 1139–41

[15] Xue H and Khawaja H 2016 Review of the phenomenon of ice shedding from wind turbine blades *Int. J. Multiphys.* **10** 265–76

[16] Gouni R 2011 A new technique to study temperature effects on ice adhesion strength for wind turbine materials Case Western Reserve University, Cleveland, OH

[17] Meuler A J, Smith J D, Varanasi K K, Mabry J M, McKinley G H and Cohen R E 2010 Relationships between water wettability and ice adhesion *ACS Appl. Mater. Interfaces* **2** 3100–10

[18] Zou M, Beckford S, Wei R, Ellis C, Hatton G and Miller M A 2011 Effects of surface roughness and energy on ice adhesion strength *Appl. Surf. Sci.* **257** 3786–92

[19] Emelyanenko K A, Emelyanenko A M and Boinovich L B 2020 Water and ice adhesion to solid surfaces: common and specific, the impact of temperature and surface wettability *Coatings* **10** 648

[20] Rønneberg S, Laforte C, Volat C, He J and Zhang Z 2019 The effect of ice type on ice adhesion *AIP Adv.* **9** 055304

[21] Wang Y, Zhu C, Xiong K and Zhu C 2024 Experimental investigation on ice–aluminum interface adhesion strength under heating conditions *Aerospace* **11** 152

[22] Laforte C and Beisswenger A 2005 Icephobic material centrifuge adhesion test *Proc. of the 11th Int. Workshop on Atmospheric Icing of Structures, IWAIS (Montreal, QC, Canada)* pp 12–6

[23] Fortin G, Beisswenger A and Perron J 2010 Centrifuge adhesion test to evaluate icephobic coatings *AIAA Atmospheric and Space Environments Conf.* p 7837

[24] Douglass R G, Palacios J and Schneeberger G 2018 Design, fabrication, calibration, and testing of a centrifugal ice adhesion test rig with strain rate control capability *2018 Atmospheric and Space Environments Conf.* p 3342

[25] Sonwalkar N, Sunder S S and Sharma S K 1993 Ice/solid adhesion analysis using low-temperature Raman microprobe shear apparatus *Appl. Spectrosc.* **47** 1585–93

[26] Huang W, Huang J, Guo Z and Liu W 2022 Icephobic/anti-icing properties of superhydrophobic surfaces *Adv. Colloid Interface Sci.* **304** 102658

[27] Cao L, Jones A K, Sikka V K, Wu J and Gao D 2009 Anti-icing superhydrophobic coatings *Langmuir* **25** 12444–8

[28] Shaikh A R, Qu J, Zhou M and Mulbah C K Y 2021 Recent progress on the surface finishing of metals and alloys to achieve superhydrophobic surfaces: a critical review *Trans. IMF* **99** 61–72

[29] Menini R and Farzaneh M 2011 Advanced icephobic coatings *J. Adhes. Sci. Technol.* **25** 971–92

[30] Menini R, Ghalmi Z and Farzaneh M 2011 Highly resistant icephobic coatings on aluminum alloys *Cold Reg. Sci. Technol.* **65** 65–9

[31] Fu D, Zheng H, Sheng W, Hao X, Zhang X, Chang S and Song M 2024 An experimental study on the influence of humidity on ice adhesion strength on superhydrophobic surfaces with microstructures *Appl. Therm. Eng.* **244** 122732

[32] He Z, Vågenes E T, Delabahan C, He J and Zhang Z 2017 Room temperature characteristics of polymer-based low ice adhesion surfaces *Sci. Rep.* **7** 42181

[33] Janjua Z A 2017 The influence of freezing and ambient temperature on the adhesion strength of ice *Cold Reg. Sci. Technol.* **140** 14–9

[34] Yeong Y H, Sokhey J and Loth E 2019 Ice adhesion on superhydrophobic coatings in an icing wind tunnel *Contamination Mitigating Polymeric Coatings for Extreme Environments* (Cham: Springer) pp 99–121

[35] Zhu C X, Zhu C L, Zhao W W and Tao M J 2018 Experimental study on the shear adhesion strength between the ice and substrate in icing wind tunnel *J. Mech.* **34** 209–16

[36] Wang Y, Xiong K, Zhu C, Zhu C, Guo R and Chen L 2023 Research on normal ice adhesion strength in icing wind tunnel *Proc. Inst. Mech. Eng.* G **237** 3272–84

[37] Anderson D, Reich A, Anderson D and Reich A 1997 Tests of the performance of coatings for low ice adhesion *35th Aerospace Sciences Meeting and Exhibit* p 303

[38] Kraj A G and Bibeau E L 2010 Measurement method and results of ice adhesion force on the curved surface of a wind turbine blade *Renew. Energy* **35** 741–6

[39] Work A and Lian Y 2018 A critical review of the measurement of ice adhesion to solid substrates *Prog. Aerosp. Sci.* **98** 1–26

[40] Mizan T I, Savage P E and Ziff R M 1996 Temperature dependence of hydrogen bonding in supercritical water *J. Phys. Chem.* **100** 403–8

[41] Silverstein K A T, Haymet A D J and Dill K A 2000 The strength of hydrogen bonds in liquid water and around nonpolar solutes *J. Am. Chem. Soc.* **122** 8037–41

[42] Cai Y Q, Mao H-K, Chow P C, Tse J S, Ma Y, Patchkovskii S, Shu J F, Struzhkin V, Hemley R J and Ishii H 2005 Ordering of hydrogen bonds in high-pressure low-temperature H_2O *Phys. Rev. Lett.* **94** 25502

[43] Kalifa P, Jones S J and Slade T D 1991 Microcrack nucleation in granular ice under uniaxial compression: effect of grain-size and temperature *Ann. Glaciol.* **15** 222–9

[44] Chaplin M F 2007 Water's hydrogen bond strength arXiv: 0706.1355

[45] Ignatov I and Mosin O 2014 Hydrogen bonds among molecules in liquid and solid state of water. Modifications of ice crystals *J. Heal. Med. Nurs.* **5** 56–79

[46] Steiner T 2002 The hydrogen bond in the solid state *Angew. Chem. Int. Ed.* **41** 48–76

[47] Gragson D E and Richmond G L 1998 Investigations of the structure and hydrogen bonding of water molecules at liquid surfaces by vibrational sum frequency spectroscopy *J. Phys. Chem.* B **102** 3847–61

[48] Smith T R 1991 The effect of stress state on the brittle compressive failure of columnar ice *Doctoral Dissertation* Dartmouth College, Hanover, NH

[49] Picu R C, Gupta V and Frost H J 1994 Crack nucleation mechanism in saline ice *J. Geophys. Res. Solid Earth* **99** 11775–86

[50] Frost H J 2001 Mechanisms of crack nucleation in ice *Eng. Fract. Mech.* **68** 1823–37

[51] Cui J, Wang T and Che Z 2024 Freezing–melting mediated dewetting transition for droplets on superhydrophobic surfaces with condensation *Langmuir* **40** 14685–96

[52] Wiedemann S, Plettl A, Walther P and Ziemann P 2013 Freeze fracture approach to directly visualize wetting transitions on nanopatterned superhydrophobic silicon surfaces: more than a proof of principle *Langmuir* **29** 913–9

[53] Chen R, Jiao L, Zhu X, Liao Q, Ye D, Zhang B, Li W, Lei Y and Li D 2018 Cassie-to-Wenzel transition of droplet on the superhydrophobic surface caused by light induced evaporation *Appl. Therm. Eng.* **144** 945–59

[54] Zhang B, Chen X, Dobnikar J, Wang Z and Zhang X 2016 Spontaneous Wenzel to Cassie dewetting transition on structured surfaces *Phys. Rev. Fluids* **1** 73904

[55] Karmouch R and Ross G G 2010 Experimental study on the evolution of contact angles with temperature near the freezing point *J. Phys. Chem.* C **114** 4063–6

[56] Wang Y and Zhao P 2022 Temperature-based analysis of droplet cooling and freezing on femtosecond laser textured surfaces *Appl. Therm. Eng.* **206** 118046
[57] Ruiz-Cabello F J M, Bermúdez-Romero S, Ibáñez-Ibáñez P F, Cabrerizo-Vílchez M A and Rodríguez-Valverde M A 2021 Freezing delay of sessile drops: probing the impact of contact angle, surface roughness and thermal conductivity *Appl. Surf. Sci.* **537** 147964
[58] Rico V J, López-Santos C, Villagrá M, Espinós J P, de la Fuente G F, Angurel L A, Borrás A and González-Elipe A R 2019 Hydrophobicity, freezing delay, and morphology of laser-treated aluminum surfaces *Langmuir* **35** 6483–91
[59] Liu Y, Zhang Z, Hu H, Hu H, Samanta A, Wang Q and Ding H 2019 An experimental study to characterize a surface treated with a novel laser surface texturing technique: water repellency and reduced ice adhesion *Surf. Coat. Technol.* **374** 634–44
[60] Zarasvand K A, Mohseni M and Golovin K 2021 Cohesive zone analysis of cylindrical ice adhesion: determining whether interfacial toughness or strength controls fracture *Cold Reg. Sci. Technol.* **183** 103219
[61] Bleszynski M and Clark E 2021 Current ice adhesion testing methods and the need for a standard: a concise review *Standards* **1** 117–33
[62] Petrenko V F and Whitworth R W 1999 *Physics of Ice* (Oxford: Oxford University Press)
[63] Shaikh A R 2026 *PhD Thesis* Technische Universität Darmastadt

Chapter 5

Fracture mechanisms in electromechanical resonant de-icing systems

Giulia Gastaldo, Valérie Pommier-Budinger and Marc Budinger

Icing is a significant challenge for aviation, affecting both safety and efficiency. While current ice protection systems can already provide effective protection, they require a significant amount of power and maintenance. Moreover, as the aviation industry shifts toward electrically powered aircraft, systems that rely on combustion engines are becoming obsolete, driving the need for new electric systems. In recent years, electromechanical resonant de-icing systems have shown significant benefits in terms of energy efficiency and reduced onboard weight, and they present a promising alternative for protecting aircraft surfaces from icing.

This chapter is a critical first step in advancing research on electromechanical resonant de-icing systems, as it explores the fracture mechanisms associated with different resonant modes and dimensions of the ice layer, to have a better understanding of how ice detaches from a surface. Using modal analysis and fracture analysis tools, the different fracture mechanisms are identified, and the triggering conditions are defined for bare plates equipped with piezoelectric actuators.

5.1 Principle of electromechanical resonant de-icing systems

Electromechanical resonant de-icing systems use actuators to induce vibrations on iced surfaces by exploiting the phenomenon of resonance, which occurs when a system is excited at one of its natural frequencies. At resonance, the amplitude of the vibrations increases significantly for the same input signal magnitude, creating high stresses and facilitating energy release in the ice. This results in the failure of the ice layer and its delamination from the surface [1, 2].

One potential technology for this kind of de-icing system is based on piezoelectric materials, which have the ability to mechanically deform in the presence of an electric field (a phenomenon known as the reverse piezoelectric effect). By applying a

doi:10.1088/978-0-7503-6009-8ch5

sinusoidal voltage signal to the electrodes of the piezoelectric ceramic, the resulting electrical charge induces sinusoidal mechanical vibrations in the structure.

Various resonant modes can be excited. The detection of these modes is typically achieved using one of the piezoelectric transducers as a sensor, while the other (or others) actuates the plate. Figure 5.1 presents an example of the frequency spectrum for a plate under free boundary conditions, covering the range from 200 Hz to 3 kHz. To de-ice a certain structure, a first low-voltage frequency sweep is performed to retrieve the frequency spectrum of the system. Once the desired resonance peak is identified, the de-icing is performed at higher voltages, targeting frequencies near the resonance.

Three distinct types of resonant modes can be used to mechanically break the ice when an electromechanical resonant de-icing system is used.

The first type of mode is characterized by significant out-of-plane bending and is known as the out-of-plane flexural mode (figure 5.2(a)). Out-of-plane flexural modes can occur at relatively low frequencies (a few hundred Hz).

The second type of mode is known as the extensional mode and is characterized by primarily in-plane vibrations involving a stretching movement (figure 5.2(b)). Extensional modes typically occur around 10–20 kHz. A comparative performance assessment of these two modes was conducted by Budinger *et al* [4, 5], highlighting that while extensional modes consume more energy than out-of-plane flexural modes, they facilitate ice detachment over larger areas, as the energy release rate is more stable throughout the propagation.

A third type of mode called in-plane flexural mode can also be identified. As the name suggests, this type of mode is characterized by bending within the plane of a

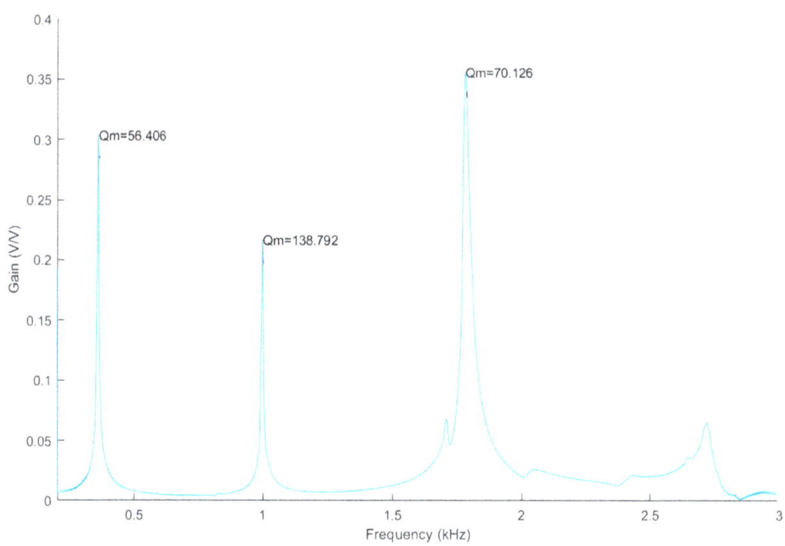

Figure 5.1. Example of a natural frequency spectrum for the first three out-of-plane flexural modes. The gain (voltage of the sensor/voltage of the actuator) is plotted against the frequency. Q_m represents the quality factor of the mode. (Adapted from [3].)

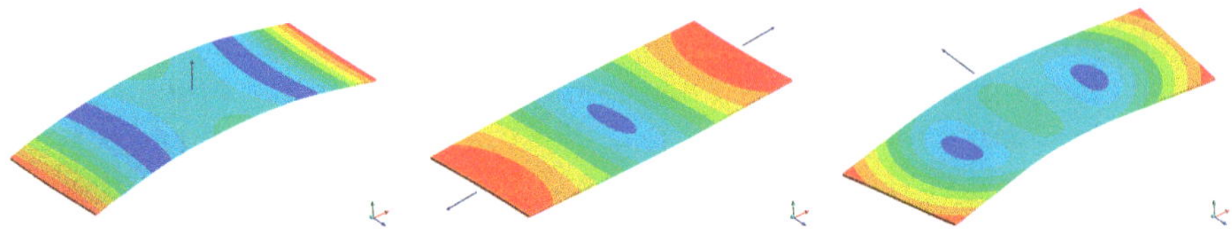

(a) First out-of-plane flexural mode. (b) First extensional mode. (c) First in-plane flexural mode.

Figure 5.2. Types of resonant modes for a plate sample ($130 \times 50 \times 1$ mm^3) in free boundary conditions. (Adapted from [3].)

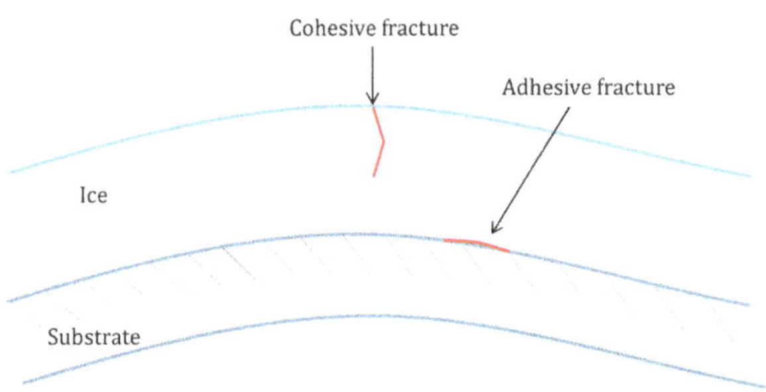

Figure 5.3. Cohesive and adhesive fractures. (Adapted from [3].)

surface (figure 5.2(c)). The first in-plane flexural mode appears at a frequency of a few kilohertz (approximately 5–7 kHz).

Two different types of fractures can appear: cohesive and adhesive fractures. Cohesive fractures occur within the ice bulk, while adhesive fractures occur at the interface between the ice and the substrate (see figure 5.3). Cohesive fractures do not necessarily occur during the de-icing process. In the absence of accompanying adhesive fractures, the ice remains adhered to the surface, preventing complete ice removal. Adhesive fractures are essential for the shedding of ice.

As mentioned above, the modes exhibit significant differences in their resulting fracture mechanisms. To assess these differences, an aluminum sample measuring $130 \times 50 \times 1$ mm^3, uniformly covered with a 3 mm thick ice layer, is analysed under free boundary conditions.

The fracture mechanism for the first out-of-plane flexural mode is illustrated in figure 5.4. In this case, a cohesive crack initially forms at the antinode of the mode, where the deflection achieves its maximum. Subsequently, adhesive fractures propagate symmetrically from the cohesive crack toward the sample's outer edges. This propagation appears in the form of a whitish region surrounding the crack; however, no significant de-icing effect is observed. The de-icing was conducted in three distinct tests due to the substantial frequency loss associated with out-of-plane flexural modes, which makes it impossible to achieve complete crack propagation in a single high-voltage de-icing sweep.

Figure 5.5 illustrates the fracture mechanism obtained through de-icing using the first extensional mode. Given the rapid nature of the de-icing process, a high-speed camera (operating at 2000 frames per second) was used to capture the mechanism in detail. In the initial frames, cohesive cracks begin to form in the transverse direction. Subsequently, a combination of adhesive fracture propagation and the formation of additional cohesive cracks in both longitudinal and transverse directions occurs. These events may take place either simultaneously or sequentially, following the propagation of the adhesive fracture in a specific region. This process ultimately results in the complete de-icing of the sample, with a total de-icing time of 10 seconds. The frequency sweep was repeated multiple times during the test.

Finally, figure 5.6 shows the fracture mechanism for the first in-plane flexural mode, which exhibits a fracture process that falls between the behaviors observed in the out-of-plane flexural and extensional modes. A cohesive crack initially develops

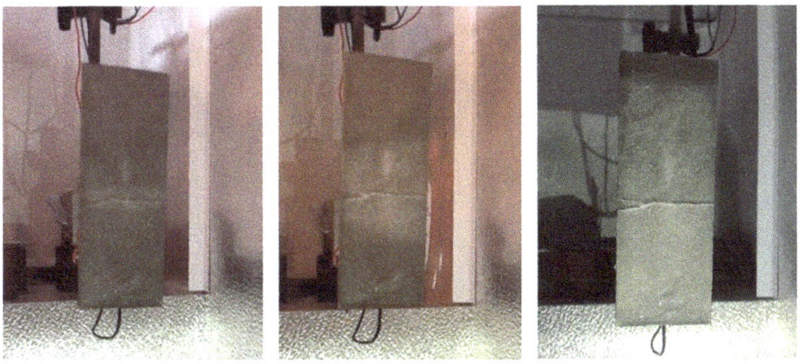

Figure 5.4. De-icing fracture mechanism for the first out-of-plane flexural mode (figure 5.2(a)). (Adapted from [3].)

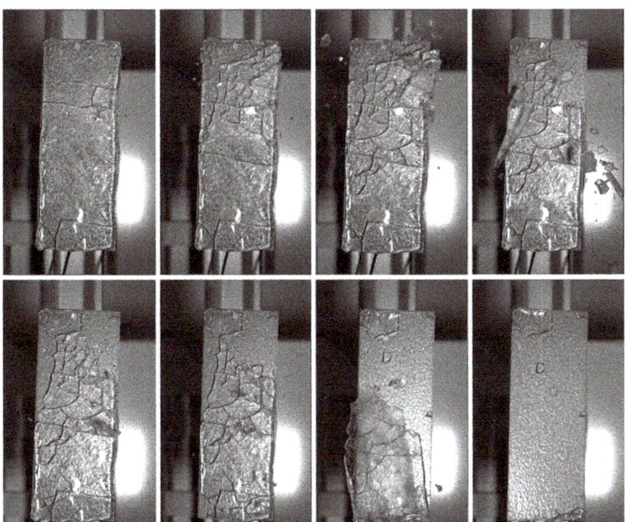

Figure 5.5. De-icing fracture mechanism for the first extensional mode (figure 5.2(b)). Total de-icing time = 10 s. (Adapted from [3].)

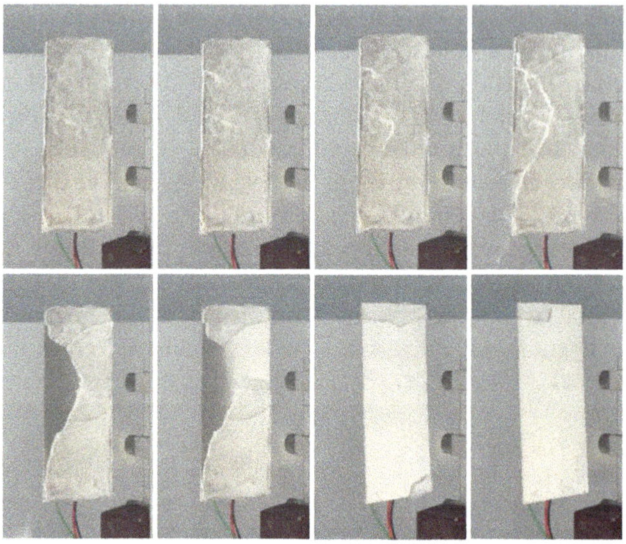

Figure 5.6. De-icing fracture mechanism for the first in-plane flexural mode (figure 5.2(c)). Total de-icing time = 10 s. (Adapted from [3].)

at the antinode of the mode, located at the center of the sample. This is followed by a combination of cohesive fractures and adhesive propagation until complete de-icing is achieved. Similarly to the extensional mode, the total de-icing time for this test was also 10 s.

A brief overview of the fracture mechanisms for the fully ice-covered sample has been presented. Hereafter, a more detailed analysis of the fracture mechanisms will be conducted, focusing on the out-of-plane flexural mode and the extensional mode. A theoretical framework, reinforced by numerical simulations, will be developed to explain the fracture mechanisms described earlier and to predict the fracture behavior under varying ice coverage conditions.

5.2 Fracture criteria and mechanical properties of ice

5.2.1 Fracture criteria

For fracture initiation, a coupled criterion is typically employed [6, 7]. This criterion considers both the stress and energy conditions at the interface or within the material. Assuming the influence of kinetic energies and the work of external forces is negligible, a fracture over a surface ΔS, smaller than the total surface S, can occur if the following two conditions are respected:

$$\tau \text{ (or } \sigma) > \tau_c \text{ (or } \sigma_c), \quad \forall \Delta S \in S \tag{5.1}$$

$$G = -\frac{\Delta E_p}{\Delta S} \geqslant G_c, \tag{5.2}$$

where $\Delta E_p = E_p(0) - E_p(i)$ represents the variation in potential strain energy between the undamaged state and the damaged state for a cracked surface ΔS. This coupled criterion is traditionally used for interfacial fractures but can also be applied to thick ice layers.

Following the initiation phase, which results in the formation of a sharp crack within the system, the stress concentration at the crack tip ensures that the stress criterion is satisfied. Consequently, the fracture propagation can be governed exclusively by the energy release rate criterion. The energy release rate criterion for propagation is expressed as

$$-\frac{\Delta E_{pi}}{\Delta S} \geqslant G_c, \tag{5.3}$$

where $\Delta E_{pi} = E_p(i) - E_p\big(i + 1\big)$ represents the difference in potential energy between two consecutive propagation steps i and $i + 1$. A schematic of the propagation steps is illustrated in figure 5.7 for a cohesive crack.

A methodology for evaluating these fracture criteria, which can be applied to both fracture initiation and propagation, is developed. The method relies on the knowledge of the critical mechanical properties of bulk ice and the ice–substrate interface, as well as some modal quantities obtained through finite element analysis. It consists in calculating the vibration amplitudes required to initiate or propagate a fracture by exceeding the tensile and shear strengths and those required to exceed the toughness of the ice–substrate interface.

Based on the methodology previously introduced in [5, 8], the amplitude required to generate a cohesive fracture based on the tensile stress criterion is defined by

$$x_{r\sigma} = \frac{\sigma_c}{\sigma_{mod}} x_{mod}, \tag{5.4}$$

where σ_c is the tensile strength of ice, σ_{mod} is the modal tensile stress in the ice bulk, and x_{mod} the maximum modal displacement of the system. If the flexural mode is considered, x_{mod} is an out-of-plane displacement, while for the extensional mode, x_{mod} is the lateral in-plane displacement. Similarly, the amplitude required to generate an adhesive fracture based on the shear stress criterion is given by

$$x_{r\tau} = \frac{\tau_c}{\tau_{mod}} x_{mod}, \tag{5.5}$$

where τ_c the shear strength of ice, and τ_{mod} denotes the modal shear stress of the ice–substrate interface and x_{mod} the modal displacement of the system. Lastly, the amplitude necessary to generate fracture based on the energy release rate criterion is given by

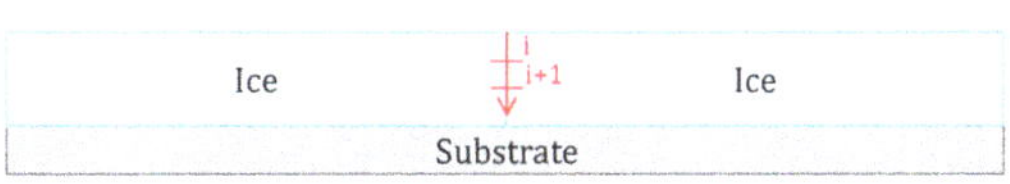

Figure 5.7. Example two consecutive steps of cohesive crack propagation.

$$x_{rG} = \sqrt{\frac{G_c}{G_{mod}}} x_{mod}, \quad (5.6)$$

where G_c represents the toughness of the interface, G_{mod} is the modal energy release rate, and x_{mod} is the modal displacement of the studied system.

For fracture initiation, the amplitudes are used to quantify the initiation fracture length. This is found when the amplitudes associated with the stress-based and energy-based mechanisms are equal. For fracture propagation, only x_{rG} is analysed.

5.2.2 Mechanical properties of ice

For an accurate assessment of the fracture mechanism, it is essential to attribute the appropriate mechanical properties to the ice. Many experiments have been performed to measure the shear strength [9–11], but the results show significant discrepancies. These variations can be attributed to the variability in the mechanical properties of ice depending on its type, as well as the challenges associated with applying pure shear stress to the interface [12]. Considering glaze ice on a bare metallic surface, an interfacial shear strength value of 0.5 MPa is selected for this study, as it is commonly mentioned in the literature for these testing conditions [13].

In [14], Hassan *et al* measured the interfacial tensile strength using a method with forced vibrations on a metal–ice interface, finding values between 0.14 MPa and 2.28 MPa for a surface roughness varying between 0.7 μm and 1.65 μm. Due to the high variability in the data, the critical value of 0.5 MPa that is considered for shear strength will also be applied to tensile strength for consistency.

The tensile strength of bulk ice ranges from 0.7 MPa to 3.1 MPa [15–19]. This parameter is significantly influenced by factors such as temperature, strain rate, tested volume, and ice grain size. A conservative tensile strength value of 3 MPa has been adopted for this study.

According to the literature, the toughness at the ice interface is highly dependent on interfacial properties, including material, roughness, and the properties of the ice itself [20]. Dawood *et al* [21] estimated that the mean value of G_c is approximately around 0.5–0.6 J m^{-2} on a bare metallic interface for both mode-I and mode-II fractures, while Palanque *et al* estimated that the value of G_{Ic} for the interface is around 0.38 J m^{-2}. In this study, a critical G_c of 0.5 J m^{-2} is used.

n his review, Petrovic reports values of bulk ice toughness between 0.3 J m^{-2} and 2.5 J m^{-2} [22]. In a less recent paper by Goodman and Tabor [23], different studies are performed (three-point bending test and indenter test) and the toughness is found to be very scattered, ranging from 0.5 J m^{-2} to 5 J m^{-2}. In this study, the critical value for bulk ice is considered slightly higher than that of the interface, $G_c = 1$ J m^{-2}.

5.3 Fracture mechanisms study

Three resonant modes are studied here: the first out-of-plane flexural mode, the first extensional mode, and the second out-of-plane flexural mode. The amplitude-based methodology is used to assess whether the fracture initiates cohesively or adhesively.

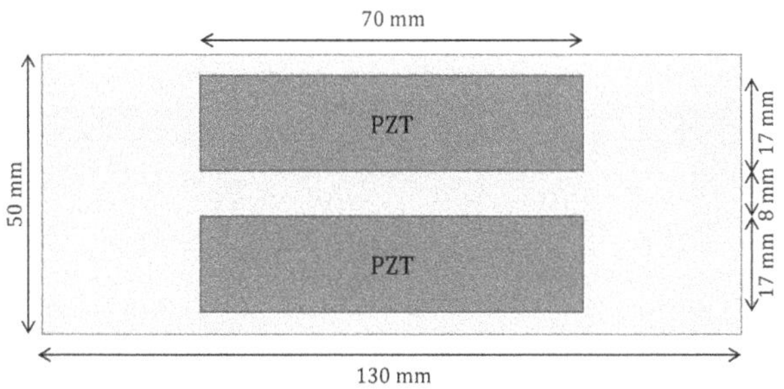

Figure 5.8. Sample dimensions for the first out-of plane flexural mode and first extensional mode—sample dimensions. (Adapted from [24]. CC BY 4.0.)

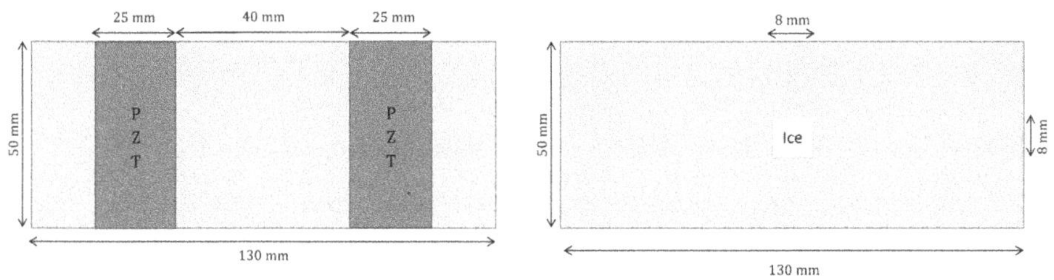

(a) Rear view schematic. (b) Front view schematic.

Figure 5.9. Sample dimensions for the second out-of plane flexural mode. (Adapted with permission from [27].)

Then, it is adopted to evaluate the propagation curves. Finally, tests are performed to verify the accuracy of numerical predictions.

For the first out-of-plane flexural mode and the first extensional mode, the study is conducted on a titanium plate, having dimensions of 130 × 50 mm × 1 mm^3, in free boundary conditions. For actuation, hard piezoceramic transducers (PIC 181 of dimensions 70 × 17 × 1 mm^3) are employed. These actuators are bonded to the titanium plate using a conductive adhesive (EPO-TEK® E4110) to ensure effective mechanical and electrical coupling. Figure 5.8 illustrates the plate's dimensions and the placement of the transducers on the substrate. For the first in-plane flexural mode, two different ice coverage configurations will be investigated: one in which the ice covers the entire length of the sample (130 mm), and another with a shorter ice block (25 mm). This comparison aims to evaluate whether the fracture mechanism differs between these two cases.

The second out-of-plane flexural mode is also studied to verify whether the size of the ice layer and its position with respect to the antinodes influence the fracture mechanism. Previous studies using a horizontal shear test set-up [25, 26] show that shear stress-dominated detachment (manifested as instantaneous delamination) occurs in small ice blocks instead of an energy-dominated mechanism. The study with the second out-of-plane flexural mode aims at evaluating if this phenomenon also occurs with electromechanical resonant de-icing systems. For this mode, the 130 × 50 × 1 mm^3 titanium sample was equipped with soft piezoceramic transducers (PIC 255, 25 × 50 × 1 mm^3), as illustrated in figure 5.9. The piezoceramic

Table 5.1. Material mechanical properties for the finite element analysis. (Adapted from [24]. CC BY 4.0.)

	Density (kg m^{-3})	Young's modulus (GPa)	Thickness (mm)
Titanium	4370.0	110.0	1.0
Ice	900.0	9.0	2.0

transducers were placed at the antinodes of the mode. A small ice block, measuring $8 \times 8 \times 5$ mm^3 is located at the center of the plate (figure 5.9(b)).

Freezer ice is used, as it produces a consistent and uniform ice layer. The mechanical properties of the materials involved in this study are summarized in table 5.1.

5.3.1 First out-of-plane flexural mode: plate entirely covered with ice

5.3.1.1 Numerical study: cohesive fracture initiation and propagation

A 3D numerical study of the sample covered by a uniform 2 mm thick ice layer is performed to understand whether a cohesive or adhesive fracture will occur first. To this end, two different numerical models are done. In the first model, a cohesive fracture within the bulk of the ice is generated at the antinode of the mode, which is located at the center of the plate. In this case, two separate ice blocks are formed, and the contact between them is gradually removed until full detachment is achieved across the thickness of ice.

In the second case, the ice layer is progressively debonded at the interface between the ice and the substrate, symmetrically on both sides. A schematic of the two processes is reported in figures 5.10(a) and 5.10(b). The stresses are retrieved at each node of the mesh in the fully bonded configuration. The distribution is uniform across the width; therefore, the results are presented only for the nodes in the center of the width. The energy release rate G is calculated at each debonding step i as $G = -\frac{\Delta E_p}{\Delta S}$ where $\Delta E_p = E_p(0) - E_p(i)$, which is the difference between the undamaged configuration and the debonding step i.

A straightforward comparison can be made between the two plots. The mechanism that requires a lower amplitude to initiate fracture will be the one that occurs. In figure 5.10(c), the amplitudes are plotted as a function of the 2 mm ice thickness, demonstrating that the coupled criterion is satisfied at an amplitude of approximately 60 μm. In figure 5.10(d), the coupled $\sigma - G$ criterion to initiate an adhesive crack from the edges of the ice layer is not met, as the amplitudes necessary to respect the tensile strength criterion are significantly higher and the two curves do not intersect. Therefore, cohesive initiation is expected, with an initiation crack length of approximately 0.6 mm.

The amplitude required to propagate the fracture within the bulk ice, x_G, is also calculated. As shown in figure 5.11, the displacement necessary to propagate the

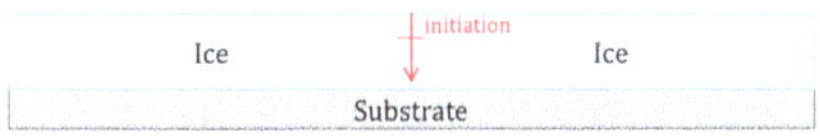

(a) 2D schematic of cohesive initiation.

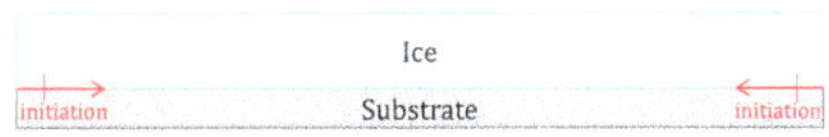

(b) 2D schematic of adhesive initiation.

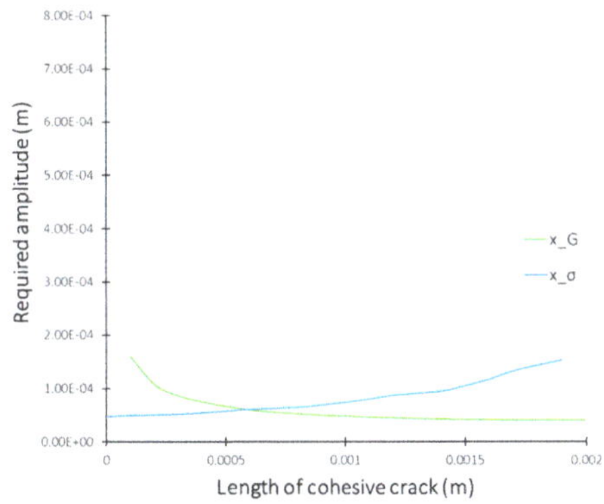

(c) Cohesive initiation.

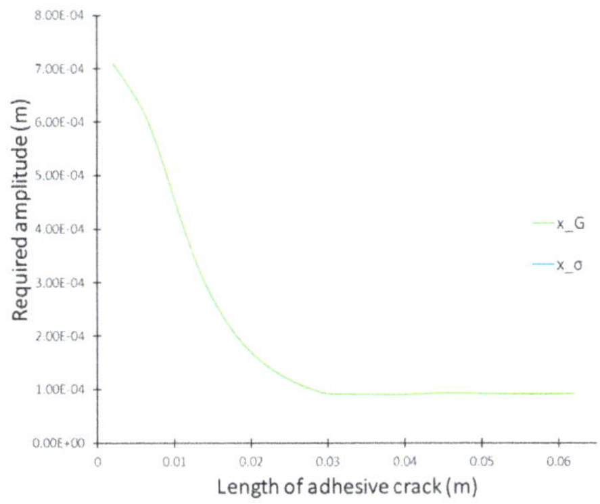

(d) Adhesive initiation.

Figure 5.10. Amplitudes required to initiate fractures with a long ice layer (130 mm total length). Half the plate is represented. (Adapted from [3].)

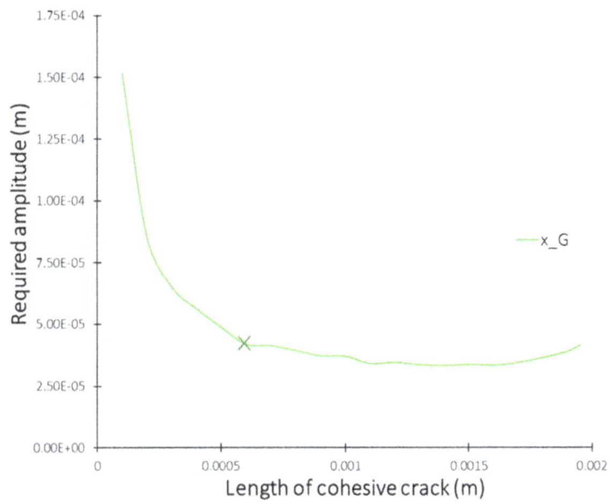

Figure 5.11. Amplitude required to propagate a cohesive fracture within a 2 mm thick ice block. The ~ indicates the fracture initiation location. (Adapted from [3].)

fracture decreases to a relatively low and uniform value, indicating that the propagation within the material is sudden.

5.3.1.2 Numerical study: adhesive fracture initiation and propagation

Once the cohesive fracture is formed, it is necessary to study the adhesive fracture. The 3D model for initiation presents a cohesive crack at the center of the ice layer in the so-called undamaged configuration or state 0 (see figure 5.12(a)). Then, the ice is

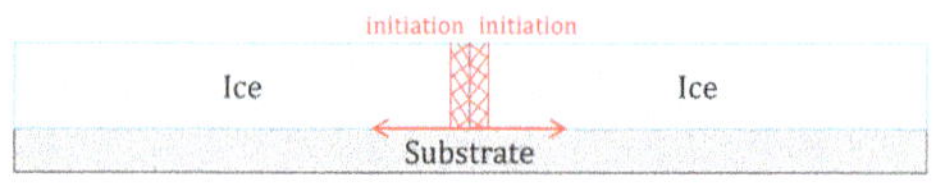

(a) 2D schematic of adhesive initiation from the center.

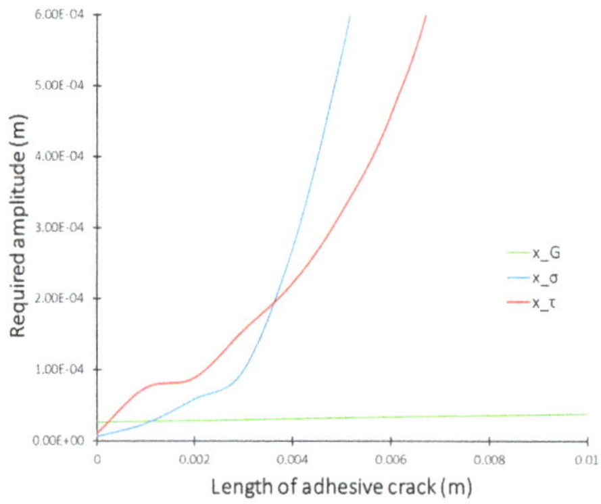

(b) Adhesive initiation from the center.

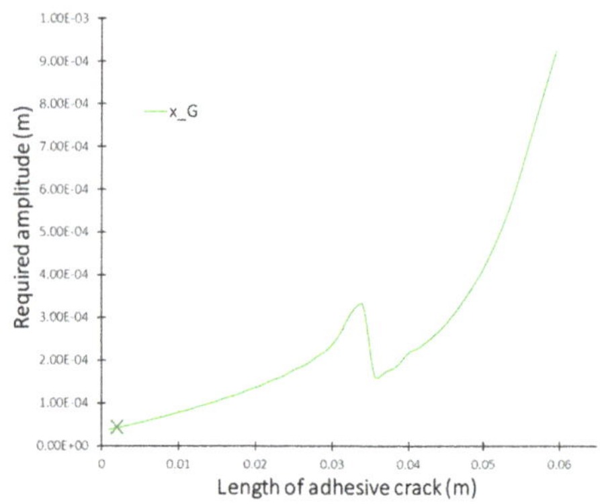

(c) Adhesive propagation from the center.

Figure 5.12. Amplitudes required to initiate and propagate adhesive fractures with a long ice layer (130 mm total length). (Adapted from [3].)

progressively removed from the center of the crack, going toward the sides. To avoid penetration problems between the ice layer and the substrate, instead of removing the contact, the ice is removed. This was done after verifying that the trend and values of the *G*-curves prior penetration were coherent. Figure 5.12(b) shows the initiation amplitudes calculated for the 3D model. In this case, both the shear stress τ and the tensile stress σ are evaluated. It is possible to observe that the coupled σ–G criterion occurs for lower amplitudes, generating a crack length of 1.2 mm.

The *G*-curve for fracture propagation is evaluated. In figure 5.12(c), the trend of the amplitudes required to ensure fracture propagation is shown. Overall, the amplitude increases with propagation, although a small decrease is observed around 35 mm, corresponding to the end of the piezoelectric transducers. The increase in amplitude is quite steep, suggesting that complete de-icing may not be achievable, as the required displacement might exceed practical limits.

5.3.1.3 Experimental validation

Experimental tests are performed to verify whether they can confirm numerical predictions. In the configuration where a 2 mm thick ice layer covers the sample, it is predicted that an initial cohesive fracture will appear, from which an adhesive fracture will gradually propagate.

The experimental tests confirm the fracture mechanism. Initially, a low-voltage sweep (20 Vpk) is performed to identify the first out-of-plane resonant frequency. Subsequently, by exciting the plate at 150 Vpk using both piezoelectric actuators, a cohesive crack appears at the center of the ice block. The voltage is then incrementally increased to 200 Vpk and once the first out-of-plane flexural mode

Figure 5.13. De-icing of a 130 mm long ice layer with the first out-of-plane resonant mode (150 Vpk, 200 Vpk and 250 Vpk). (Adapted from [24]. CC BY 4.0.)

is excited, the adhesive crack propagates from the previously formed cohesive crack (indicated by the whitish area). This process is repeated at 250 Vpk. Figure 5.13 illustrates the various testing stages. In the final image, the ice remains adhered to the surface, and complete de-icing is not achieved. Increasing the voltage to further propagate the adhesive fracture was not feasible, as it would risk depolarizing the piezoelectric transducers.

5.3.2 First out-of-plane flexural mode: plate covered with a short ice layer

5.3.2.1 Numerical study: adhesive fracture initiation and propagation

The same numerical study is performed to evaluate initiation on a short ice layer for the first out-of-plane flexural mode, positioned at the center of the plate and having a total length of 25 mm. The goal is again to understand whether the fracture will initiate cohesively or adhesively (figures 5.14(a) and 5.14(b)). As before, two 3D numerical models are studied: one with ice debonded within its thickness at the center of the plate and the other with ice debonded on the sides, at the interface between the ice layer and the substrate. In figure 5.14(c), the amplitudes are plotted against the 2 mm ice thickness, and it is possible to see that the coupled criterion is respected for an amplitude of approximately 0.16 mm. In figure 5.14(d), the amplitudes are plotted against the half-length of the ice block; it is possible to observe that the amplitudes for shear stress and tensile stress are quite similar, with the σ curve being slightly below the τ curve. The $\sigma - G$ coupled criterion is respected when the plate vibrates with a displacement of 80 μm. From these two plots, it is expected that adhesive initiation will occur, with an initiation length of approximately 1.25 mm.

The propagation behavior, i.e. the amplitude x_G required to continue propagating the adhesive fracture, is reported in figure 5.15. The figure demonstrates that an increase in displacement is required to further propagate the fracture.

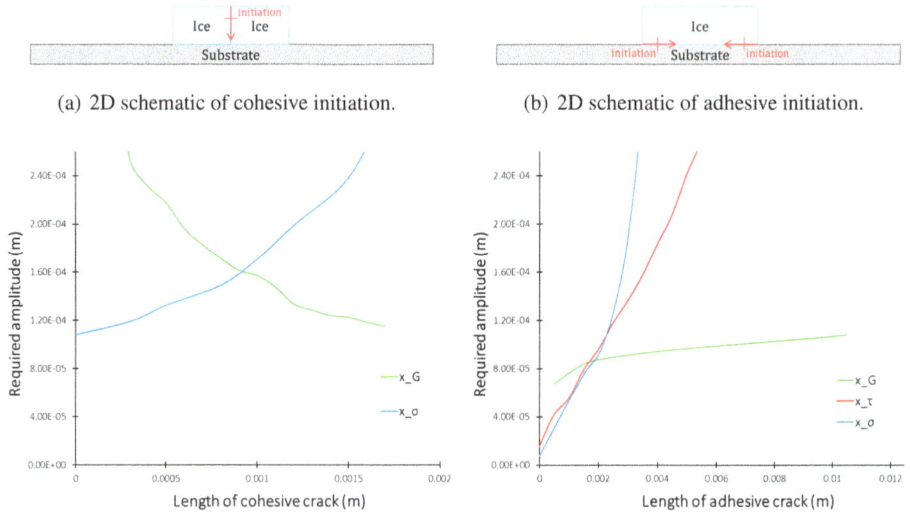

(a) 2D schematic of cohesive initiation.

(b) 2D schematic of adhesive initiation.

(c) Cohesive initiation.

(d) Adhesive initiation.

Figure 5.14. Amplitudes required to initiate fracture with a short ice layer (25 mm total length). (Adapted from [3].)

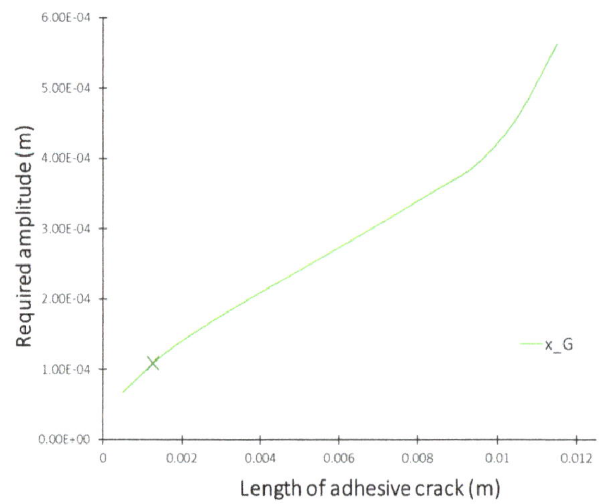

Figure 5.15. x_G amplitude required to propagate fracture with a short ice layer (25 mm total length). The ~ indicates the fracture initiation location. (Reproduced from [3].)

5.3.2.2 Experimental validation

In this configuration, debonding is predicted to begin at the edges of the interface and progress toward the center of the plate without the formation of cohesive fractures. The ice layer, having a 25 mm length and a thickness of 2 mm, is formed in a freezer by pouring supercooled demineralized water on a limited area of the plate. After detecting the resonance peak, the de-icing test is performed at 200 Vpk. As presented in figure 5.16, the propagation begins at the plate edges, and the ice

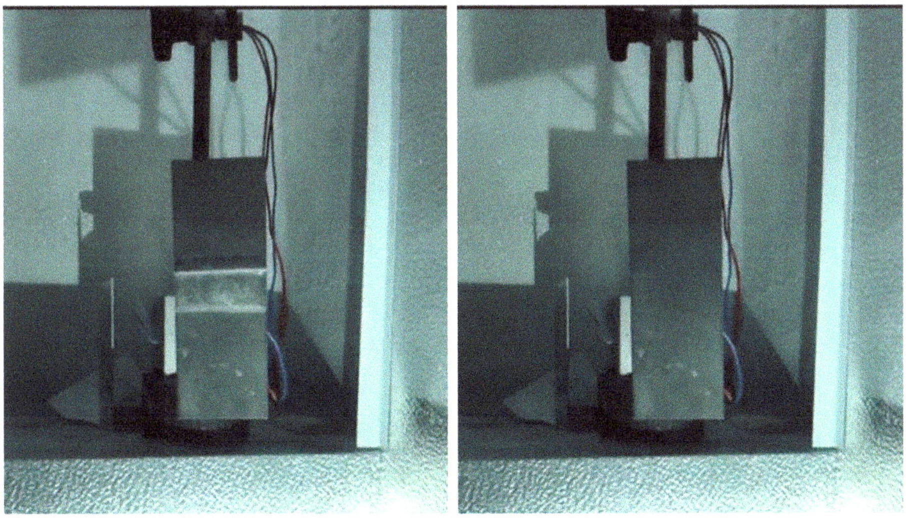

Figure 5.16. De-icing of a 25 mm long ice layer (200 Vpk). (Adapted from [24]. CC BY 4.0.)

detaches without the formation of any cohesive fracture. Therefore, the experimental tests confirm what was predicted by the numerical analyses.

5.3.3 First extensional mode: plate entirely covered with ice

5.3.3.1 Numerical study: cohesive fracture initiation and propagation

Finally, a numerical study is performed for the first extensional mode. Similarly to the study of the first out-of-plane flexural mode, the initiation of the cohesive fracture is studied at the center of the ice block (see figure 5.17(a)) and the initiation of the adhesive fracture is studied from the sides of the ice block.

The adhesive initiation study did not result in any measurable energy release rate. This outcome can be attributed to the observation, as demonstrated in numerical simulations, that for an extensional mode, the energy distribution is minimally affected by an adhesive fracture originating at the sides. Significant variation in the energy distribution requires the presence of a cohesive fracture within the system.

Figure 5.17(b) illustrates the plot corresponding to cohesive crack initiation, which requires an amplitude of approximately 0.15 μm to generate a crack of 0.7 mm, while figure 5.17(c) reports the curve for cohesive fracture propagation. Similarly to cohesive fracture propagation for the first out-of-plane flexural mode, the displacement required to propagate the adhesive fracture after its initiation is relatively constant, indicating that the fracture is propagating suddenly.

5.3.3.2 Numerical study: adhesive fracture initiation and propagation

The adhesive fracture initiation and propagation once the cohesive fracture is formed are also evaluated. For simplicity, the appearance of cohesive fractures subsequent to the first fracture is not evaluated. Still, based on the experimental results obtained with the high-speed camera shown in figure 5.5, they seem to appear either simultaneously

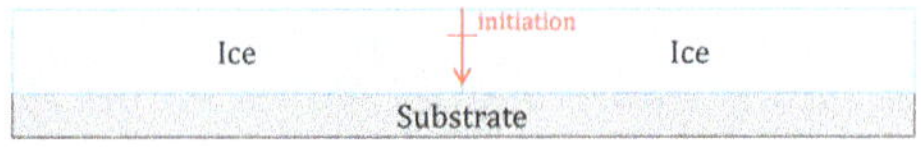

(a) 2D schematic of cohesive initiation.

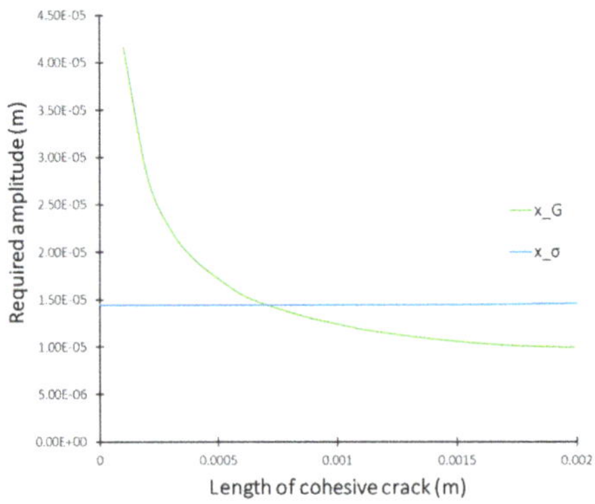

(b) Cohesive initiation.

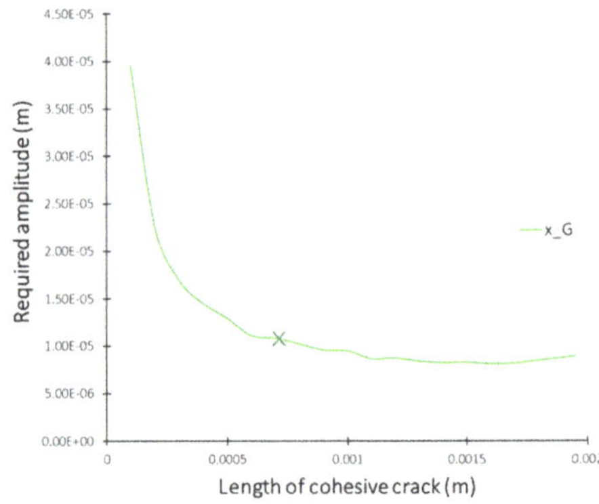

(c) Cohesive propagation.

Figure 5.17. Amplitudes required to initiate and propagate cohesive fractures with a long ice layer (130 mm total length). (Adapted from [3].)

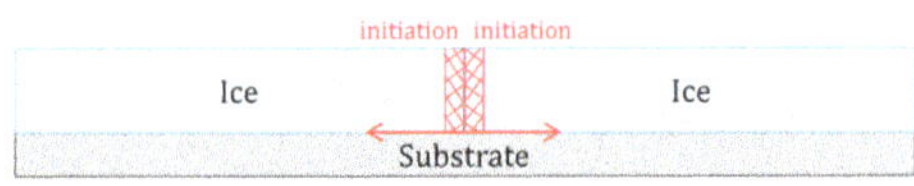

(a) 2D schematic of adhesive initiation from the center.

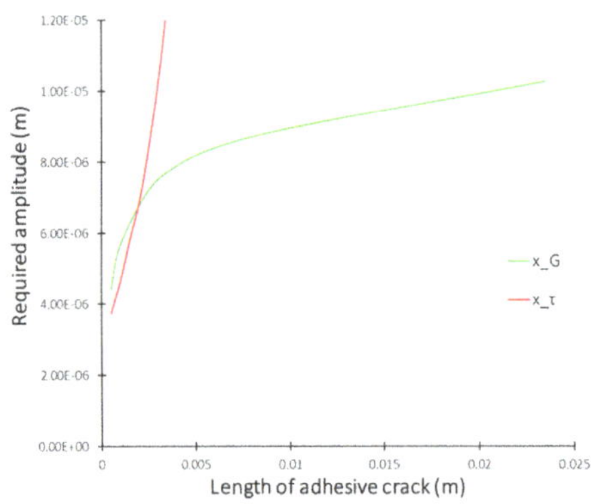

(b) Adhesive initiation from the center.

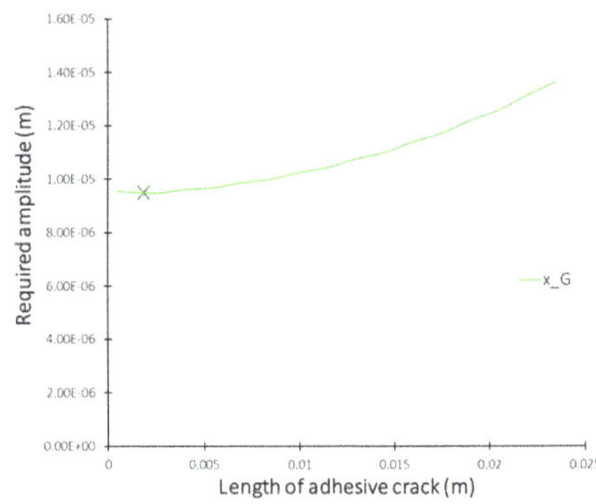

(c) Adhesive propagation from the center.

Figure 5.18. Amplitudes required to initiate and propagate adhesive fracture with a long ice layer (130 mm total length). (Adapted from [3].)

or slightly after adhesive fracture propagation. It seemed reasonable to consider only the adhesive mechanism starting from the central cohesive crack.

Again, the 3D model at state 0 for studying the adhesive fracture initiation presents a cohesive crack at the center of the ice layer and is shown in figure 5.18(a).

Then, the ice is progressively removed from the center of the crack, going toward the sides. Figure 5.18(b) shows the amplitudes calculated for the 3D models. In this case, only the shear stress τ is plotted, as the amplitudes required for the tensile stress mechanism are too big. The coupled $\tau - G$ criterion generates a crack length of approximately 2 mm. The required amplitudes for propagating the adhesive fractures are also reported in figure 5.18(c). It is possible to notice that the curve is again quite uniform, thus explaining the sudden detachment from the surface.

5.3.3.3 Experimental validation

In the configuration where a 2 mm thick ice layer uniformly covers the entire sample, the fracture is first supposed to appear cohesively. To identify the frequency of the first extensional mode, a low-voltage sweep (20 Vpk) is performed. The de-icing is then performed at 200 Vpk using both piezoelectric actuators, generating several cohesive cracks within the ice and a simultaneous adhesive propagation that leads to the detachment of the ice. Snapshots of the de-icing process are presented in figure 5.19, and the ice completely sheds 1 s after activation of the piezoelectric actuators.

Figure 5.19. De-icing of a 130 mm long ice layer (200 Vpk). (Adapted from [24].)

5.3.4 Considerations on the de-icing performance of resonant modes

It is important to observe that certain resonant modes exhibit better de-icing performance than others. This difference can be attributed to the more uniform energy release rate during propagation in extensional modes as already mentioned in [5], but also to the contribution of kinetic energy. In fact, for a material volume, the energy balance can be expressed as follows:

$$\frac{\mathrm{d}W_{\mathrm{m}}}{\mathrm{d}S} = \frac{\mathrm{d}E_{\mathrm{p}}}{\mathrm{d}S} + \frac{\mathrm{d}E_{\mathrm{a}}}{\mathrm{d}S} + \frac{\mathrm{d}E_{\mathrm{d}}}{\mathrm{d}S} + \frac{\mathrm{d}E_{\mathrm{k}}}{\mathrm{d}S}, \tag{5.7}$$

where W_{m} is the mechanical work that is supplied to the material by the external forces, E_{p} is the potential energy, E_{a} is the surface energy, E_{d} is the dissipated energy and E_{k} is the kinetic energy.

Conventionally, the dissipation energy is neglected, which results in Griffith's fracture criterion

$$G > G_c, \tag{5.8}$$

with G the energy release rate defined as

$$G = \frac{dW_m}{dS} - \frac{dE_p}{dS} - \frac{dE_k}{dS} \tag{5.9}$$

and G_c the critical energy release rate defined as

$$G_c = \frac{dE_a}{dS}. \tag{5.10}$$

In an undamped vibratory system, the mechanical energy W_m supplied to the system is constant, and the term $\frac{dW_m}{dS}$ can be neglected, simplifying equation (5.9).

For out-of-plane flexural modes, the contribution of kinetic energy is minimal (see figure 5.20 for the distribution of kinetic and potential energies within the ice layer). Therefore, the quasi-static assumption is valid and the variation of the kinetic energy can be neglected. Equation (5.9) is even more simplified and leads to Griffith's fracture criterion $G > G_c$ expressed as

$$-\frac{dE_p}{dS} > \frac{dE_a}{dS}. \tag{5.11}$$

For other types of modes, such as extensional and in-plane flexural modes, the quasi-static assumption is valid only in the central area of the plate, where kinetic energy is negligible. However, for these types of modes, when the fracture propagates toward the edges, small ice blocks remain attached to the surface and the variation of the kinetic energy can no longer be neglected, as illustrated in figure 5.21. This leads to Griffith's fracture criterion $G > G_c$ expressed as

$$-\frac{dE_k}{dS} - \frac{dE_p}{dS} > \frac{dE_a}{dS}. \tag{5.12}$$

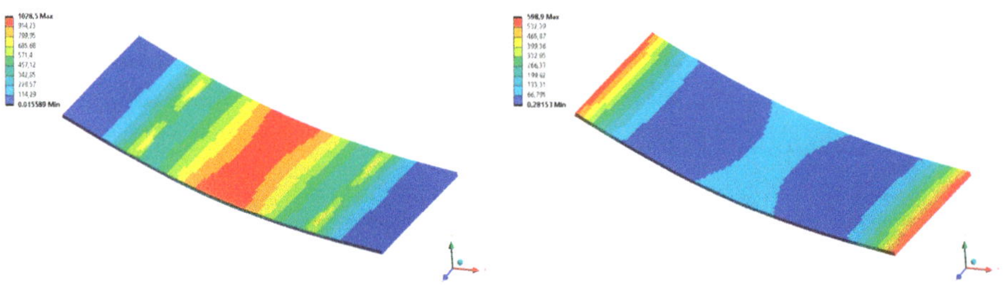

(a) Potential energy. (b) Kinetic energy.

Figure 5.20. Energy distributions in the ice layer for the first out-of-plane flexural mode. (Adapted from [3].)

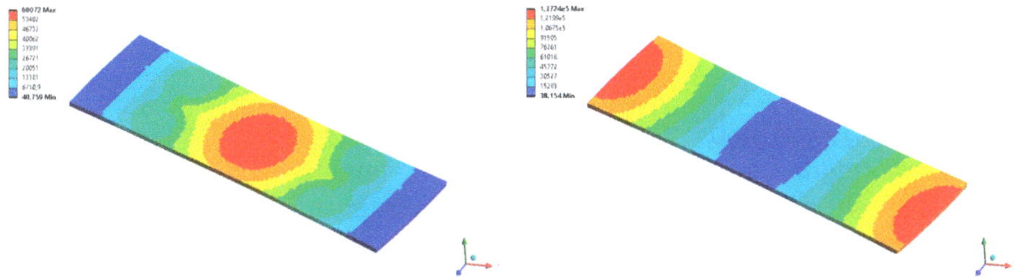

(a) Potential energy. (b) Kinetic energy.

Figure 5.21. Energy distributions in the ice layer for the first extensional mode. (Adapted from [3].)

When considering the coupled criterion, it is possible that the energy release rate criterion is consistently satisfied at the edges due to the combined contributions of the two energy components previously discussed. Consequently, the propagation might be controlled exclusively by the shear stress criterion. This hypothesis offers a potential explanation for the observed behavior; however, additional numerical validations are required to confirm its validity.

5.3.5 Origin of fracture mechanism in the case of an electromechanical de-icing system

The fracture mechanism originated by electromechanical de-icing systems became a subject of inquiry following Golovin's study [25] who found that, for a horizontal shear test, the fracture mechanism depends on the size of the ice sample. For long interfaces, above a certain critical length l_c, the mechanism is energy-dominated, while, below l_c, the mechanism transitions to stress-dominated. Later, Stendardo *et al* [26] stated that for the fracture mechanism to be stress-dominated, the shear stress must exceed the shear strength of the interface at each point along the interface. The energy-dominated mechanism for long ice blocks is evident from the studies of Golovin and Stendardo and is confirmed by the present study. However, for short ice blocks, the results differ and deserve a specific analysis.

5.3.5.1 Numerical study

It is necessary to identify a stress distribution that is sufficiently uniform for an ice block of small dimensions. For this study, an ice block with a surface area of $8 \times 8\ \mathrm{mm}^2$ and a height of 5 mm is chosen *a priori* and placed at the center of a titanium plate with dimensions $130 \times 50 \times 1\ \mathrm{mm}^3$. A 2D modal analysis is performed under free boundary conditions. The first four out-of-plane flexural modes are observed within the frequency range of 281 Hz to 2450 Hz.

The 2D stress distributions for these modes are compared by analysing the ratio $\tau/\sqrt{P}$, which is a normalized indicator. Here, τ represents the shear stress along the interface, and P denotes the mechanical power per unit area required to vibrate the plate, calculated as in [28]:

$$P = \frac{1}{2A} F\dot{x} = \frac{1}{2A}\frac{K_{\mathrm{eq}}x^2\omega}{Q_{\mathrm{m}}} = \frac{1}{2A}\frac{E_{\mathrm{p}}\omega}{Q_{\mathrm{m}}}. \tag{5.13}$$

In equation (5.13), A is the surface area of the sample, E_{p} is the potential strain energy of the mode under consideration, ω is its angular velocity, and Q_{m} is its quality factor, set to a value of 50. For the first four flexural modes, the ratio $\tau/\sqrt{P}$ is plotted in figure 5.22. This ratio is evaluated at the ice–substrate interface, specifically around 65 mm, since the ice is centrally located on the plate.

For modes 1 and 3, having an antinode at the plate center, the stress distribution is not uniform, with a zero value at the center. In contrast, the modes 2 and 4 with a node at the center exhibit a more uniform stress distribution, but the sides are still characterized by a sharp decrease in value that passes by 0. Therefore, the pure shear mechanism is by default not possible in any of these configurations. The second out-of-plane flexural mode was selected for further studies, to confirm that an energy-dominated mechanism could also occur for a small ice block positioned on the node of a resonant mode.

It is assumed that fracture initiation occurs at the center of the sample, with subsequent propagation toward the edges. Both scenarios (initiation and propagation from the edges, as well as from the center) were analysed, and the results indicate that only the center-initiated fracture mechanism is practically achievable. In the case of edge-initiated fracture, the amplitude required to propagate the fracture is excessively high, preventing effective detachment of the central area of the ice block.

In figure 5.23(c), the coupled τ–G criterion for fracture initiation is presented. Following the formation of the initial crack, the fracture propagation behavior is assessed. Figure 5.23(d) illustrates the propagation curve, where the x mark represents the initial crack length. The final part of the propagation curve, which corresponds to the phase of effective propagation, is quite stable, and eventually

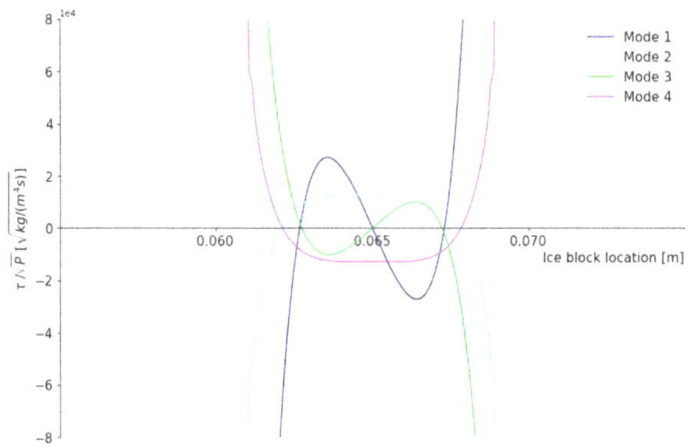

Figure 5.22. $\tau/\sqrt{P}$ at the ice–substrate interface for the first four resonant flexural modes. The ice block is located at the center of the plate (65 mm). (Adapted with permission from [27].)

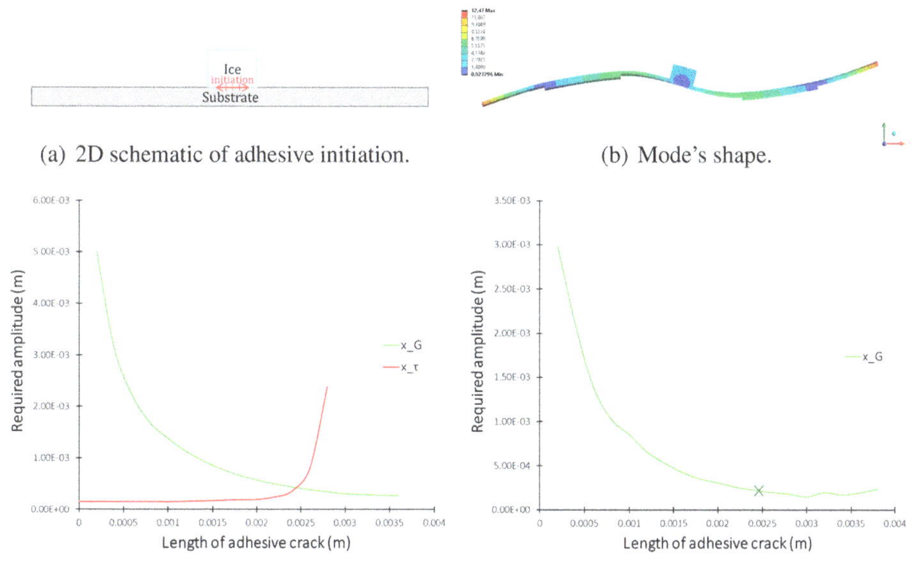

(a) 2D schematic of adhesive initiation. (b) Mode's shape.

(c) Adhesive initiation from the center. (d) Adhesive propagation from the center.

Figure 5.23. Amplitudes required to initiate and propagate adhesive fracture with a short ice layer (8 mm). (Adapted from [3].)

leads to a rapid detachment of the small ice block. The study shows that, even for small ice blocks, the fracture propagates with an energy-dominated detachment.

5.3.5.2 Experimental verification

The ice block, measuring $8 \times 8 \times 5$ mm^3, is prepared on the sample in a freezer set at −20 °C using demineralized water and placed at the center of the plate using a flexible mold (figure 5.9(b)). Initially, a low-voltage sweep (20 Vpk) is employed to identify the second out-of-plane resonant frequency. Following this, the system is subjected to a 200 Vpk excitation with a downward frequency sweep, transitioning from the highest to the lowest frequency. Both actuators are used during this process. Given the rapid nature of the detachment, a laser vibrometer is used to measure the system's speed at one of the mode's antinodes and observe its behavior near the resonant peak. As illustrated in figure 5.24, the oscillation speed progressively increases as the frequency approaches the resonance at approximately 1050 Hz. At this point, the ice block suddenly detaches from the titanium plate, evidenced by a sharp drop in oscillatory speed. After the detachment, the mechanical properties of the system are slightly altered due to the mass reduction, resulting in a shift in the resonant frequency. Continuing the frequency sweep reveals a second, shifted peak at around 1025 Hz.

Several tests were conducted, and in all cases, the detachment mechanism occurred abruptly, with the fracture propagation path remaining undetectable in the recorded videos. The support provided by numerical analyses proved essential in identifying the fracture mechanism.

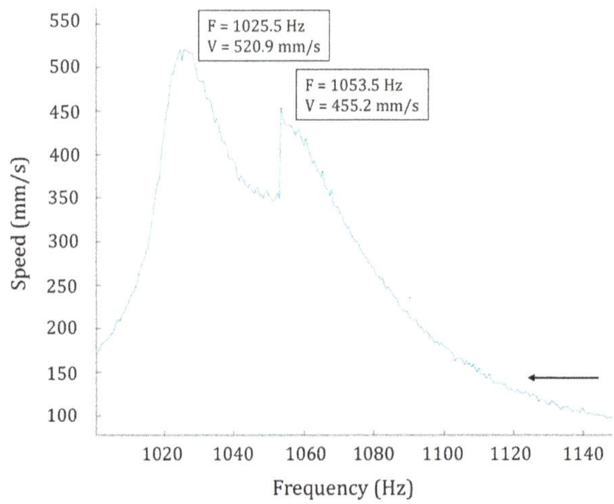

Figure 5.24. Oscillation speed versus excitation frequency. A downward frequency sweep was performed until the ice block detached (first drop of V at about 1050 Hz). (Adapted with permission from [27].)

5.3.5.3 Considerations on the de-icing mechanism: resonant system versus horizontal shear test

As previously discussed, the results of the works on horizontal shear tests and electromechanical resonant de-icing systems differ for small ice blocks. In a horizontal shear test, the mechanism is stress-dominated, while for an electromechanical resonant de-icing system it is energy-dominated. It was previously demonstrated that it is important for the stress distribution to be quite uniform along the interface, but another explanation relies on the amount of energy that can be supplied to the system.

As already mentioned, for electromechanical resonant de-icing systems, the energy release rate can be expressed as

$$G = -\frac{\mathrm{d}E_\mathrm{p}}{\mathrm{d}S} \tag{5.14}$$

or

$$G = -\frac{\mathrm{d}E_\mathrm{k}}{\mathrm{d}S} - \frac{\mathrm{d}E_\mathrm{p}}{\mathrm{d}S}. \tag{5.15}$$

However, in a horizontal shear test, as the actuator pushes against the ice block, the mechanical work input to the system can increase continuously. In this case, $\frac{\mathrm{d}W_\mathrm{m}}{\mathrm{d}S} \neq 0$ and, considering equation (5.9), the energy release rate is expressed as

$$G = \frac{\mathrm{d}W_\mathrm{m}}{\mathrm{d}S} - \frac{\mathrm{d}E_\mathrm{p}}{\mathrm{d}S}. \tag{5.16}$$

It is possible to conclude that, for horizontal shear tests, the energy release rate criterion can be consistently satisfied as the supplied energy increases. Under these conditions, fracture propagation is stress-dominated. In contrast, for electromechanical resonant de-icing systems, where the mechanical energy input is fixed, such continuous energy supply is not possible, making the mechanism energy-dominated.

5.4 Conclusions

This chapter examines the mechanisms of fracture initiation and propagation in electromechanical resonant de-icing systems. It starts by introducing the working principles of electromechanical resonant de-icing systems and types of resonant modes that can be adopted. A hybrid analytical/numerical study has been performed to assess whether the adopted fracture theory would be able to predict how fracture initiates and propagates for different modes. The method involves calculating and comparing the vibration amplitudes necessary to attain the critical mechanical properties of the ice (tensile strength, shear strength, or toughness) and using them to assess whether the cohesive or adhesive coupled criterion is satisfied first.

This technique has proven to accurately predict the de-icing mechanisms for both out-of-plane flexural and extensional modes. Additionally, this study demonstrates that the fracture mechanism is significantly influenced by the size of the ice layer.

Further analysis of the second out-of-plane flexural mode examined the effects of both the dimensions of the ice block and its position relative to the antinodes of the resonant mode. The results show that, even for ice blocks with very small dimensions, the de-icing mechanism is likely to be energy-dominated rather than stress-dominated. This is primarily due to the uneven stress distribution, which prevents ice detachment by uniformly exceeding the shear strength across the entire interface.

It is important to observe that certain resonant modes exhibit better de-icing performance. In this study, it was observed that extensional modes can remove large portions of ice from a surface, whereas flexural modes are unable to achieve the same result, even with increased voltage. This difference can be attributed not only to the more uniform energy release rate during propagation in extensional modes but also to the contribution of kinetic energy, which plays an important role in helping to remove residual ice along the edges of the surface.

The study focused on simple geometries and low-order modes; however, the methodology can be effectively adapted to more complex structures and higher-order modal shapes, providing a useful tool for predicting de-icing mechanisms in resonant ice protection systems.

Acknowledgements

This project has received funding from the European Union's Horizon 2020 research and innovation program under the Marie Skłodowska-Curie Grant Agreement No. 956 703 (SURFICE Smart surface design for efficient ice protection and control).

References

[1] Bai T, Zhu C, Miao B, Li K and Zhu C 2015 Vibration de-icing method with piezoelectric actuators *J. Vibroeng.* **17** 61–73

[2] Villeneuve E, Harvey D, Zimcik D, Aubert R and Perron J 2015 Piezoelectric deicing system for rotorcraft *J. Am. Helicopter Soc.* **60** 1–12

[3] Gastaldo G 2025 Optimization of hybrid ice protection systems based on coatings and electromechanical actuators *PhD Thesis* ISAE-SUPAERO (submitted)
[4] Budinger M, Pommier-Budinger V, Bennani L, Rouset P, Bonaccurso E and Dezitter F 2018 Electromechanical resonant ice protection systems: analysis of fracture propagation mechanisms *AIAA J.* **56** 4412–22
[5] Budinger M, Pommier-Budinger V, Reysset A and Palanque V 2021 Electromechanical resonant ice protection systems: energetic and power considerations *AIAA J.* **59** 2590–602
[6] Leguillon D 2002 Strength or toughness? A criterion for crack onset at a notch *Eur. J. Mech. A* **21** 61–72
[7] Martin E, Vandellos T, Leguillon D and Carrère N 2016 Initiation of edge debonding: coupled criterion versus cohesive zone model *Int. J. Fract.* **199** 157–68
[8] Palanque V, Marbœuf A, Budinger M, Pommier-Budinger V and Bennani L 2022 Improving mechanical ice protection systems with substrate thickness and topology optimization *Cold Regions Sci. Technol.* **202** 103641
[9] Liu Y, Xi N, Zhang X and Liu N 2020 Study of atmospheric ice adhesion properties of superhydrophobic surface by *in-situ* shear system *Mater. Express* **10** 1704–10
[10] Chu M C and Scavuzzo R J 1991 Adhesive shear strength of impact ice *AIAA J.* **29** 1921–6
[11] Tryde P 1980 Physics and mechanics of ice *Naturwissenschaften* **67** 556–9
[12] Frederking R M W, Svec O J and Timco G W 1988 On measuring the shear strength of ice National Research Council of Canada, Institute for Research in Construction pp 76–88
[13] Rønneberg S, Laforte C, Volat C, He J and Zhang Z 2019 The effect of ice type on ice adhesion *AIP Adv.* **9** 055304
[14] Hassan M F, Lee H P and Lim S P 2010 The variation of ice adhesion strength with substrate surface roughness *Meas. Sci. Technol.* **21** 075701
[15] Andrews E H and Lockington N A 1983 The cohesive and adhesive strength of ice *J. Mater. Sci.* **18** 1455–65
[16] Wenyuan M and Yingkui G 2015 Experimental study on mechanical properties of ice *AASRI International Conference on Industrial Electronics and Applications (IEA 2015) Proceedings of the AASRI International Conf. on Industrial Electronics and Applications* (Atlantis Press) pp 192–96
[17] Petrovic J J 2003 Review mechanical properties of ice and snow *J. Mater. Sci.* **38** 1–6
[18] Druez J, Nguyen D D and Lavoie Y 1986 Mechanical properties of atmospheric ice *Cold Reg. Sci. Technol.* **13** 67–74
[19] Reich A, Scavuzzo R and Chu M 1994 Survey of mechanical properties of impact ice *32nd Aerospace Sciences Meeting and Exhibit* (Reno, NV: American Institute of Aeronautics and Astronautics) p 712
[20] Huang X, Tepylo N, Pommier-Budinger V, Budinger M, Bonaccurso E, Villedieu P and Bennani L 2019 A survey of icephobic coatings and their potential use in a hybrid coating/active ice protection system for aerospace applications *Prog. Aerosp. Sci.* **105** 74–97
[21] Dawood B, Yavas D and Bastawros A 2023 Ice adhesion characterization using mode-I and mode-II fracture configurations *J. Appl. Mech.* **90** 121015
[22] Petrovic J J 2003 Review mechanical properties of ice and snow *J. Mater. Sci.* **38** 1–6
[23] Goodman D J and Tabor D 1978 Fracture toughness of ice: a preliminary account of some new experiments *J. Glaciol.* **21** 651–60
[24] Gastaldo G, Palanque V, Budinger M and Pommier-Budinger V 2022 Stress and energy release rate influence on ice shedding with resonant electro-mechanical de-icing systems *ICAS 2022–33rd Congress of the Int. Council of the Aeronautical Sciences*

[25] Golovin K, Dhyani A, Thouless M D and Tuteja A 2019 Low-interfacial toughness materials for effective large-scale deicing *Science* **364** 371–5

[26] Stendardo L, Gastaldo G, Budinger M, Pommier-Budinger V, Tagliaro I, Ibáñez-Ibáñez P F and Antonini C 2023 Reframing ice adhesion mechanisms on a solid surface *Appl. Surf. Sci.* **641** 158462

[27] Stendardo L, Gastaldo G, Budinger M, Antonini C, Pommier-Budinger V and Patiño A C O 2023 Dynamic and static test methods: quantifying the shear strength at the interface of iced substrates *Technical Report* SAE Technical Paper

[28] Palanque V, Budinger M, Pommier-Budinger V, Bennani L and Delsart D 2021 Electro-mechanical resonant ice protection systems: power requirements for fractures initiation and propagation *AIAA Aviation 2021 Forum, Session: Ice Protection System Design and Analysis*

IOP Publishing

Smart Surface Design for Efficient Ice Protection and Control

Carlo Antonini and Irene Tagliaro

Chapter 6

The promise of reticular materials: scalable icephobicity and beyond

Simrandeep Bahal, Vikramjeet Singh, Jianhui Zhang, Priya Mandal, Prasenjit Kabi and Manish K Tiwari

Parts of this chapter have been reproduced from [23]. CC BY 4.0.

Reticular materials, metal–organic frameworks (MOFs) and covalent organic frameworks (COFs) are constructed by linking molecular building blocks (organic and/or organic) into extended, periodic structures, at the molecular level. Among others, these materials have shown to be very useful for their surface applications which also include anti-icing and anti-scaling. Surfaces that can prevent the formation of ice and scale often require a multiplicity of additional features such as optical transparency, robust impact resistance, and the ability to prevent contamination from low surface energy liquids. Most of the surface solutions have relied on using perfluoro compounds, which are bio-persistent and/or highly toxic, unfortunately. We have recently shown that MOFs/COFs can be utilized for these applications and possess necessary functionalities. Surface-grown MOFs resulted in transparent coatings with excellent repellent properties and low adhesion towards ice. A simple strategy to exploit the nanoconfinement effect, which remarkably delays the nucleation of ice and scale formation on surfaces, was observed on COFs. Ice nucleation is suppressed down to −28 °C, scale formation is avoided for >2 weeks in supersaturated conditions, and jets of organic solvents impacting at Weber numbers $>10^5$ are resisted with surfaces that also offer optical transparency (>92%). We believe that these coatings have potential and should be explored at larger scale for anti-icing and other features.

6.1 Introduction

Controlling undesirable nucleation, specifically ice nucleation, is an issue of broad implications, especially in the case of scaling and icing [1–4]. Preventing/delaying ice formation on surfaces is a challenge, with a major impact in a variety of

doi:10.1088/978-0-7503-6009-8ch6

applications, such as heat exchangers, oil and gas extraction, water treatment systems [3, 5], etc. In fact, nucleation control has broader implications. For example, in a number of these systems nucleation control needs to extend to other undesirable deposits such as either inorganic or organic or hybrid fouling. As a specific example, the scaling of $CaCO_3$ on pipelines adds to high energy consumption and dangerous microbial growth. There are several active and passive scale inhibition methods, such as mechanical cleaning, acid–base cleaning and the addition of scale inhibitors (phosphates and polymeric) that are being practiced at the commercial level [6, 7]. However, they are either expensive and/or inefficient and, often, pose a direct threat to the environment by releasing hazardous chemicals [8]. There have been extensive efforts to understand the effect of ions, pH, temperature, medium flow rate, impurities, suspended solids, substrate materials, and surface physicochemical properties on the nucleation and crystal growth of ice and foulants such as $CaCO_3$ [3]. For example, Zhao *et al* [9] showed that a thin hydrophobic polymer coating significantly reduced the heterogenous nucleation of $CaCO_3$. However, these coatings often face durability challenges and are unsuited for conditions where other contaminants (e.g. microbes, dirt, oils, etc) are present. Similar to scaling, undesirable ice formation on infrastructure and transport components (e.g. power lines, ground transport vehicles, aircrafts, etc) can impair safety and may cause life-threatening accidents [10]. Therefore, passive prevention of ice and scale formation on surfaces, by using liquid-repellent, superhydrophobic and slippery liquid infused surfaces (SLIPSs) has received due attention [11, 12]. However, most of these surface treatments have relied on the use of perfluoro compounds (PFCs) due to their very low surface energy (<10 mN m^{-1}), among others, which has serious health and environmental concerns [13–15]. In particular, the use of per- and polyfluoroalkyl substances (PFASs) has emerged as a major environmental and human health concern. Additionally, these surfaces tend to lose their non-wetting characteristics under harsh conditions and are not suitable for long-term use [16, 17]. Recent efforts to replace the PFAS with safer compounds and/or their decomposition into benign inorganic fluoride ions and oxygenated organic molecules explain the seriousness of the situation [14, 18, 19]. Due to their high surface energy (>20 mN m^{-1}), fluorine-free surfaces are susceptible to contamination from oils and low surface tension liquids which are commonplace in everyday usage. Strategies, such as re-entrant surface texture to achieve superoleophobicity are impractical due to their complicated design and low impact resistance [20, 21].

Reticular materials, including metal–organic framework (MOF) and covalent organic framework (COF) based slippery nanohierarchical surfaces, were recently introduced to overcome some of these problems [22, 23]. Reticular chemistry is the study and synthesis of extended structures by the directional bonding of molecular building units. This area of chemistry primarily focuses on the design and creation of frameworks with pre-defined and tunable structures and functionalities. The most well-known outcomes of reticular chemistry are MOFs and COFs. MOFs consist of metal ions or clusters coordinated to organic ligands to form one-, two-, or three-dimensional structures [24].

6.2 Metal–organic frameworks

MOFs are two-dimensional or three-dimensional porous crystalline materials constructed of metal nodes and organic linkers through coordination bonds. They can be adjusted by altering the nature of the metal cations and linkers as well as post-synthetic modifications (PSMs), hence their flexible structures can be specifically designed according to applications [54]. MOFs possess an ultrahigh specific surface area, reaching up to 10 000 m^2 g^{-1}, which is significantly higher than that of zeolites and activated carbon [25]. Because of these excellent characteristics, MOF materials have a wealth of applications in fields including gas storage, energy conversion, chemical sensing, drug delivery, proton conduction, and catalysis [26, 27]. Among others, UiOs (UiO stands for University of Oslo), MOF-5 and ZIF-8 have been popular among researchers due to their associated structural advantages [28–30]. For example, due to the high oxidation state of Zr(IV) and its strong coordination bond with hard bases carboxylates, UiO-type MOFs possess excellent chemical stability, thermal stability (stable up to 500 °C), and mechanical stability. MOF-5 is constructed by connecting Zn_4O clusters with terephthalic acid (BDC) to form a cubic framework. MOF-5 has a robust structure, with a porosity of 61%, and a BET surface area of 2320 m^2 g^{-1} [30]. ZIF-8 is a type of small-pore zeolitic imidazolate framework (MOF) composed of zinc ions coordinated with 2-methylimidazole ligands [31]. The sizes of its connection windows and the internal pore diameters are 3.4 and 11.6 Å, respectively. ZIF-8 is easy to synthesize and possesses good chemical stability. It remains unchanged after immersion in boiling alkaline water and organic solvents [32].

An MOF based nanocoating, prepared through layer-by-layer technique on glass, was tested for its ice adhesion capabilities (figure 6.1). Bare glass and glass functionalized with the same silane used to functionalize the MOF were used as

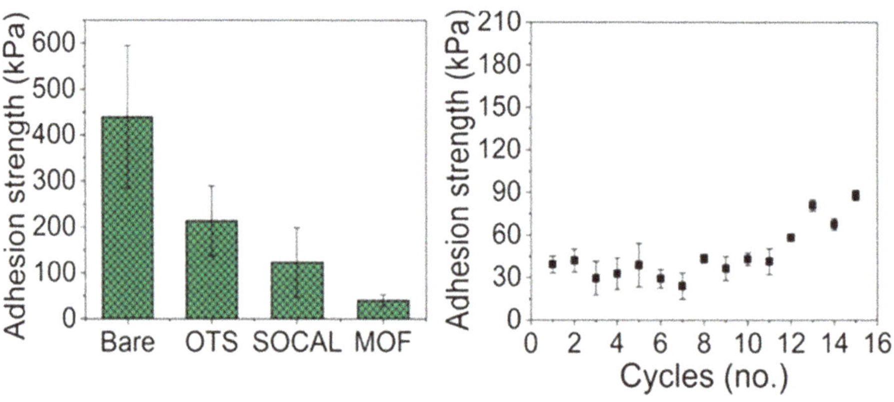

Figure 6.1. Ice adhesion on MOF surfaces. (A) Ice adhesion strength of different treatments on glass substrate including bare glass as a control, glass functionalized with trichloro-octadodecyl silane (OTS), an ultrasmooth surface based on polydimethylsiloxane oligomers (4) (SOCAL), and our nanohierarchical MOF surfaces. (B) Change in the ice adhesion strength of the MOF surface with up to 15 repeated icing/deicing cycles. (Adapted with permission from [22]. Copyright (2021) American Chemical Society.)

Figure 6.2. SEM image recorded after 15 icing–deicing cycles showing cracks. (Adapted with permission from [22]. Copyright (2021) American Chemical Society.)

references. Additionally, ultrasmooth polydimethylsiloxane oligomers (SOCAL)-treated glass was also used. The lowest ice adhesion strength of 35 ± 10 kPa was recorded on the nanohierarchical MOF surface, whereas the ice adhesion strength on hydrophilic MOFs (prior to functionalization) was measured to be 100.5 ± 5 kPa. The surface robustness was tested by repeating icing–deicing cycles; no change in the ice adhesion strength was observed up to 11 cycles. The morphology changed noticeably after 15 icing–deicing cycles and showed some cracks in the coating (figure 6.2).

6.3 Covalent organic frameworks

COFs, another type of reticular structure, are ordered, porous, crystalline polymers composed of organic small molecule ligands connected by strong covalent bonds [33, 34]. Unlike MOFs, COFs are primarily made up of light elements, hence they possess a low mass density [35]. The strong covalent bonds endow COFs with high thermal stability, chemical stability, and permanent porosity [36]. The covalent bonding in COFs occurs only within the 2D sheets. These sheets form an ordered, periodic layered structure through π–π stacking interactions. And 3D COFs utilize sp^3 carbon, silanes, or boron atoms to extend the network into three-dimensional space. Due to characteristics such as porosity, high stability, and π stacking, COFs also have a wide range of applications in gas storage, gas separation, superhydrophobic interfaces, catalysis, and optoelectronics, among others. A commonly used COF synthesized through the Schiff base reaction, TpPa, has high chemical stability [37]. The chemically stable COF connected by β-ketoenamine linkage is constructed by a one-pot reaction between 1,3,5-triformylphloroglucinol (Tp) and p-phenylenediamine (Pa-1). The irreversible proton tautomerism leads to the formation of COF linked by β-ketoenamine. Due to its irrevversible nature, the β-ketoenamine form does not revert to the starting material upon contact with water and exhibits unprecedented chemical stability in water as well as in acidic (9 M HCl) and basic (9 M NaOH) media [38].

Overall, because of the precise control of their pore size, shape and chemistry, both MOFs and COFs have been exploited for their potential in surface applications including anti-icing. Understanding the nucleation and crystal growth mechanisms of ice and limescales have motivated works on suitable surfaces [39–41]. For example, Bai *et al* investigated the critical nucleus size of ice formation and showed that beyond 8 nm size, the presence of graphene oxide sheets increased the nucleation temperature (T_{IN}) to around 10 °C [42]. With substantial improvements in classic nucleation theory, some of us previously showed that surfaces with a specific concave nanotexture can significantly suppress ice nucleation through the interface nanoconfinement effect [43]; however, the issue of producing such surface treatments at scale is a major challenge. The fabrication of surfaces with mechanically robust nanotextures with manufacturing precision of a few-nanometre is extremely challenging, even when compromising the scalability. The use of anti-freezing proteins is also an attractive approach, albeit somewhat limited by the highly susceptible nature of the peptide linkages that degrade under mild acidic/basic conditions [44]. Similarly, nucleation investigations on the formation of $CaCO_3$ in fine pores also indicate that nanoscale confinement may be a general strategy that also works for scale inhibition [45]. Despite having a clear theoretical steer, potentially scalable strategies to make surfaces with <10 nm precise nanocavities remains a challenge. The advances in classical nucleation theory suggest that a surface concave nanotexture smaller than the critical nucleus radii (often below 10 nm) could potentially inhibit the nucleation of ice and minerals [43, 46]. Therefore, a O(1) nm precise texture, which is naturally accessible through porous materials such as COFs, MOFs or graphene, etc, offers a tremendous opportunity to realise nanocoatings for nucleation prevention. Top-down nanofabrication approaches are limited in their ability to achieve such high precision. Specifically, COFs have advantages over graphene and MOFs, being mechanochemically more robust and defect-free due to inherent structural characteristics [47, 48]. In this work an imine-linked COF, with 1.8 nm pores, was selected for application as a simple nanocoating for surface nucleation control. Scalable solid–vapour interfacial polymerization (SV-IP) of COFs and subsequent post-functionalisation with silane (to reduce surface energy) is used to obtain surfaces that show excellent anti-icing and anti-scaling characteristics (figure 6.3).

6.4 Surface fabrication and characterization

The concentration of the organic linkers (p-phenylenediamine and 2,4,6-triformylphloroglyucinol) was tuned to maintain the transparency and mechanical robustness by controlling thickness of the coating. As synthesized COF coating was functionalised with trichloro-octadecylsilane to alter the surface wettability from hydrophilic to hydrophobic. The reactant concentrations and polymerisation times were systematically varied to obtain a surface that exhibited $\geqslant 90\%$ transmittance throughout the visible spectrum and the chemical structures of the COF was confirmed via using FTIR, Raman and powder *x*-ray spectroscopies.

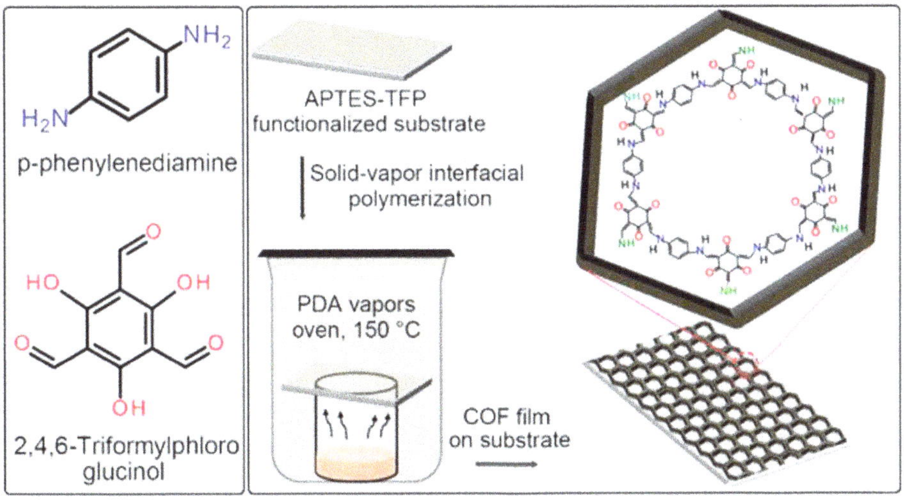

Figure 6.3. (A) Chemical structures of the organic linkers, p-phenylenediamine and 2,4,6-triformylphloroglyucinol used in COF synthesis. (B) Schematic showing solid–vapour interfacial polymerisation method of COF synthesis on 3-aminopropyl-triethoxysilane (APTES) functionalized substrates (glass, aluminium, copper and silicon) through covalent linkage. (Adapted from [23]. CC BY 4.0.)

The silane-treated surface was slippery (figure 6.4(A)) with water drop advancing contact angle (θ_{adv}) 110°, contact angle hysteresis ($\Delta\theta$) 7° and sliding angle 14° (tested using a ~15 μl water droplet). Carbon powder was easily removed by water drops sliding off the surface, confirming the self-cleaning characteristics. A smooth surface with shallow roughness is observed at higher magnifications. The thickness of the COF film, checked by imaging the cross-section under a scanning electron microscope (SEM), was ~231 nm (figure 6.4(B)). The porosity of the COF film was analysed using a transmission electron microscope (TEM) (see figure 6.4(C)). The smooth topology of the COF coating was also confirmed using atomic force microscopy (AFM) (figures 6.4(D)–(F)). The 3D AFM image further confirmed the smoothness and the root-mean-square roughness (R_a) of the coating to be 4.5 nm. Height analysis by AFM confirmed the average coating thickness to be 300 nm.

The scalability potential of the method is shown by coating half of the 20 cm × 15 cm glass substrate. Half of the substrate is left uncoated for a clear presentation of the wettability difference. The functionalised COF surface offered a smooth amphiphobic surface on which low surface tension liquids including alcohols and ketones slid off easily at ⩽15° angles (figure 6.5(A)). Drop impacts at 1.4 and 4.2 m s^{-1} (drop diameter = ~2.7 mm), with relatively low and high impact kinetic energy offered first the insight into liquid impacts on the COF surfaces. The key stages in impact dynamics are shown in figure 6.5(B) and 5(C). A fraction of liquid detached from the droplet at higher speed due to hydrodynamic instability of the rim [49]. Next, high-speed jet impact tests were performed (figure 6.5(D) and (E)). The surface resisted water jet at maximum speed attainable in our set-up (~35 m s^{-1}, liquid Weber number ~42 500), even after repeated tests. We also tested the surface by

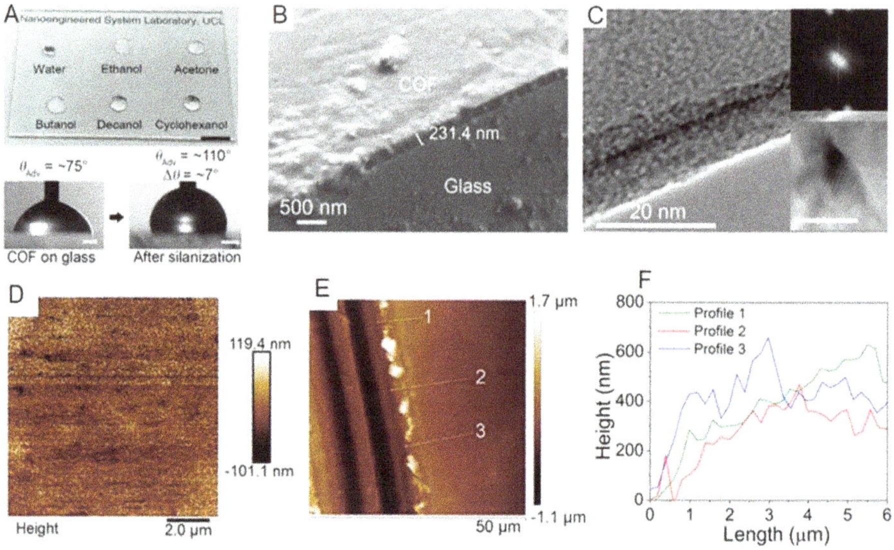

Figure 6.4. (A) Image showing droplets of water, ethanol, acetone, butanol, 1-decanol and cyclohexanol on the transparent COF coated glass substrate (scale bar = 1 cm). The amphiphobicity is achieved without any halogen chemistry. Advancing contact angle (θ_{adv}) and contact angle hysteresis ($\Delta\theta$) for a water droplet on a pristine COF on glass and after functionalisation with trichloro-octadodecyl silane (scale bar = 1 cm). (B) Cross-section of a COF coated on glass confirming nanometre thickness. (C) TEM image showing the porous nature of the COF film. Inset showing the FET pattern and the corresponding zoomed in image. (D) AFM image of the COF surface. (E) AFM scan of a scratched COF coating for height analysis and (F) height profiles of the scratched section on the glass substrate. (Adapted from [23]. CC BY 4.0.)

impacting low surface tension liquids such as jets of ethanol and butanol tested at $\sim$35 m s^{-1}, resulting in a liquid Weber number (We_l) of $\sim$109 300 and $\sim$99 200, respectively [50]. After repeated jet impacts (three times), the water droplet sliding was repeated to confirm the retention of the slippery behaviour of the coating. The effect of substrate topology and composition was checked by measuring the contact angle hysteresis $\Delta\theta$ and sliding angles for water droplets on different substrates coated with COFs (figure 6.5(F)). The coating showed low $\Delta\theta$ ($\leq 10°$) and sliding angles ($\leq 15°$) on all the substrates, being lowest on aluminium (mirror finish). The thermal stability of the COF coating was assessed by heating at various temperatures (20 °C–200 °C) for 60 min followed by cooling down to room temperature and measuring the wetting angles. The surface maintained the slipperiness ($\Delta\theta \leq 10°$) up to 200 °C (figure 6.5(G)). The stability of the coating against corrosion was confirmed by immersing in strongly acidic (pH = 1–2) and basic media (pH = 12–13) for one week each (figure 6.5(H)). Next, the mechanical robustness was tested using the standard tape-peel test. A high-tack tape (3 M VHBTM 5952 with a strong adhesive peel strength of 3900 N m^{-1}) was pressed on the surface with a 2 kg roller, before peeling off to complete one peel cycle (figure 6.6). No significant change was observed even after 50 repetitive cycles (figures 6.5 and 6.7).

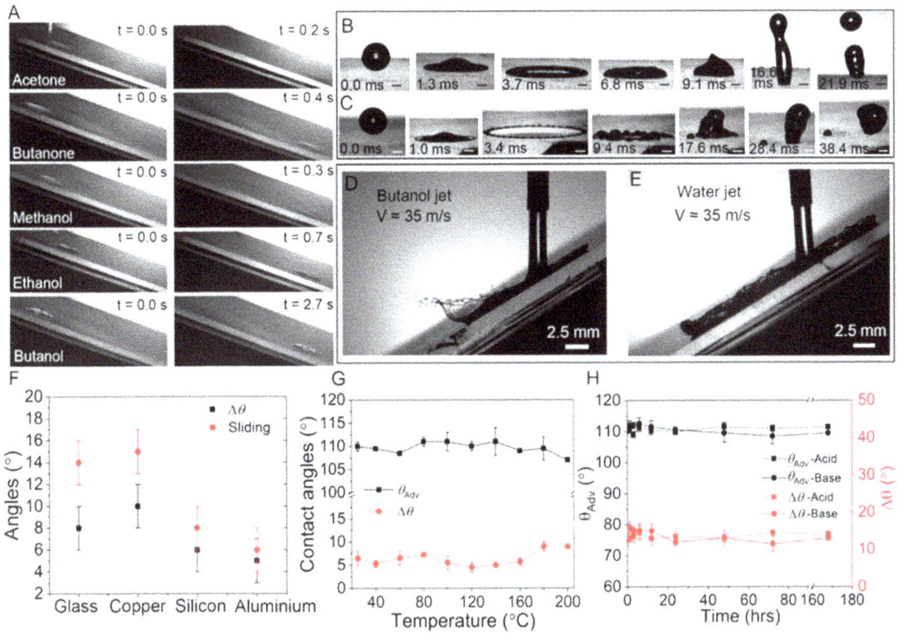

Figure 6.5. Amphiphobicity and robustness testing. (A) Image sequence showing the sliding of low surface tension liquids at 15° tilt angles. The images are presented according to the drop speed (high to low from top to bottom). (B) Time lapse snapshots of drop impact on the COF surfaces at 1.4 $m\,s^{-1}$ and (C) 4.2 $m\,s^{-1}$ impact velocities, respectively (scale = 2 mm). (D) Butanol and (E) water jets (nozzle diameter = 2.5 mm) impacting on COF surfaces at ~35 $m\,s^{-1}$. (F) Sliding angles and on amphiphobic COF film coated on different substrates. (G) Thermal stability test of the surface showing change in and advancing angles of water droplets in a temperature range from 20 °C to 200 °C. (H) Wetting angles (θ_{adv} and $\Delta\theta$) representing the stability of the COF coated on a glass surface in acid (pH = 1–2) and base (pH = 12–13) tested for one week. (Adapted from [23]. CC BY 4.0.)

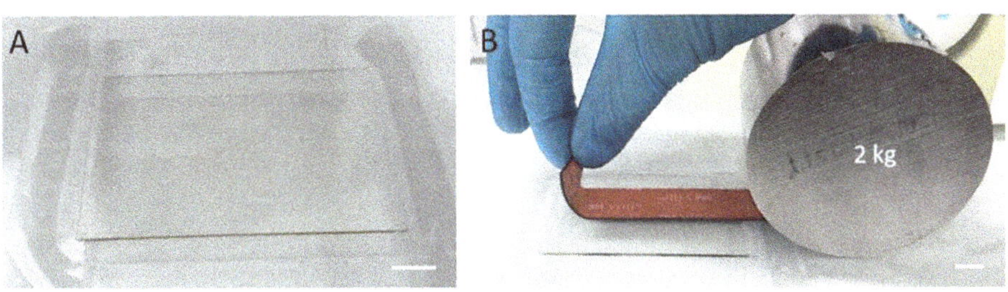

Figure 6.6. Details of the tape-peel test to check the adhesion of the COF coating on the glass. (A) A COF coated glass slide fixed with transparent tape and (B) the tape-peel test featuring the application of strong adhesive tape on the coated substrate and a 2 kg roller used to press the tape. (Adapted from [23]. CC BY 4.0.)

6.5 Icephobicity

The anti-icing capabilities of the COF surface were determined by measuring the delay in ice nucleation in supercooled droplets and ice adhesion. For the ice

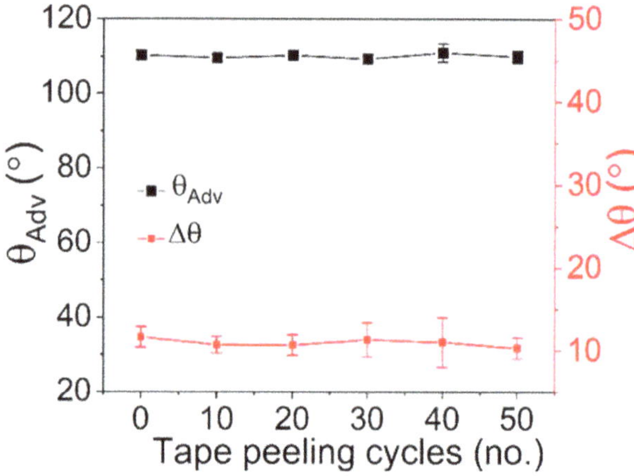

Figure 6.7. The effect of tape-peel cycles on θ_{Adv} and for 50 repeated cycles. (Adapted from [23]. CC BY 4.0.)

adhesion measurement, an extension rod connected to a force gauge (M4-50, MARK-10) was mounted on a custom-made driving system with a stepper motor (17HS19-2004S1). This enabled deflecting the cuvettes with frozen liquid laterally and measuring the required forces. The extension rod is 127 mm long, and its flattened end has a diameter of 25 mm. LabView software was used to operate and record the forces and determine the corresponding adhesion strength. Vaseline petroleum jelly was used as control due to its well-known inertness towards ice nucleation [51] and compared against a functionalised COF coated on glass substrate (figure 6.8(A)). Clearly, the COFs are excellent at suppressing ice nucleation, with the last few droplets freezing at ~−28 °C as opposed to ~−23 °C on Vaseline. Furthermore, unfunctionalised COF surfaces (without silane functionalisation) were also tested and are compared in figure 6.8(B) to check the impact of silane (hydrophobicity) on ice nucleation. An almost similar freezing trend was observed on both the surfaces, unfunctionalised COFs and Vaseline, possibly due to perfectly arranged small nanopores which are akin to the nanocavities reported elsewhere [43]. The ice adhesion test results are shown in figures 6.8(C)–(F). The adhesion on our functionalised COF films is compared against bare glass and various nanoscale thick coatings such as a bare glass functionalised with trichloro-octadodecyl silane (referred as OTS), covalently attached dimethyl siloxane oligomer chains (SOCAL) and recently reported amphiphobic MOFs [22] (figure 6.8 (C)). The lowest ice adhesion strength of 15 ± 5 kPa was recorded on the COF coated on a glass substrate which is comparable to the value obtained on lubricant-infused slippery surfaces [52, 53]. In terms of functionalisation, tricholorooctadecyl silane functionalised COFs showed the lowest ice adhesion, possibly due to the greater flexibility and larger chain length when compared to SOCAL oligomers which are relatively shorter in length [54]. Figure 6.8(E) compares the substrates and the decrease in ice adhesion strength is quite clear in each case. The surface robustness was also tested by repeating icing–de-icing cycles on a COF coated glass substrate; within experimental repeatability error, no change in the ice adhesion

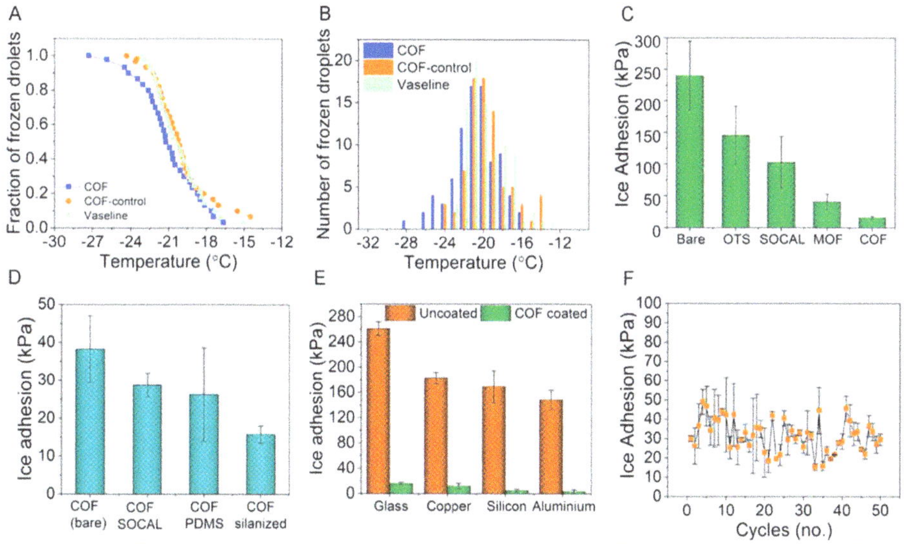

Figure 6.8. Icephobicity of the COF coating. (A) Droplet freezing delay test showing the fraction of droplets frozen against temperature range comparing the functionalised COF surface with unfunctionalised COFs (COF-control) and Vaseline petroleum jelly and (B) histogram showing the number of droplets frozen against temperature on three different surfaces including Vaseline as a control. The data plotted here are the average of the three measurements. (C) Ice adhesion strength of different treatments on a glass substrate including bare glass as a control, glass functionalized with trichloro-octadodecyl silane (OTS), an ultrasmooth surface based on polydimethylsiloxane oligomers [54] (SOCAL), our previously published nanohierarchical MOF surface [22] and silane functionalised COF surfaces. (D) Change in ice adhesion strength on bare COFs (without functionalisation) and COFs with different functionalisation, with flexible SOCAL, PDMS polymer and long alkyl chains (trichloro-octadecylsilane). (E) Ice adhesion strength comparison on glass, metals (copper and aluminium) and silicon substrates before and after being coated with COFs functionalised with silane. (F) Unchanged ice adhesion strength on the COF surface for 50 repeated icing–de-icing cycles. (Adapted from [23]. CC BY 4.0.)

strength was observed after 50 icing–de-icing cycles. Unlike MOF based surfaces, following these cyclic assessments, the surface showed no sign of mechanical damage (figure 6.9).

The nucleation for ice and scales on a COF surface can be understood through classical nucleation theory. The critical radii to overcome the free energy barrier during crystal growth were calculated for both types of nuclei. The critical nucleus radii for the ice and $CaCO_3$ were calculated as ~2.3 nm (at −20 °C) and 3.89 nm, respectively, which are much larger than the radius of the COF pore itself (0.9 nm). The results suggest that it is almost not possible for water molecules and mineral ions to form stable nuclei inside the COF pores which might explain the experimental observations above. Further to the COF work, the effect of silane chain length on ice nucleation was also assessed. Figure 6.10 shows the results of ice nucleation temperature measurements on the reticular COF surfaces functionalized with the three flexible alkyl molecular chains of various lengths and rigid fluroroalkyl chain. The four different surfaces were characterized to confirm the chemical and

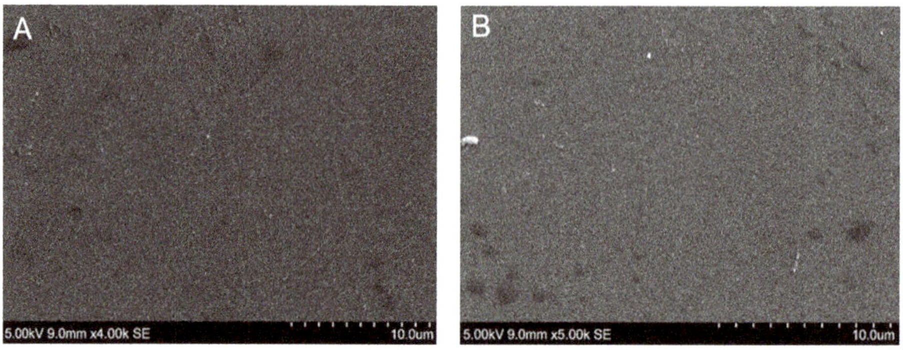

Figure 6.9. Surface topology of the COF coating (A) before and (B) after 50 repeated icing–de-icing cycles. (Adapted from [23]. CC BY 4.0.)

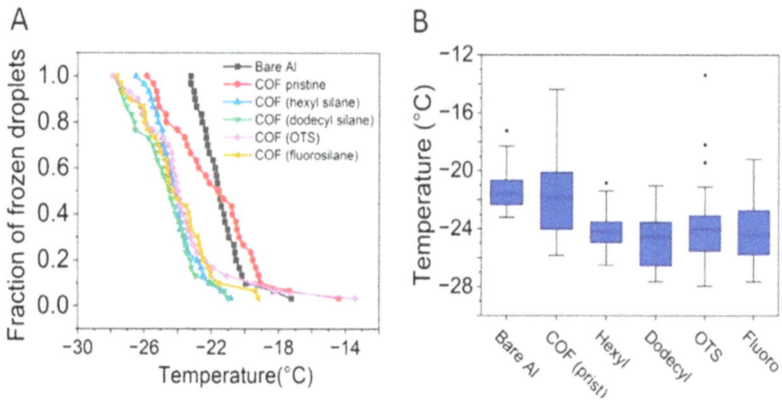

Figure 6.10. (A) Ice nucleation graph showing the fraction of frozen droplets against temperature range comparing different surfaces. (B) Box plot graph showing the median nucleation temperature for different surfaces. (Adapted from [61]. CC BY 4.0.)

morphology of COF surfaces and then subjected to ice nucleation experiments. Pristine (unfunctionalised) COF surface, bare aluminium substrate were used as a control to compare against functionalised regime. The obtained data are shown in figures 6.10(A) and (B). Clearly, pristine COF is excellent in suppressing the ice nucleation compared to bare aluminium substrate as the last few drops freeze at ≈−26 °C compared to ≈−23 °C on bare aluminium. Furthermore, silanized COF substrates are even better at delaying the ice nucleation compared to the pristine COF substrate, with the last few drops freezing at ≈−26.5 °C, −27.6 °C, −28 °C and −27 °C on hexyl silane, dodecyl silane, octadecylsilane and fluoro silane, respectively.

6.6 Anti-scaling

The COF coated surfaces performed exceptionally well in the prevention of $CaCO_3$ crystal growth; after two weeks only ≈ 0.5% and ≈ 5% for coated glass and aluminium substrates, respectively, were covered with $CaCO_3$. The corresponding uncoated substrates had $\approx$ 85% and $\approx$ 75% coverage, respectively (figures 6.11(A)–(C)). The silane functionalized hydrophobic glass (control) showed some resistance against crystal growth, with $\approx$ 38% of the surface covered with $CaCO_3$ crystals after two weeks. The uncoated samples demonstrate thin patch-like depositions as well as large micrometre-sized crystals (20–50 μm) of $CaCO_3$. Amorphous $CaCO_3$ is

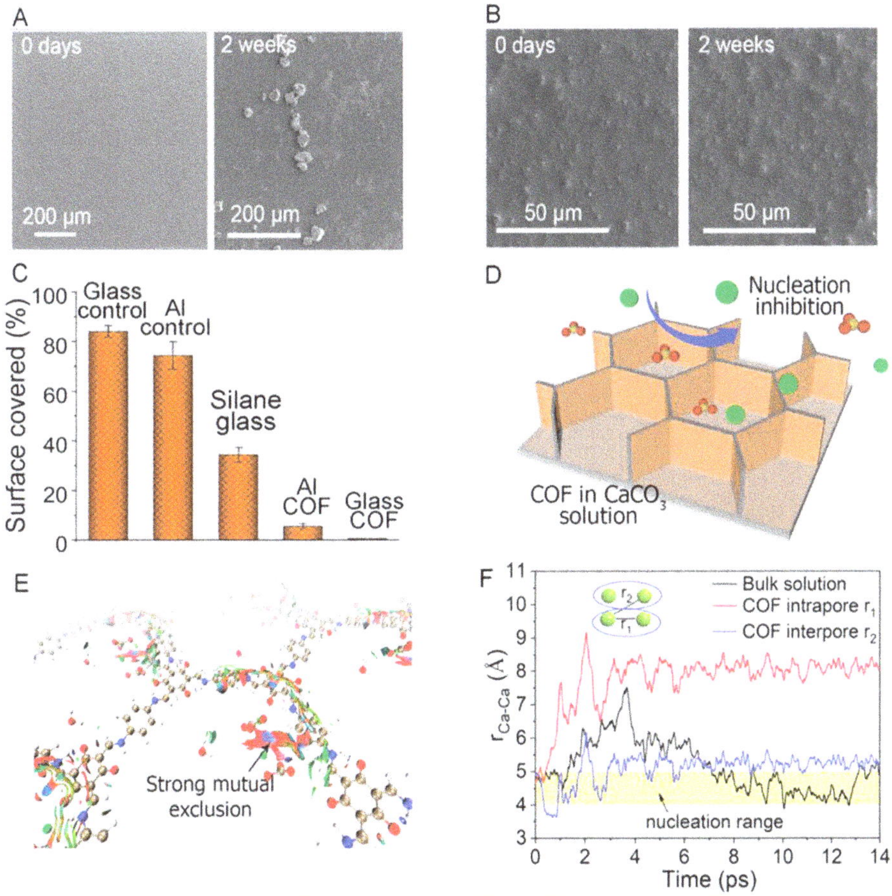

Figure 6.11. Scaling inhibition on COF coated substrates. (A) SEM images showing glass control and (B) COF coated glass before (0 days) and after (2 weeks) immersion in the supersaturated solution of $CaCl_2$ and $NaHCO_3$. (C) The percentage of control and COF surfaces covered (on glass and aluminium substrates) with $CaCO_3$ deposition after two weeks of the scaling test. (D) Schematic showing the nucleation inhibition of Ca^{2+} and CO_3^{2-} ions in COF pores. (E) Strong mutual exclusion (repulsion) observed between Ca^{2+} and CO_3^{2-} ions in simulated COF pores. (F) Evolution of the distance between two Ca^{2+} ions r_{Ca-Ca} over time in bulk and COF pores calculated in *ab initio* simulation. r_{Ca-Ca} reduced to the nucleation range in the bulk solution, whereas the Ca^{2+} ions remain well above the nucleation range due to strong repulsion in the COF pores. In the schematic insets, the blue circles and green balls represent COFs and Ca^{2+} ions, respectively. r_1 is the distance between two Ca^{2+} ions inside the same COF pore while r_2 is the longer distance between two Ca^{2+} ions in two adjacent COF pores. (Adapted from [23]. CC BY 4.0.)

involved in the nucleation and growth of scales, and it usually follows a non-classical pathway [55]. A Born–Oppenheimer molecular dynamics (BOMD) simulation based on density functional theory (DFT) was used [56] to further understand the nucleation process inside COF pores from a non-classical perspective (figures 6.11 (D)–(F)). The three-dimensional COF crystal structure was obtained from the Cambridge Structural Database [57, 58]. The formation of prenucleation clusters in the COF pores was studied at an atomistic level and two possible modes, monodentate (η^1) and bidentate (η^2) were considered based on the number of oxygen atoms coordinate with the Ca^{2+} ions. A strong repulsion in between Ca^{2+} and CO_3^{2-} ions was observed in the COF pore, possibly due to nanoconfinement effects [59, 60], whereas ion pairs were observed to interact strongly via van der Waals forces in the bulk solution during *ab initio* simulation. The time evolutions of the distances between two Ca^{2+} ions in bulk solution, in COF pores, within a single pore (intra, r_1), and from an adjacent pore along a 2D channel (inter, r_2) was studied by comparing the nucleation range of a typical $CaCO_3$ crystal structure. Ions in bulk solution start to form nuclei after 7 ps, while Ca^{2+} ions inside the COF pores remain separated, well above the nucleation range in both inter and intra pores, confirming strong repulsion. The time evolutions of the distances between Ca^{2+} and the oxygen (O_c) of the carbonate ions ($r_{Ca\text{-}Oc}$) in bulk solution and the COF pores are shown figure 6.12,

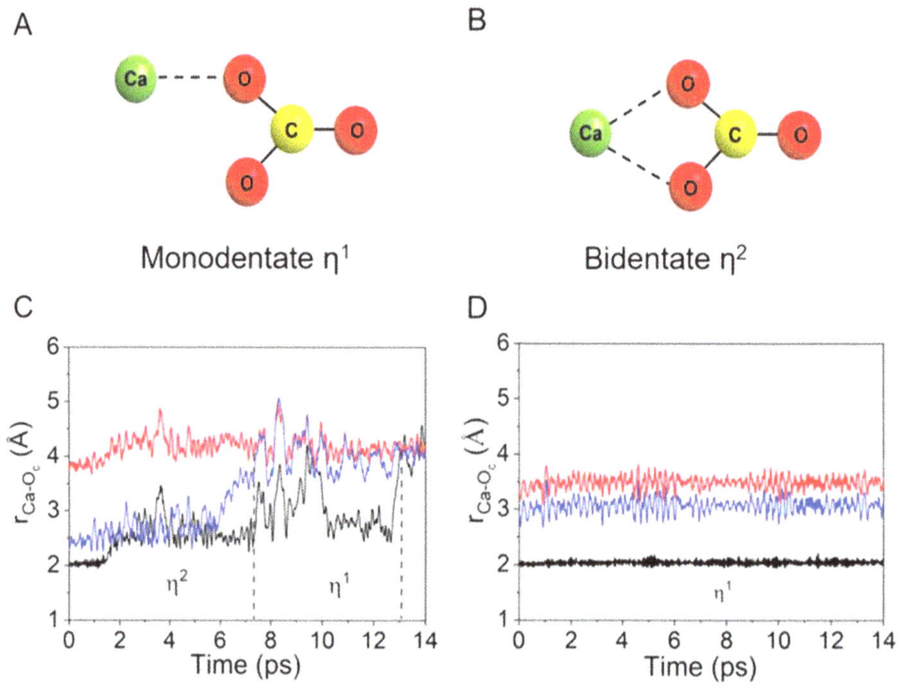

Figure 6.12. Two coordination modes of the carbonate ion, (A) monodentate and (B) bidentate. (C) Time evolution of the distance between Ca^{2+} and O_c (oxygen in carbonate ion) for bulk solution simulation and (D) COF pore simulation. (Adapted from [23]. CC BY 4.0.)

respectively. The $r_{Ca\text{-}Oc}$ of the bulk solution show a facile change between the two coordination modes (η^1 and η^2), which is in agreement with previous reports [61].

6.7 Molecular simulations

This association mechanism is unstable thermodynamically, which helps in the formation of prenucleation amorphous $CaCO_3$ clusters by lowering the free energy barrier [55]. Additionally, the coordination modes inside COF pores remain unchanged for over 14 ps, suggesting that the confinement of ions inside the COF pores might help in stabilizing the $CaCO_3$ monomer. After equilibrium, the second hydration shells for both the calcium ions were occupied by oxygen from water molecules in the bulk solution, whereas the second hydration shells for Ca^{2+} ions were empty in the COF pores. All these observations, including the strong mutual exclusion and stable coordination mode for the cation–anion pair in COF pores and the distance between the cation ions above nucleation range, suggest that the COF pores with nanoscale precision should be highly efficient in inhibiting the nucleation of $CaCO_3$, in agreement with our experiments.

6.8 Conclusion

In summary, the surface-grown reticular materials, specifically, MOF/COF films with post-synthetic alkylation, offer exceptional properties as thin and multifunctional coatings due to their inherent precise nanoporous and robust structure. Fluorine-free functionalisation of COF with flexible alkyl chains enables the surface to resist the impact of high-speed jets of low surface tension liquids (such as ethanol and butanol). Additionally, the miniature and precision COF pores (1.8 nm), can effectively inhibit nucleation in both cases, icing and scaling. With potentially scalable interfacial polymerisation-based synthesis of defect-free surfaces, excellent anti-icing and anti-scaling characteristics, the COF based surface presented may elicit widespread usage.

References

[1] Atkinson J D *et al* 2013 The importance of feldspar for ice nucleation by mineral dust in mixed-phase clouds *Nature* **498** 355–8

[2] Kawasaki T and Tanaka H 2010 Formation of a crystal nucleus from liquid *Proc. Natl Acad. Sci.* **107** 14036–41

[3] Lin L *et al* 2020 A critical review of the application of electromagnetic fields for scaling control in water systems: mechanisms, characterization, and operation *npj Clean Water* **3** 25

[4] Demichelis R *et al* 2011 Stable prenucleation mineral clusters are liquid-like ionic polymers *Nat. Commun.* **2** 590

[5] Kohler F *et al* 2022 Crystal growth in confinement *Nat. Commun.* **13** 6990

[6] Shen Z *et al* 2012 The effect of synthesized hydrolyzed polymaleic anhydride (HPMA) on the crystal of calcium carbonate *Desalination* **284** 238–44

[7] Macedo R G M D A *et al* 2019 Water-soluble carboxymethylchitosan as green scale inhibitor in oil wells *Carbohydr. Polym.* **215** 137–42

[8] Li A *et al* 2022 Effects of chemical inhibitors on the scaling behaviors of calcite and the associated surface interaction mechanisms *J. Colloid Interface Sci.* **618** 507–17

[9] Zhao J *et al* 2019 Fundamental nanoscale surface strategies for robustly controlling heterogeneous nucleation of calcium carbonate *J. Mater. Chem.* A **7** 17242–7

[10] Jung S *et al* 2012 Mechanism of supercooled droplet freezing on surfaces *Nat. Commun.* **3** 615

[11] Kim P *et al* 2012 Liquid-infused nanostructured surfaces with extreme anti-ice and anti-frost performance *ACS Nano* **6** 6569–77

[12] Kreder M J *et al* 2016 Design of anti-icing surfaces: smooth, textured or slippery? *Nat. Rev. Mater.* **1** 15003

[13] Shabanian S *et al* 2020 Rational design of perfluorocarbon-free oleophobic textiles *Nat. Sustain.* **3** 1059–66

[14] Trang B *et al* 2022 Low-temperature mineralization of perfluorocarboxylic acids *Science* **377** 839–45

[15] Jokinen V *et al* 2018 Superhydrophobic blood-repellent surfaces *Adv. Mater.* **30** 1705104

[16] Peppou-Chapman S *et al* 2020 Life and death of liquid-infused surfaces: a review on the choice, analysis and fate of the infused liquid layer *Chem. Soc. Rev.* **49** 3688–715

[17] Liu J *et al* 2021 One-step synthesis of a durable and liquid-repellent poly(dimethylsiloxane) coating *Adv. Mater.* **33** 2100237

[18] Joudan S and Lundgren R J 2022 Taking the "F" out of forever chemicals *Science* **377** 816–7

[19] Lahiri S K *et al* 2023 Polydimethylsiloxane-coated textiles with minimized microplastic pollution *Nat. Sustain.* **6** 559–67

[20] Tuteja A *et al* 2007 Designing superoleophobic surfaces *Science* **318** 1618–22

[21] Tuteja A *et al* 2008 Robust omniphobic surfaces *Proc. Natl Acad. Sci.* **105** 18200–5

[22] Singh V *et al* 2021 Transparent and robust amphiphobic surfaces exploiting nanohierarchical surface-grown metal–organic frameworks *Nano Lett.* **21** 3480–6

[23] Singh V *et al* 2023 Precision covalent organic frameworks for surface nucleation control *Adv. Mater.* **35** 2302466

[24] Jiang H *et al* 2021 A reticular chemistry guide for the design of periodic solids *Nat. Rev. Mater.* **6** 466–87

[25] Khandelwal G *et al* 2019 Metal–organic framework: a novel material for triboelectric nanogenerator-based self-powered sensors and systems *Adv. Energy Mater.* **9** 1803581

[26] Hajra S *et al* 2021 A green metal–organic framework-cyclodextrin MOF: a novel multifunctional material based triboelectric nanogenerator for highly efficient mechanical energy harvesting *Adv. Funct. Mater.* **31** 2101829

[27] Khandelwal G *et al* 2020 Zeolitic imidazole framework: metal–organic framework subfamily members for triboelectric nanogenerators *Adv. Funct. Mater.* **30** 1910162

[28] Guo Y *et al* 2020 Fluorinated metal–organic framework as bifunctional filler toward highly improving output performance of triboelectric nanogenerators *Nano Energy* **70** 104517

[29] Winarta J *et al* 2020 A decade of UiO-66 research: a historic review of dynamic structure, synthesis mechanisms, and characterization techniques of an archetypal metal–organic framework *Crystal Growth Design* **20** 1347–62

[30] Gangu K K *et al* 2022 The pioneering role of metal–organic framework-5 in ever-growing contemporary applications—a review *RSC Adv.* **12** 14282–98

[31] Wei X *et al* 2025 Facile synthesis of multifunctional zeolitic imidazolate framework-8 coatings on diverse substrates using a one-step strategy *Separat. Purif. Technol.* **357** 130031

[32] Deacon A *et al* 2022 Understanding the ZIF-L to ZIF-8 transformation from fundamentals to fully costed kilogram-scale production *Commun. Chem.* **5** 18
[33] Tan K T *et al* 2023 Covalent organic frameworks *Nat. Rev. Methods Primers* **3** 1
[34] Chang J *et al* 2025 Synthesis of three-dimensional covalent organic frameworks through a symmetry reduction strategy *Nat. Chem.*
[35] Lohse M S and Bein T 2018 Covalent organic frameworks: structures, synthesis, and applications *Adv. Funct. Mater.* **28** 1705553
[36] Yin Y *et al* 2024 Ultrahigh-surface area covalent organic frameworks for methane adsorption *Science* **386** 693–6
[37] Natraj A *et al* 2022 Single-crystalline imine-linked two-dimensional covalent organic frameworks separate benzene and cyclohexane efficiently *J. Am. Chem. Soc.* **144** 19813–24
[38] Wu X *et al* 2018 Control interlayer stacking and chemical stability of two-dimensional covalent organic frameworks via steric tuning *J. Am. Chem. Soc.* **140** 16124–33
[39] Dobberschütz S *et al* 2018 The mechanisms of crystal growth inhibition by organic and inorganic inhibitors *Nat. Commun.* **9** 1578
[40] Deng H *et al* 2015 Two competitive nucleation mechanisms of calcium carbonate biomineralization in response to surface functionality in low calcium ion concentration solution *Regen. Biomater.* **2** 187–95
[41] Salvati Manni L *et al* 2019 Soft biomimetic nanoconfinement promotes amorphous water over ice *Nat. Nanotechnol.* **14** 609–15
[42] Bai G *et al* 2019 Probing the critical nucleus size for ice formation with graphene oxide nanosheets *Nature* **576** 437–41
[43] Eberle P *et al* 2014 Rational nanostructuring of surfaces for extraordinary icephobicity *Nanoscale* **6** 4874–81
[44] Eskandari A *et al* 2020 Antifreeze proteins and their practical utilization in industry, medicine, and agriculture *Biomolecules* **10** 1649
[45] Zeng M *et al* 2018 Confinement generates single-crystal aragonite rods at room temperature *Proc. Natl Acad. Sci.* **115** 7670–5
[46] Zhou X *et al* 2021 Designing anti-icing surfaces by controlling ice formation *Adv. Mater. Interfaces* **8** 2100327
[47] Ding M *et al* 2019 Improving MOF stability: approaches and applications *Chem. Sci.* **10** 10209–30
[48] Côté A P *et al* 2005 Porous, crystalline, covalent organic frameworks *Science* **310** 1166–70
[49] Yarin A L 2006 Drop impact dynamics: splashing, spreading, receding, bouncing *Annu. Rev. Fluid Mechan.* **38** 159–92
[50] Peng C *et al* 2018 All-organic superhydrophobic coatings with mechanochemical robustness and liquid impalement resistance *Nat. Mater.* **17** 355–60
[51] Polen M *et al* 2018 Cleaning up our water: reducing interferences from nonhomogeneous freezing of 'pure' water in droplet freezing assays of ice-nucleating particles *Atmos. Meas. Tech.* **11** 5315–34
[52] Sandhu A *et al* 2019 Perfluoroalkane wax infused gels for effective, regenerating, anti-icing surfaces *Chem. Commun.* **55** 3215–8
[53] Liu B *et al* 2016 Strategies for anti-icing: low surface energy or liquid-infused? *RSC Adv.* **6** 70251–60
[54] Wang L and McCarthy T J 2016 Covalently attached liquids: instant omniphobic surfaces with unprecedented repellency *Angew. Chem. Int. Ed. Engl.* **55** 244–8

[55] Raiteri P and Gale J D 2010 Water is the key to nonclassical nucleation of amorphous calcium carbonate *J. Am. Chem. Soc.* **132** 17623–34
[56] Hutter J *et al* 2014 CP2K: atomistic simulations of condensed matter systems *WIREs Comput. Mol. Sci.* **4** 15–25
[57] Biswal B P *et al* 2015 Pore surface engineering in porous, chemically stable covalent organic frameworks for water adsorption *J. Mater. Chem.* A **3** 23664–9
[58] Zhang K *et al* 2017 Computational design of 2D functional covalent-organic framework membranes for water desalination *Environ. Sci.: Water Res. Technol.* **3** 735–43
[59] Hedges L O and Whitelam S 2012 Patterning a surface so as to speed nucleation from solution *Soft Matter* **8** 8624–35
[60] Katsman A *et al* 2022 On the mechanism of calcium carbonate polymorph selection via confinement *Faraday Discuss.* **235** 433–45
[61] Tommaso D D and de Leeuw N H 2008 The onset of calcium carbonate nucleation: a density functional theory molecular dynamics and hybrid microsolvation/continuum study *J. Phys. Chem.* B **112** 6965–75
[62] Bahal S, Singh V and Tiwari M K 2024 Functionalized covalent organic frameworks for ice nucleation inhibition *Proc. of the 18th UK Heat Transfer Conf. (Birmingham, 9–11 Sept.)*

Chapter 7

Rigid and soft materials for icephobicity: the role of surface and mechanical properties

Catalina Ospina, Irene Tagliaro, Luca Stendardo, Pablo F Ibáñez Ibáñez and Carlo Antonini

Understanding the mechanical and wetting properties that lead to icephobic performance is a challenging task due to the complex dynamic phenomena occurring at the surface during ice adhesion tests. Moreover, due to the variety of possible de-icing mechanisms, it is necessary to evaluate the traditional empirical relations useful for predicting the icephobic performance based on the work of adhesion on liquid-water–substrate interactions and on the mechanical properties of elastic materials.

This chapter focuses on the different de-icing mechanisms on rigid surfaces with different hydrophobicity, soft materials with different lubrication degrees, and systems composed of soft and rigid materials, assessing them by means of a horizontal shear ice adhesion set-up.

The existence of liquid-like behavior in the ice adhesion tests is discussed for rigid materials. For soft materials, the contact angle evolution depending on the humidity as well as the viscoelastic properties are confirmed. Additionally, we discuss how the integration of rigid and soft components at the substrate surface allows the improvement of icephobicity.

7.1 Introduction

In chapter 2, various strategies for developing icephobic surfaces have been presented. Here, we briefly recap the key points relevant to the discussion of results.

It is important to highlight that ice adhesion strength (τ_{ice}), a key parameter in the current icephobic models, is heavily influenced by experimental variables such as nucleation temperature, stress contributions, and ice geometry. Additionally, the current models for understanding icephobic performance often assume that fracture is always governed by the critical stress. However, this assumption does not hold true in all cases [1, 2], frequently resulting in an incomplete account of the fracture

doi:10.1088/978-0-7503-6009-8ch7

mechanics involved in ice detachment, particularly on complex surfaces. Therefore, in this work, although τ_{ice} is calculated and reported, it is used solely for comparing samples tested under consistent experimental conditions.

7.1.1 Icephobicity based on wetting performance

Non-wetting surfaces are commonly used for icephobicity, assuming that weak interactions with water translate to weak ice adhesion. Superhydrophobic surfaces effectively reduce ice adhesion by promoting liquid shedding and easing ice removal. However, under high humidity and dynamic conditions such as droplet impact, water can infiltrate surface cavities, transitioning from the Cassie–Baxter to the Wenzel state. This transition increases ice adhesion through mechanical interlocking.

Alternatively, slippery liquid infused surfaces (SLIPSs) or liquid-iInfused surfaces (LISs) use lubricants to create low-adhesion interfaces, facilitating ice release. However, the durability of these systems depends on lubricant stability, which is influenced by surface morphology and texture density [3]. While SLIPSs and LISs demonstrate superior performance, they require excess lubricant, limiting their long-term applicability.

Recent advancements include the development of systems described as 'liquid-like', 'quasi-liquid', or 'pseudo-slippery', which retain liquid-like properties. These systems consist of functionalized solid surfaces that mimic liquid behavior due to flexible, mobile chains. Such surfaces, referred to as slippery covalently attached liquid surfaces (SCALSs), including polymer brushes, achieve low contact angle hysteresis, allowing various organic and aqueous liquids to slide, regardless of their surface tension [4]. This performance is driven by the surface chain mobility, which is influenced by molecular structure, bond characteristics (e.g. length and angle), and the electronegativity differences between constituent atoms, therefore, by tuning the molecular structure it is possible to design systems that exhibit liquid-like behavior.

The ability of these surfaces to prevent the aggregation of solid matter, such as ice, has been explored in the context of icephobicity using methylated PDMS brushes [5]. Macroscopic evidence of liquid-like behavior has been observed in the sliding of ice, the shear-dependent τ_{ice} and the similarity in surface energy to the liquid PDMS. These findings highlight the need for improved models and comprehensive experimental descriptions to advance icephobic technology.

7.1.2 Icephobicity based on mechanical properties

7.1.2.1 Rigid materials

For rigid, non-deformable materials, ice adhesion strength depends on both surface energy and mechanical properties. The adhesion to materials with different mechanical properties has been derived considering a tensile experiment [6], where the force required to separate a solid with a contact area A from thick adhesive layers can be expressed as

$$\tau_{ice} \propto \sqrt{\frac{W_{adh} K}{t}}. \tag{7.1}$$

In this equation W_{adh} denotes the interfacial surface energy, t represents the thickness of the adhesive layer (ranging from 0.4 to 4.8 mm), and K is the bulk modulus of the adhesive. The study referenced indicated that K was approximately ten times greater than the Young's modulus (E). Equation (7.1) has frequently been applied in research to compare ice adhesion strength across various materials, particularly for films with thickness under 500 μm in shear tests, using without distinction E or shear modulus, G [7–10], despite the initial relation being proposed for tensile tests and considering the bulk modulus, K.

Other authors prefer to use the fracture mechanics approach [11–13], estimating ice adhesion strength by

$$\tau_{ice} \propto \sqrt{\frac{K^* W_{adh}}{\pi a \Lambda}}, \tag{7.2}$$

where K^* is the 'apparent' elastic modulus, W_{adh} is the surface energy of the crack, usually calculated considering the receding contact angle [11], a is the length of the interface crack, and Λ is a non-dimensional parameter determined by the geometric configuration of the crack.

Those models suggest that reducing the elastic modulus and interfacial energy can lower τ_{ice}. It is important to mention that the available expressions mentioned here are developed for rigid materials, where, if deformation occurs, it is permanent, therefore, they do not account for the viscoelastic recovery of deformable substrates.

Nevertheless, the data available in the literature agree that when a block of ice is separated from a softer material (typically $E < 10$ MPa, according to [14]), τ_{ice} is reduced due to the mismatch in the mechanical properties, which generates micro-fractures at the interface, promoting the easier detachment of ice blocks [8].

7.1.2.2 Soft-deformable materials

In contrast, soft materials present a different paradigm. Their ability to dissipate applied energy through bulk deformation reduces ice adhesion strength, diverging from the brittle fracture mechanisms observed in rigid materials [1].

The mechanical behavior of elastic materials is analysed by measuring their strain responses to various types of stress. Young's modulus (E) quantifies the strain response to tensile stress, the shear modulus (G) represents the response to shear stress, and the bulk modulus (K) characterizes the response to compressive stress. For isotropic and homogeneous materials, the theoretical relation among the tensile and shear response is approximated in the literature as $E = 2G(1 + \mu)$, where μ is Poisson's ratio.

In addition, soft polymers are intrinsically lubricated deformable materials. They are composed of mobile chains with dangling ends embedded in the polymer bulk, providing interesting features such as the combination of elastic and viscous response, evidenced in wetting experiments where the coupling between elastic and capillary effects has been reported [15, 16].

Considering viscoelastic materials, polydimethylsiloxane (PDMS) is commonly included among icephobic materials due to its hydrophobicity and tunable mechanical properties [10, 14, 17–20]. For PDMS, μ is generally taken as 0.5 [21], giving a ratio of 3 between the E and G moduli. This indicates that the material offers greater resistance to stretching or compression compared to shearing. This distinction highlights the varied performance a material can demonstrate in shear versus tensile ice adhesion tests [10], which depend on both the film thickness and the strain rate during ice removal [9].

However, the reported mechanical properties for PDMS obtained experimentally in different studies give E values obtained by dynamical mechanical analysis [9, 22], AFM indentation [23], and tensile [24, 25] and compressive tests [10], in the range of thousands of kPa, with small variations that may depend on the specific curing procedure. On the other hand, the reported shear moduli, G^* [7, 17, 25–27] are in the range of some hundreds of kPa. This difference of about one order of magnitude between E and G suggest that the above relationship ($E = 2G\,(1 + \mu) = 3\,G$) does not hold for PDMS, indicating that, based on experimental data, it does not behave as an ideal rubber [28]. Such non-ideal rubber behavior also affects ice adhesion strength predictions based on traditional models.

Moreover, the time-dependency of its contact angle triggered upon contact with water has been recently discovered [29, 30]. The adaptive wetting behavior is explained by the migration of uncrosslinked chains to the surface, decreasing the surface energy in wet conditions [30], further challenging the application of conventional models on equations (7.1) and (7.2).

7.1.3 Icephobicity on discontinuous mechanical properties

Materials with discontinuous properties, such as variations in chemical nature, surface morphology, thermal conductivity or mechanical properties, can promote the reduction of ice adhesive force by different mechanisms, exploiting mismatches in local properties to initiate cracks at multiple length scales at the ice–substrate interface [12, 31–33].

For example, combining soft polymers with rigid structures introduces stress concentration sites that enhance de-icing efficiency. This approach has demonstrated significant reductions in τ_{ice} by fostering crack propagation [32].

Within this framework, several composite materials have been produced to increase such stress concentration sites by inducing sub-structure voids in rigid or viscoelastic materials [12, 31–33], or combining materials with different mechanical properties [34–36].

Previous studies proposed that in surfaces composed of hard and soft components, fractures may start in the softer material [31, 32, 34]. Nonetheless, it has been recently observed in composite materials containing a rigid component, that the fracture initiates over the harder parts [36, 37]. Although the concept of surface discontinuity has been previously visualized and used to explain experimental results, the effect of discontinuity dimensions and geometrical parameters in the icephobic performance has not been demonstrated quantitatively.

As of today, discontinuous surface elasticity can also be combined with other approaches, such as including a quasi-liquid lubricating layer [38], lubricant reservoirs in the form of aerogels [39] or in more complex formulations such as including solid grease and silicon oil [8], where the interplay between surface morphology, elasticity gradients, and thermal conductivity discontinuities offers a versatile framework for icephobic design [34–36].

7.2 Rigid materials

7.2.1 Experimental details

The functionalization of rigid glass substrates was achieved through condensation reactions between a silane precursor and hydroxyl units on the glass substrate [40]. In figure 7.1 the schematic of the process is shown.

To do so, the glass substrates are first cleaned by soaking them in ethanol for five minutes and dried with fluxed air. In order to expose the hydroxyl groups at the surface, the glasses are soaked in freshly made piranha solution consisting of a 3:2 v/v mixture of concentrated H_2SO_4 and H_2O_2 (30%). The oxidizing solution increases the temperature and generates bubbles while removing the impurities and exposing the active hydroxyl groups in the glass. After the oxidation stops, the glass is submerged in KOH for 5 min to further ensure the availability of oxygen atoms for the condensation reaction. Then, the glass substrates are thoroughly rinsed with deionized water and dried under air flow right before the functionalization.

7.2.1.1 Perfluorododecyl-trichlorosilane functionalization

In this approach, the activated glasses are submerged in a solution of perfluorododecyl-trichlorosilane (figure 7.2), a white solid, in a concentration of 110 mmol l^{-1} in

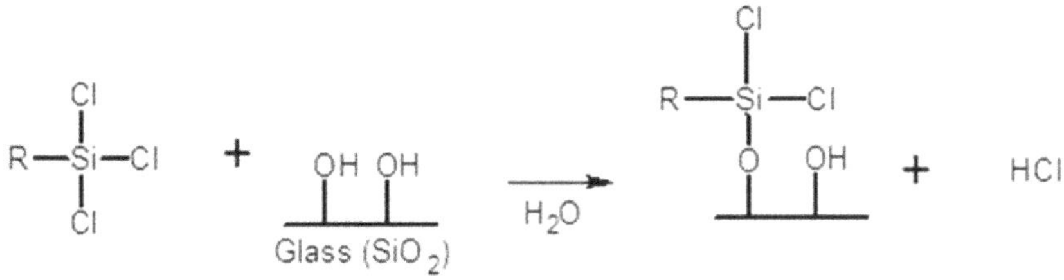

Figure 7.1. Schematic of the functionalization of glass substrates. Condensation between the precursor and the activated glass. The further condensation mechanism depends on the nature of the silane precursor and the relative humidity. (Adapted from [41].)

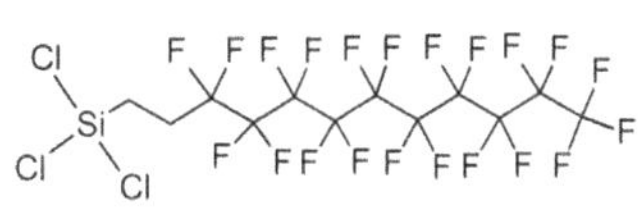

Figure 7.2. Perfluorododecyl-trichlorosilane used to obtain F-silane functionalized glass. (Adapted from [41].)

hexane for different times between 5 min and 1 h. The experiments are performed at ambient temperature, 23 °C. After the immersion, the coatings are dried for 1 h under a hood.

The chemisorption of the surface-active head group at the substrate's surface allows the *in situ* chemical bond condensation to proceed. In figure 7.2, the hydrophilic (tri-chloro) heads undergo the condensation reaction with the hydroxyl groups in the substrate.

7.2.1.2 Methyl-trichlorosilane functionalization

Using methyl-trichlorosilane (figure 7.3), the functionalization of the glass slides is performed by two different methods, as follows.

The first method consists in a chemical vapor deposition (CVD) process inside a desiccator (6 l volume) as the reaction chamber. The environmental temperature and relative humidity (outside the desiccator) are registered and the activated glass slides are placed inside the dessicator. Then, a container with a solution 1 M of methyl-trichlorosilane in toluene is placed at the bottom and the dessicator is closed to start the reaction. No other gases are flushed in or withdrawn until the reaction time finishes; then, the samples are taken out and brought into an oven for thermal treatment at 80 °C for 1 h.

The second method consists in the direct deposition of 1 ml of pure methyl-trichlorosilane in activated glass slides of 2.5 × 2.5 cm, allowing the immediate condensation of the reactant while covering the substrate with a Petri dish under a hood.

7.2.2 Surface characterization

7.2.2.1 Surface morphology

The surface morphology is observed using atomic force microscopy (AFM) and scanning electron microscopy (SEM).

The AFM (Nanosurf, Core AFM) is used for samples with thickness in the nanometer range, since the nanosized tip cannot follow microscopic features. The images are obtained in tapping mode (tip from Nanosensors, model PPP-NCR), with a 10 nm radius, a resonance frequency of around 280 kHz and a spring constant of 25–30 N m^{-1}.

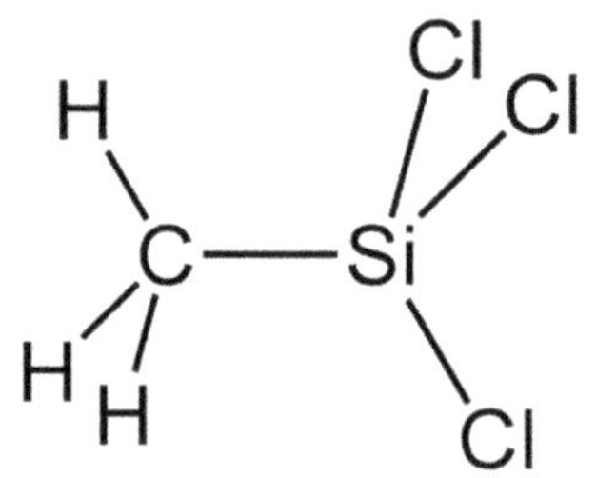

Figure 7.3. Methyl-trichlorosilane (M-silane) used to obtain nano- and micro-functionalized glass. (Adapted from [41].)

The SEM, from Zeiss, uses an accelerating voltage of 5 kV. To enhance the conductivity of the samples before the analysis, they are coated with a thin gold layer, varying the magnification to observe different features.

7.2.2.2 Wetting properties

To assess the wettability of the samples, sessile drop experiments are performed using a custom-built set-up composed of a high-speed camera (Photron NOVA S6) and an automatic syringe pump.

The experiment involves several steps: (i) deposition of an initial 3 μl water droplet, (ii) a 3 s pause for droplet stabilization and damping of oscillations, (iii) water inflation at a constant rate of 3 μl min^{-1} until the droplet reaches a maximum volume of 8 μl, and (iv) droplet deflation at the same constant rate of 3 μl min^{-1}. The video of the experiments is recorded at 10 fps and then the advancing, θ_A, and receding, θ_R, contact angles are calculated using Dropen software [42], developed in the MatLab environment using circle fitting. The reported data include a minimum of three independent measurements.

7.2.2.3 Drop impact tests

The drop impact tests are performed by dropping 6 μl droplets, in the range of Weber number $11 < \mathrm{We} < 161$ (corresponding to drop fall heights $20 < h < 250$ mm from the surface, and impact speed $0.7 < V < 2.2$ m s^{-1}) to verify the differences in drop impact outcomes between the samples. The experiments are recorded using a high-speed camera, controlled through a specific software supplied with the camera, PFV4, with a typical frame rate of 2000 to 4000 fps and sufficient exposure time to optimize contrast. A backlight lamp is used to provide sufficient illumination at a reduced exposure time. The configuration of the camera gives a resolution of 20 μm/pixel. Every experiment is replicated three times, where the drop, air and surface temperature are equal to room temperature.

7.2.2.4 Ice adhesion tests

To evaluate the icephobic properties of the various samples, ice adhesion push tests are performed using a custom-built set-up, as illustrated in figure 7.4. The set-up consists of a horizontal shear test, which has been extensively detailed in previous studies [43, 44] and in chapter 3. In short, the system includes an environmental chamber that maintains humidity at approximately 3% through a low nitrogen flow, a cooling system regulated by a Peltier element, and a motion stage connected to a force transducer to apply a load to the ice block. The position of the components in the system is seen from the top view in figure 7.4(a). Figure 7.4(b), a schematic from a lateral view, shows the ice formed by using cylindrical nylon molds of different diameter D and distilled water. The necessary amount of liquid water to exceed the pushing height (h) is added to the mold (typically 600–1000 μl) when the surface reaches 0 °C; then, the temperature is decreased until −10 °C. After freezing, the ice is allowed to stabilize for 20 min while maintaining the low humidity. The force probe is then pushed against the mold containing the ice at a 1 mm distance from the substrate (h in figure 7.4(b)) with a constant velocity of 0.01 mm s^{-1}, so

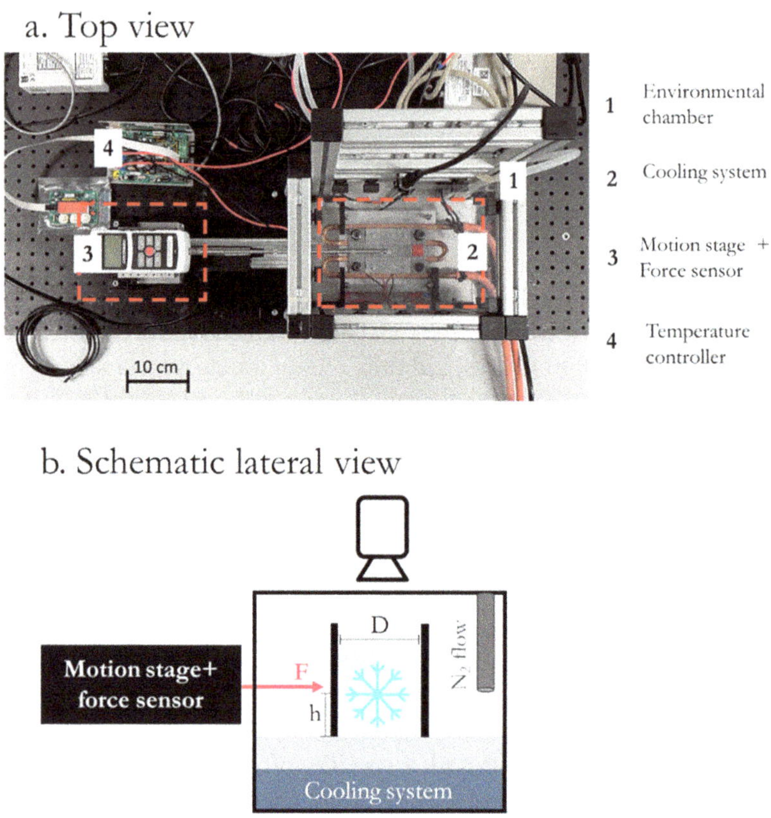

Figure 7.4. (a) Top view and (b) schematic lateral view of the horizontal ice adhesion push test. (Adapted from [43].)

the applied force on the ice increases until an ice detachment events occurrs. The ice block diameter used in this set of experiments is 8 mm.

7.2.3 Results for rigid materials

The exact sequence and kinetics of the condensation and hydrolysis reaction shown in figure 7.1 depend on many factors, such as the nature of the hydrolysable groups and their availability. Depending on experimental procedure and the substituents -R in the silane precursor, the reaction could produce a variety of arrangements. In figure 7.5, the morphology of the samples is visualized using different techniques, and the estimated roughness is reported. The bare glass used as a substrate has an average mean roughness of 0.8 nm (figure 7.5(a)). F-silane functionalization produces a smoother surface compared to the methyl-trichlorosilane functionalization (M-nano and M-micro).

The polymerization of the M-silane generates different morphologies at the nanoscale, depending on the relative humidity. The M-nano samples in figure 7.5(c) are obtained following a CVD process such that when the ambient humidity reaches

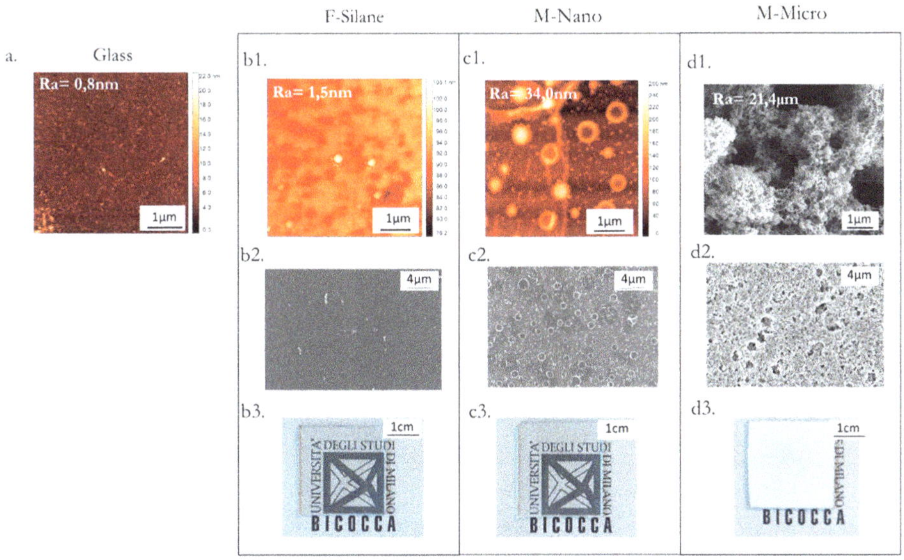

Figure 7.5. Morphology of glass and silane-based samples. (a), (b1) and (c1) are AFM images, whereas (d1), (b2), (c2) and (d2), are SEM images and (b3), (c3) and (d3) are optical images. (Adapted from [41].)

80%, its morphology consists of cylinders with homogeneous wall thickness (around 75 nm) of different dimensions in the nanometer range. The height of the structures varies in the range of 50–300 nm in the same sample depending on the analysed area.

When M-silane is deposited directly on the activated glass, the M-micro samples are obtained due to a rapid self-condensation at the interface. Since the system was not isolated from the environmental conditions, the polymeric structures obtained depend on factors beyond the relative humidity, such as the air flow under the hood.

It should be noted that alkyl trifunctional silanes undergo intermolecular condensation, growing further siloxane chains [45]. The tri-chlorine substituted precursor could give a crosslinked network when the self-condensation of the monomer occurs.

The wettability analysis through the sessile drop method is summarized in figure 7.6. In general, the functionalization procedures enhance the hydrophobic performance of glass slides. For F-silane the results were consistent for immersion times longer than 20 min, with $\theta_A = 113° \pm 1°$, $\theta_R = 95° \pm 3°$, and $\Delta H = 18° \pm 3°$. The M-nano functionalization increased the contact angle compared to the bare glass, but due to the inhomogeneous roughness, the hysteresis was high, indicating an increased drop adhesion. Lastly, the M-micro samples show averages $\theta_A = 162° \pm 1°$ and $\Delta H = 9° \pm 1°$, thus superhydrophobicity was obtained by the dual scale roughness, inducing a Cassie–Baxter wetting state.

In drop impact tests, the M-micro samples showed a high contact angle while recoiling, promoting the water drop full rebound. This agrees with previous studies [46], showing that a drop can rebound when $\theta_R > 100°$. For M-nano samples, the value of θ_R is well above the threshold. The drop rebound of M-micro samples present a promissory potential to avoid the liquid water accumulation in systems subject to

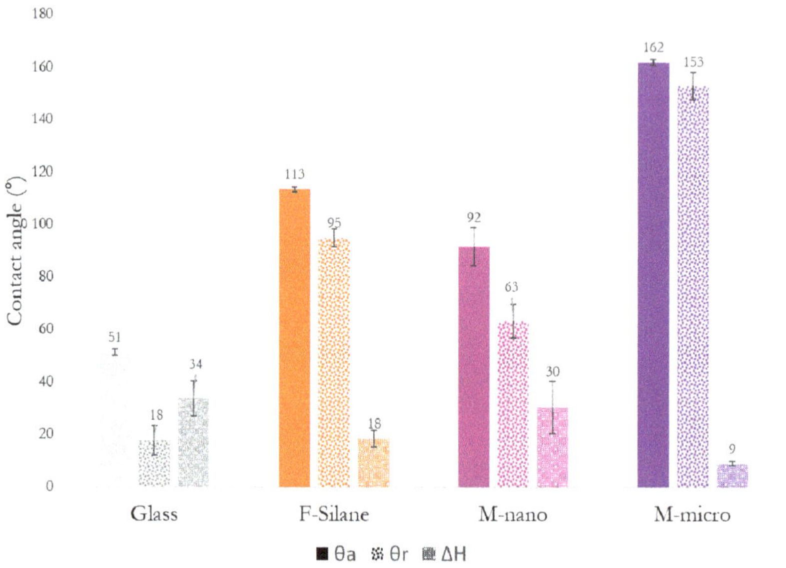

Figure 7.6. Wettability of the functionalized glass analysed by the sessile drop method. (Adapted from [41].)

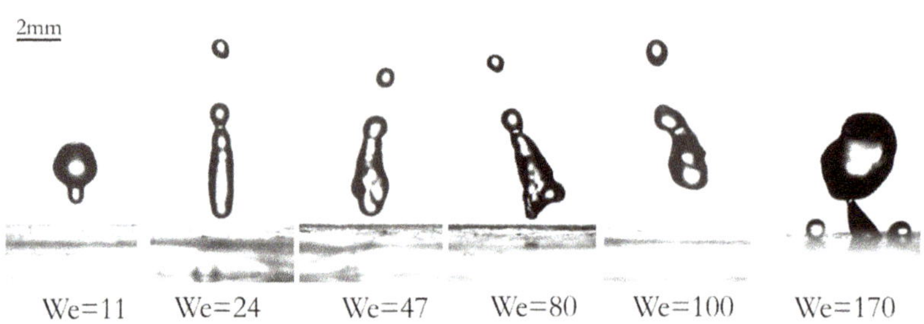

Figure 7.7. Experiments at We = 11, 24, 47, 80, 100 and 170 on M-micro samples at around 15 ms after impact. (Adapted from [41].)

vibrations. Therefore, experiments at higher We number (from 11 to 170) were made and the frames after the drop impact are shown in figure 7.7. For We < 100 there is a complete water rebound as the Cassie–Baxter state is preserved. As the We increases, drop fragmentation is observed together with the detachment of a secondary drop during the recoil phase. For We > 100, the fragmentation gives several secondary drops and, in the recoiling phase, the contact angle becomes lower than 90°, and the drop does not detach completely, giving only a partial rebound, with a smaller drop stuck on the surface (see figure 7.7, We = 170). This is an evidence of the transition from Cassie–Baxter to Wenzel wetting regime, when the impact pressure of the liquid is enough to penetrate the cavities at the surface, causing the droplet impalement.

After assessing the wettability, the ice adhesion performance was tested in a horizontal shear configuration, and the results are summarized in figure 7.8. For glass, and the M-nano and M-micro samples, the force initially increases until a maximum is reached, which decreases rapidly to zero when the ice block is

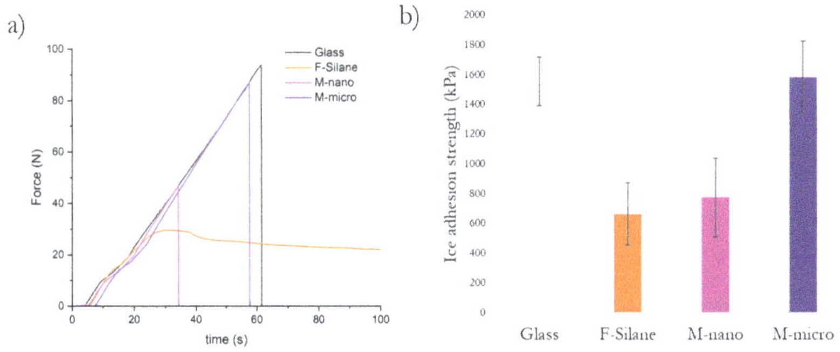

Figure 7.8. Results of the ice adhesion tests for rigid materials. (a) Force curves. (b) Ice adhesion strength. (Adapted from [41].)

completely detached. The de-icing mechanism is different for the F-silane functionalized glass, where, after a maximum is reached, the ice slides at a constant average force close to 20 N (figure 7.8).

It is important to mention that in the ice adhesion tests performed on the bare glass, a cohesive fracture was observed, as a portion of ice remain adhered to the glass. In contrast, for the M-micro sample part of the coating visibly remained adhered to the ice, indicating a cohesive fracture of the coating, which may compromise the durability of the surface over multiple cycles.

Since the icephobic performance is affected by the microstructure at the interface, the M-functionalized coatings present an important variation in the average value of the maximum force when samples obtained in different batch are compared. To check the durability of the structures on the M-nano samples, AFM microscopy was assessed after the ice adhesion tests. The images reveal that the maximum average height was reduced around 50%, suggesting a break of the structures during the de-icing tests.

The tri-chlorine substituted precursor on M-coatings gives a highly crosslinked network, that restricts the mobility and ensures the performance as a rigid material. In contrast, for perfluorododecyl-trichlorosilane the bulky aliphatic fluorinated chain presents a hysteric hindrance that limits the crosslinking of the polymer. Promoted by the presence of water in the hexane precursor solution, lateral fluorinated siloxane chains could grow under the right kinetic conditions, as schematized in figure 7.9, which contain mobile siloxane bonds with an optimal grafting density to induce the sliding behavior observed. We speculate that this may induce chain flexibility, that can lead to ice sliding on polymer brushes, as proposed by [47]. However, in the same study no sliding was observed for fluorinated self-assembled monolayers obtained by thiol chemistry. Also, such sliding has not been previously reported on alkyl-silane-based coatings, e.g. [48, 49].

The performance of the F-silane was maintained after several de-icing tests and thorough cleaning with ethanol, showing excellent durability. As a further perspective, research into the velocity dependence of this sliding behavior would be interesting to confirm the liquid-like performance of the samples.

Moreover, the differences in θ_R among the samples were not adequate to predict the icephobic behavior by using the models based on wetting, since the reduced

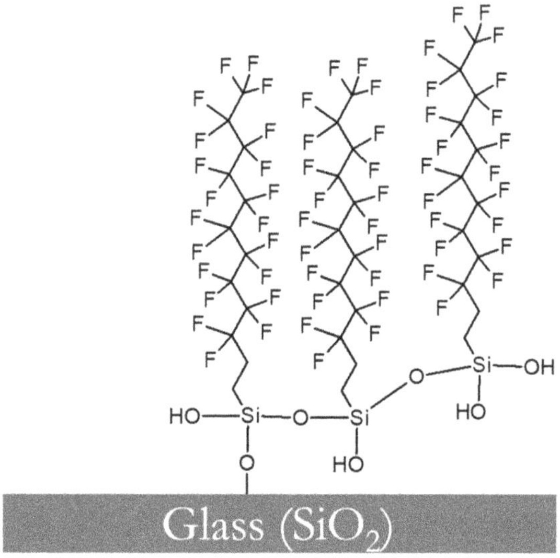

Figure 7.9. Possible schematic of siloxane lateral chains obtained when using perfluorododecyl-trichlorosilane. (Adapted from [41].)

hysteresis was roughness induced. The water could fill the cavities and promote a mechanical crosslinking among the coating and the ice, increasing the ice adhesion values expected according to the model, indicating no significant correlation with the ice adhesion strength [50].

7.3 Soft materials

7.3.1 Experimental details

The soft polymer coatings are obtained using Sylgard 184, a two-component silicone elastomer consisting of base and crosslinking fractions.

The pristine samples are fabricated by mixing the base and crosslinker in 10:1 proportion by weight. The appropriate amount of reactive mixture to reach a thickness of 1 mm is deposited on a glass slide (2.5×7.5 cm^2) followed by a curing process for 2 h at 80 °C.

Dry samples are prepared using the same procedure as for pristine samples, with the difference being that the coatings are immersed in toluene for 48 h, with the solvent changed every 24 h. This process removes the uncrosslinked chains (approximately 4% by weight) from the polymer [29]. The formulation for lubricated samples differs from both the dry and pristine ones only by the addition of 8% silicone oil (with a viscosity of 20 cSt, Sigma Aldrich) to the base formulation, which is incorporated prior to the crosslinking treatment at 80 °C, leading a total content of 12% lubricant in the sample. The resultant samples present a thickness of 1.1 ± 0.2 mm.

7.3.2 Surface characterization

The wetting properties were assessed as for rigid materials (section 7.2.2.2).

7.3.2.1 Mechanical properties

7.3.2.1.1 AFM nanoindentation

The Young's modulus of the samples is assessed by AFM indentation using PPP-CONTR tips for operating in contact mode (radius = 10 nm, Res. Freq. = 12 kHz and spring constant = 0.12 $N\,m^{-1}$).

During the indentation process, the tip is moved toward and away from the sample while its deflection is recorded. The indentation depth is set to 1 μm, and the sensitivity is determined using glass as a reference material. This allows for converting the voltage into the actual displacement of the cantilever. In the penetration region, the Johnson–Kendall–Roberts model is applied to fit the slope of the curve and calculate the material's Young's modulus (E). All experiments are conducted under low-humidity conditions. The reported results represent the average of ten independent trials.

7.3.2.1.2 Rheology analysis

The sample shear (G') and loss (G'') moduli are evaluated in an MCR92 RheoCompass™ rheometer. By bringing together parallel plates with a diameter of 25 mm, using the viscoelastic moving profile, samples of approximately 0.9 mm thickness are compressed until a safe gap is achieved, with compression kept below 12% of the sample's thickness. A preliminary amplitude sweep is conducted first, selecting a strain amplitude of 1.5% for performing a frequency sweep between 1 and 100 rad s^{-1}. The reported results are recorded at 100 rad s^{-1} and represent the average of five independent measurements.

7.3.2.2 Ice adhesion

The ice adhesion tests are conducted similarly as for rigid materials, but using ice blocks of 8, 10, 12 and 14 mm. Moreover, transparent ice that permits the visualization of fracture propagation is used. The experiments are video recorded from above the ice cylinder using a smartphone camera with a resolution of 1920 × 1080 at 30 fps to correlate the peaks in the force/time plots with the fracture events.

7.3.3 Results for soft materials

Parts of this section have been reproduced from [54]. CC BY 4.0.

The obtained soft PDMS coatings containing different lubricant content are visually smooth (figure 7.10), with an estimated roughness of 20 nm reported previously for material with similar composition and fabrication methods [51, 52].

To provide a complete mechanical characterization, contact angle, AFM indentation and rheology tests are performed on the different samples. The schematic representation and the outcome of those experiments are shown in figure 7.11, and a summary of the results is also presented in table 7.1.

The evolution of the contact angle during the inflation and deflation cycle is shown in figure 7.11(a). It can be seen that the three samples show different behaviors. The dry sample (the dark green line in figure 7.11(a)) exhibits a constant θ_A = 117°. The lubricated sample (the blue line in figure 7.11(a)) also shows a

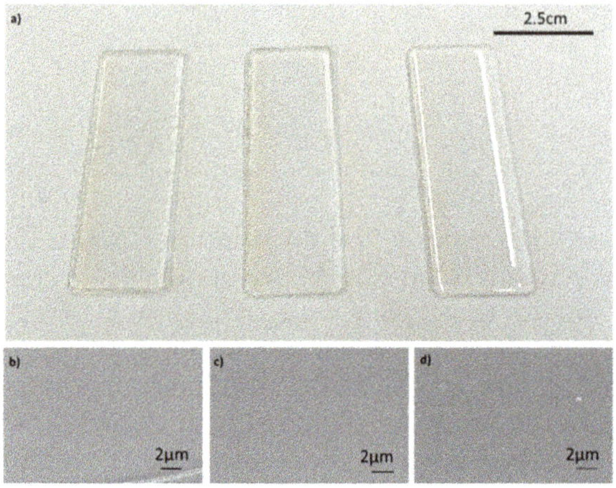

Figure 7.10. (a) Optical images of dry, pristine and lubricated samples (from left to right) and SEM micrographs of (b) dry, (c) pristine and (d) lubricated samples. (Reproduced from [53]. CC BY 4.0.)

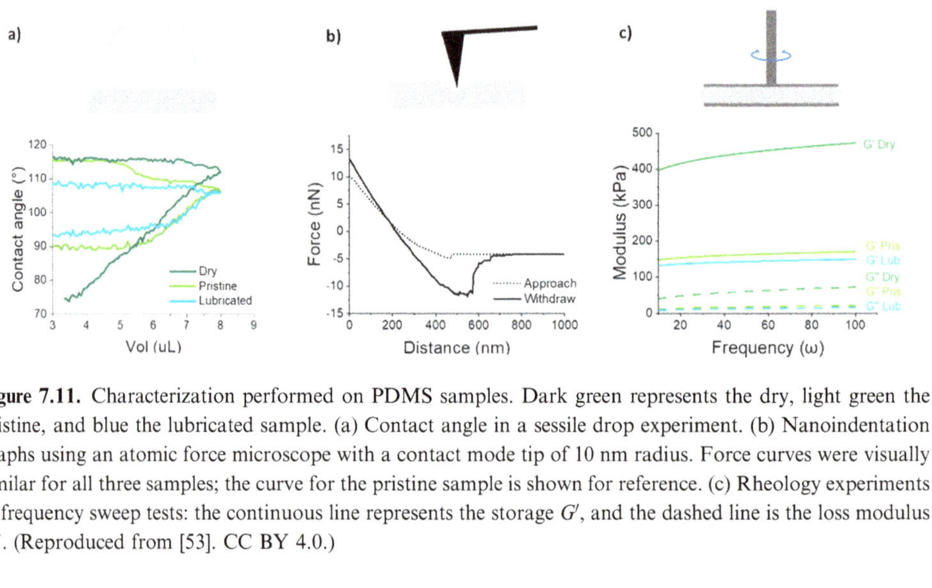

Figure 7.11. Characterization performed on PDMS samples. Dark green represents the dry, light green the pristine, and blue the lubricated sample. (a) Contact angle in a sessile drop experiment. (b) Nanoindentation graphs using an atomic force microscope with a contact mode tip of 10 nm radius. Force curves were visually similar for all three samples; the curve for the pristine sample is shown for reference. (c) Rheology experiments in frequency sweep tests: the continuous line represents the storage G', and the dashed line is the loss modulus G''. (Reproduced from [53]. CC BY 4.0.)

Table 7.1. Data from sessile drop experiments and mechanical properties (AFM indentation and rheology) for the three samples: dry, pristine and lubricated PDMS [53].

	Contact angles			AFM indentation	Rheology	
	θ_A (°)	θ_R (°)	ΔH (°)	E (kPa)	G^* (kPa)	tanδ
Dry	117 ± 1	74 ± 2	43 ± 2	3060 ± 660	477 ± 94	0.16 ± 0.002
Pristine	109 ± 3	91 ± 1	18 ± 3	2983 ± 560	171 ± 16	0.13 ± 0.005
Lubricated	107 ± 2	96 ± 1	11 ± 2	2508 ± 620	146 ± 12	0.11 ± 0.002

constant $\theta_A = 107°$. On the other hand, the pristine sample (the light green line in figure 7.11(a)) presents an intermediate behavior, evidencing a clear transition of its wetting during the experiment. At the beginning, the contact angle is 117°, which corresponds with the θ_A of the dry sample, however, after ~40 seconds from the beginning of the experiment, the contact angle decreases until it reaches a value of $\theta_A = 109$, similar to the θ_A of the lubricated sample. This property has previously been explained by the migration of uncrosslinked chains from the bulk to the interface with water. The phenomenon has been called 'adaptive wetting' [29] and results in the macroscopic modification of the wetting behavior.

For convenience, the contact angles and the calculated hysteresis for the different samples are summarized in table 7.1. Considering the ice adhesion tests protocol, where the liquid water is deposited and allowed to stabilize for some minutes before freezing occurs, the θ_A reported for the pristine material is 109°, which corresponds to the water-adapted value, once the mobile chains of lubricants have migrated to the interface.

The wetting experiments showed a decrease in ΔH when increasing the lubricant content, an indication of the presence of a movable layer at the interface. The experiments presented here demonstrate that the mobile units on the PDMS surface reorganize upon contact with water, exposing the dangling ends to the interface. This reorganization influences the wetting properties, which vary depending on prior contact with water, potentially affecting the material's mechanical and icephobic properties.

The mechanical properties were assessed by AFM nanoindentation and rheology experiments, which were tested avoiding any pre-wetting treatment of the samples to elude surface reorganization that may affect the mechanical performance of the samples.

The indentation tests evidenced a similar mechanical response of the three different samples, represented by the black line in figure 7.11(b), therefore they present similar Young's moduli, E, in the range 2.5–3.1 MPa (see table 7.1), since the differences are not statistically significant using a t-test distribution with 95% of confidence, there is no effect of the lubricant content in E values.

The mechanical behavior when shear stress is applied, similar as in the ice adhesion tests, was assessed by oscillatory rheology tests, by evaluating the storage, G', and loss, G'' moduli, as a function of the angular frequency, ω (figure 7.11(c)). The results indicate a predominant elastic response ($G' > G''$) in the frequency range considered. The ability to store and dissipate energy can be assessed by calculating the complex shear modulus, $G = \sqrt{G'^2 + G''^2}$, which considers the combined contribution of elastic and viscous response. The complex shear modulus decreases strikingly due to the inclusion of lubricants, from $G^* = 477 \pm 94$ kPa for the dry sample, which contains no lubricants, to values of 171 ± 16 and 146 ± 12 kPa for pristine and lubricated PDMS containing 4% and 12% of lubricants, respectively. Notably, G^* is an order of magnitude smaller than E.

The results on the mechanical properties of the PDMS support the idea of a segmented structure that could influence the isotropic nature of the samples. The potential phase segmentation may occur in discrete agglomerates of lubricant dispersed within the elastomeric network [54, 55]. This structure may store energy and prevent its dissipation throughout the material.

After the mechanical and wetting characterization of the samples, the ice adhesion tests were performed and the time evolution of the force curves is registered. Observing the simultaneous detachment of ice from the substrate and its progression at the ice–PDMS interface enables the correlation of fracture propagation with force measurements. Different ice adhesion experiments varying the ice block diameter (from 8 to 14 mm of diameter) were performed, allowing different detachment mechanisms to be identified depending on the dimensions of the ice blocks and the lubricant content on the samples (figure 7.12).

It is important to note that the force peaks have a rounded shape, in contrast to rigid surfaces, where a sharp peak detachment is typically observed (see figure 7.8 (a)), and the fracture propagates within milliseconds [43]. The round-shaped peaks suggest that some of the energy released during the initial detachment is absorbed by the material, contributing to its lateral deformation before the fracture propagates [7], which slows the crack speed compared to rigid materials [24, 56], hence, for soft materials, the fracture propagation occurs in the range of seconds, so it can be thus visualized with a standard camera.

The representative force curves have in common the initial force increase while the shear stress is accumulated at the interface until reaching a maximum, however, after this maximum is reached, different scenarios are observed both in the evolution of the push force and by the visualization of fracture events, thus, three different ice detachment mechanisms are identified: (i) single detachment, (ii) stick–slip and (iii) interfacial slippage (figure 7.12).

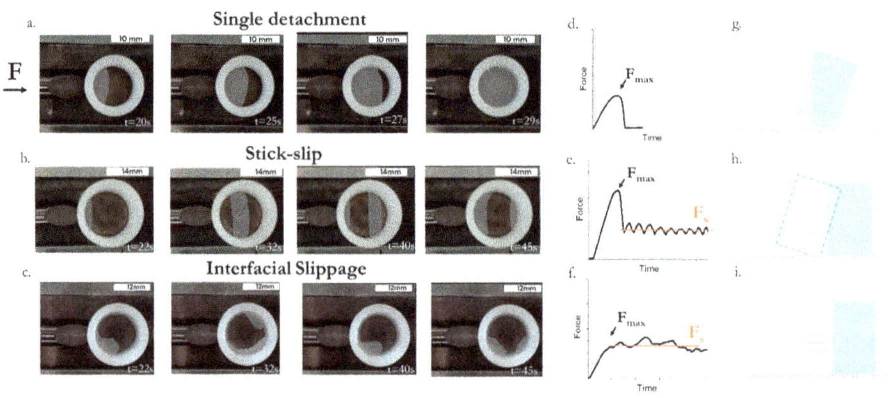

Figure 7.12. The three different mechanisms observed. Panels (a), (b) and (c) show a schematic representation of the fracture evolution from the top-view recording. The whitish areas represent the separation of the ice–coating interface observed in the videos. Panels (d), (e) and (f) show the representative force versus time curves for single detachment, stick–slip and interfacial slippage, respectively, and panels (g), (h) and (i) are schematic representations of the three different detachment mechanisms. ((a), (b), (c) adapted from [41]. (d), (e), (f), (g), (h), (i) adapted from [53]. CC BY 4.0.)

Single detachment occurs with ice blocks of 8 and 10 mm diameter, irrespective of the amount of lubricant in the coating. It is defined by the presence of a single force peak. As soon as the maximum force F_{max} (represented as a black arrow in figure 7.12) is reached, the fracture grows at the interface in the direction of the applied force and the force curve reduces to zero, revealing a complete adhesive failure between ice and PDMS.

For larger dimensions, when using ice blocks of 12 and 14 mm diameter, two phases of ice detachment can be identified: in the first 'static' phase the force reaches a peak (F_{max}) and in the second 'dynamic' phase the ice moves along the interface while remaining partially adhered, with the force oscillating around an average value F_s, depicted as an orange line in figures 7.12(e) and (f). Figures 7.13(a) and (b) graphically summarize the averaged values of F_{max} and F_s from the different experiments for the three different samples.

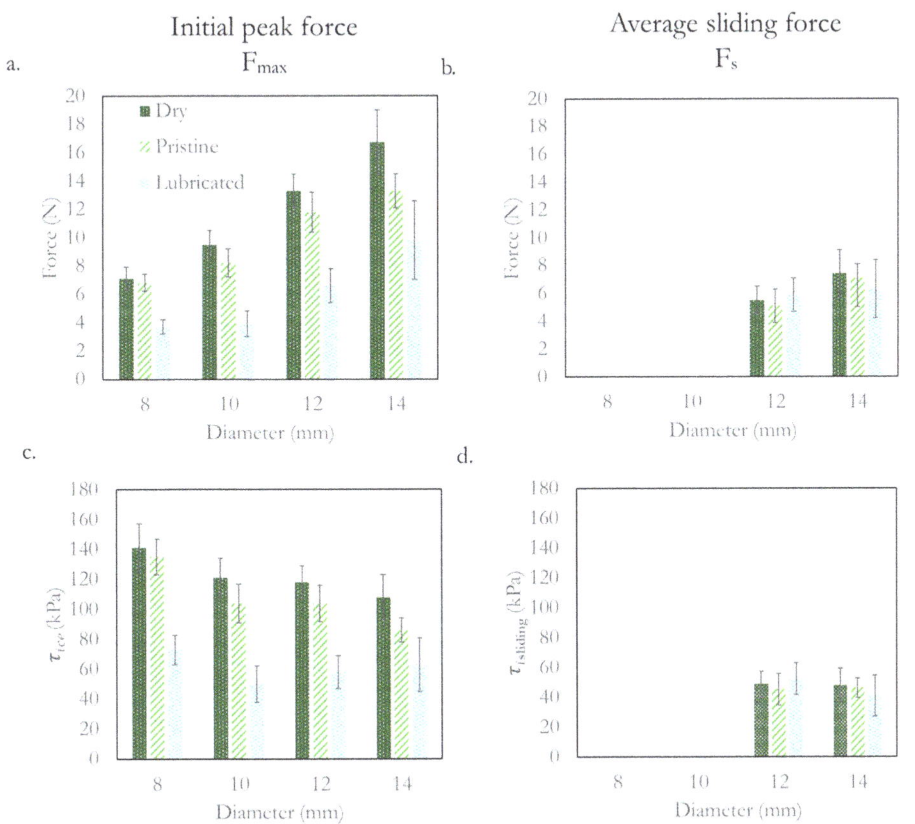

Figure 7.13. Results of ice adhesion tests for samples with different lubrication degrees and different ice dimensions. (a) Average measured force for initial detachment F_{max}. (b) Average force for sliding, F_S. (c) Ice adhesion strength τ_{ice} (F/A) for initial detachment. (d) Ice adhesion strength for sliding $\tau_{sliding}=(F_S/A)$. (Adapted from [53]. CC BY 4.0.)

The stick–slip mechanism is characterized by a high initial peak force F_{max} followed by a dynamic regime consisting on a regular pattern of fracture propagation (figure 7.12(e)) in which the force oscillates around F_s.

When the coating reaches its maximum deformation, the ice detaches from the substrate. Due to the material's elastic recovery and the dangling ends at the interface, the ice reattaches while shifting from its original position, leading to cyclic fracture propagation [57], commonly referred to in the literature as a separation pulse [58]. When the attachment–detachment or stick–slip cycles [7] present a regular pattern, the adhesion energy and the ability to store elastic energy are comparable, with the velocities of both the crack opening and closing edges being equal [59].

In contrast, the lubricated PDMS exhibits interfacial slippage: after reaching the initial peak, the force oscillates around this value, i.e. $F_{max} \approx F_s$.

As shown in figure 7.12(f), the separation pulses generated at the interface overlap with irregular slip pulses, where the material slides without a distinct separation pattern at the interface [58]. These mechanical instabilities lead to the release of shear energy, causing part of the substrate to detach from the ice but reattach after a short time. This creates a non-contact region at the interface that propagates in the direction of the applied force, resulting in an irregular pattern in the force curves.

For dry and pristine samples, the crosslinking density remains constant, whereas in the lubricated sample, the non-reactive chains of the silicone oil, added before thermal curing, may reduce the crosslinking density [60]. By increasing the lubricant content and lowering the crosslinking density, interfacial slippage is facilitated in the lubricated sample, allowing ice to slide over the surface without fully detaching from the load-bearing chains [61], as they experience relative motion compared to the static bulk solid.

The conventional method for evaluating icephobic performance, which solely focuses on comparing F_{max}, overlooks the fact that after an initial detachment, the ice can reattach and remain adhered. This is especially true for ice diameters larger than 10 mm, where reattachment is encouraged, resulting in different mechanisms depending on the lubricant content. Therefore, it is important to consider not only the first peak but also the secondary peaks, as they reveal distinct mechanisms across samples with varying lubricant contents.

In figures 7.13(c) and (d), τ is calculated using F_{max} and F_s. In the latter, there are no data for 8 and 10 mm diameter ice blocks because in those cases the ice presented a single detachment.

Figure 7.13(a) clearly shows that the lubricant content plays a crucial role in reducing ice adhesion. Therefore, the lubrication effect on icephobic performance cannot be overlooked. However, this reduction cannot be explained solely by wetting, as both pristine and lubricated samples exhibit similar θ_R, yet their ice adhesion results show different behaviors. As a result, the wetting properties, represented by the receding contact angle, are insufficient to predict the ice adhesion strength of soft materials with varying lubrication levels [61].

However, the lubricated sample with a lower shear modulus has the lowest ice adhesion values, irrespective of the dimension of the ice block. The F_{max} trend in figure 7.13(a) evidences an effect of the size of the ice block. The calculation of τ_{ice}

considers the size effects divided by the total area in contact. The data in figure 7.13 (c) suggest that τ_{ice} remains approximately constant only for lubricated PDMS. For dry and pristine PDMS, the maximum force does not scale with the area ($A \propto D^2$). This has significant implications, as it indicates that τ_{ice} is not an intrinsic material property, but instead depends on the experimental conditions, such as the ice block size used in our tests.

In this context, recent studies based on classical fracture theory for rigid materials have emphasized that τ_{ice} must be measured under appropriate conditions, as strength governs detachment for small ice block sizes (with dimension $L < L_c$, a critical length), while toughness governs detachment for larger ice block sizes, $L > L_c$. For hard surfaces, interfacial toughness Γ can be estimated by considering only the elasticity of the ice, as $\Gamma \approx \tau^2 L_c^2 / 2E_{ice} h$, where τ is the adhesion strength, E_{ice} is the modulus of ice, h is the ice block thickness, and L_c is measured from experiments [1, 43]. As observed in [1], where rectangular ice blocks with fixed width were used, in a toughness-dominated regime, above a critical length scale, the force becomes independent of the ice block length. This is because the force is required to initiate a crack, which then propagates along the entire length of the ice block. For soft polymers, applying the equation for interfacial toughness is questionable, as it does not take substrate elasticity into account.

Furthermore, in our experimental data, we observe no transition between the two regimes within the considered ice block dimensions. This suggests that toughness-mediated fracture is not achieved, as the energy released during fracture does not extend across the entire ice length. Instead, this energy is absorbed by the coating and used for its deformation [62]. Therefore, classical mechanical fracture models for hard materials cannot be directly applied to systems involving soft, deformable materials. This issue helps explain the significant variability observed in the literature regarding the average ice adhesion strength, τ_{ice}, for the same material, as well as the different ice accretion conditions [9].

On the other hand, in the dynamic regime, for larger ice blocks, it can be observed that F_s values are quite similar, regardless of the ice block dimensions or the lubricant content applied (figure 7.13(d)). Given the uniformity of the material components across different samples, the dynamic regime is more influenced by the lubricant's viscosity and its internal friction within the crosslinked network. The results suggest that when the dynamic phase occurs, it is not possible to reduce F_{max} further below F_s, as it represents the characteristic frictional force within the elastomeric network [24, 60, 61].

7.4 Discontinuity enhanced icephobic surfaces for low ice adhesion

To evaluate the if the discontinuity in mechanical properties contribute to the icephobic performance, a patterned surface containing rigid (photopolymerized resin) and elastic fractions (PDMS) was designed. Alongside experimental ice adhesion tests, the study includes numerical simulations of the ice–substrate interface stress by using a finite element method, that helps to understand the de-icing mechanism.

7.4.1 Experimental and characterization methods

7.4.1.1 Wetting performance, elastic modulus and roughness

The wetting properties were characterized for the two base materials, resin and PDMS as depicted in section 7.2.2.2, but with a bigger droplet of 15 μl, using constant rate of 10 μl min^{-1} for inflation and deflation stages. The θ_A and θ_R were obtained from the analysis of the images (over 300 for each experiment) using the automatic analysis software Dropen [42].

The elastic modulus of the PDMS was calculated from the indentation experiment explained in section 7.3.2.1.1. For the resin, the elastic modulus was calculated using the three-point bending test using rectangular slabs of 20.4 × 10.0 × 2.1 mm. The E was calculated with the expression $E = L^3F/4wh^3d$, where L is the distance between supports (18.0 ± 0.05 mm), w and h are the width and thickness of the sample, and F/d is the slope of the loading point force versus displacement curve.

The morphology at the surface was studied using a white light confocal microscope (PLμ 2300, Sensofar). A total of four measurements allowed us to calculate the arithmetic mean roughness (R_a) in the different regions.

7.4.1.2 Ice adhesion test

The ice adhesion performance was evaluated on the different substrates using the horizontal shear tests previously described in section 7.3.2.2, but considering the following modifications: a cylindrical nylon mold of 12 mm diameter was placed on the sample and filled with the necessary amount of water to exceed the pushing height of 3 mm. After reducing the temperature to −10 °C and allowing the stabilization time, the ice was pushed with a velocity of 0.01 mm s^{-1} until total detachment.

The ice adhesion tests were repeated eight times for each sample, with the ice block randomly positioned on the surface. The ice detachment process was captured from above through the clear ice using a phone camera, and some experiments were also recorded using a high-speed camera with a resolution of 1024 × 1024 pixels.

7.4.1.3 Numerical simulation

The ice–substrate interface stresses are calculated using a finite element method (FEM) numerical simulation. The model replicates the experimental set-up, and consists of a complex rigid–soft surface, an ice column, and a nylon mold. A similar numerical model has been used in previous studies [43, 63].

The force applied to the outside of the mold is simulated by imposing a displacement on a 1 × 1 mm^2 area, representing the estimated contact surface between the force application rod and the nylon mold in the experimental set-up. The materials used in the simulation are modeled as isotropic elastic, considering the density, elastic modulus, and Poisson's ratio of each material.

The interface stresses are computed through static structural analysis (Ansys Mechanical 2020 R1). After the analysis and comparison of the interface stresses of different surface geometries under an arbitrary, constant displacement value, the configuration resulting in the highest stress concentrations is identified. More details on the numerical simulations can be found in [64].

7.4.2 Results and discussion

Parts of this section have been reproduced from [64]. CC BY 4.0.

The results of the mechanical and wetting properties, as well as the mean roughness of each component, are summarized in table 7.2.

To test the hypothesis that surface elasticity discontinuities can promote ice crack initiation, various model surfaces containing different rigid fractions (11% and 25%) and different discontinuity line (DL) densities (expressed dimensionally as cm cm^{-2}) were tested. If discontinuities have no effect, the ice adhesion force would be represented by a weighted average between the adhesion on the rigid area $(F/A)_r$ = 610 ± 140 kPa, and the soft area $(F/A)_s$ = 82 ± 6 kPa, and would be independent of the pillar dimensions. In this case, adhesion could be expressed as a weighted average, e.g. $\tau_{ice} = f_r(F/A)_r + f_s(F/A)_s$, based on the rigid and soft fractions, f_r and f_s (represented as green and blue lines in figure 7.14(a)). On the other hand, if discontinuities induce interfacial cracks, then the ice adhesion force would decrease with longer DL, i.e. on surfaces with smaller pillars.

The data in figures 7.14(a) and (b) show a clear trend, with ice adhesion decreasing as pillar width decreases. A similar trend is observed for both rigid fractions (11% and 25%). For the largest pillar width of 2 mm, the adhesion value

Table 7.2. Properties of the rigid and soft components of the discontinuous surface [64].

	Rigid part	Soft part
E (MPa)	1150 ± 80	1.7 ± 0.4
θ_A (°)	83 ± 4	109 ± 2
θ_R (°)	39 ± 4	78 ± 3
R_a (nm)	224 ± 20	32 ± 3

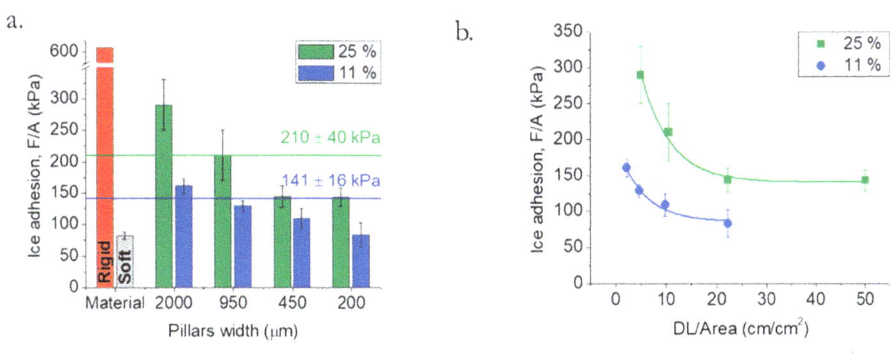

Figure 7.14. (a). Comparison of τ_{ice} of surfaces with different rigid fractions and DL densities versus pristine rigid and soft materials. The blue and green horizontal lines correspond to the weighted averages based on the rigid fraction. (b) Ice adhesion as a function of the discontinuity line density (DL/A). (Adapted from [64]. CC BY 4.0.)

exceeds the weighted average. However, the adhesion is lower for the other samples. Notably, the ice adhesion on the sample with $f_r = 11\%$ and the longest DL matches the adhesion value of the pure soft material (PDMS).

To understand the working principle of these surfaces, the ice–substrate interface was recorded. Since the crack propagation is slower and more visible on rough surfaces due to optical contrast, the surfaces were abraded to study the detachment process. The image sequence in figure 7.15(a) is representative of an ice detachment process, with a top-view perspective through the transparent ice. The whitish color that indicates the ice–substrate interface separation propagates to the back of the ice block (front-to-back). Moreover, the detachment occurs in two stages: first the ice detaches from the rigid pillars, and then from the soft areas.

Since ice adhesion is expected to be higher on the rigid areas due to their higher surface energy (indicated by lower contact angle values), and it is well known that a low elastic modulus and low surface energy promote easier ice detachment, looking at table 7.2, it is surprising that the ice detaches first from rigid areas. However, on rigid materials, the ice detachment occurs at lower strains than on soft materials, which explains the particular performance of the discontinuous surfaces.

When the fracture on the individual pillars is observed closely (figure 7.15(b)), a crack propagation opposite to the direction of the macroscopic fracture is revealed. The crack propagation on the individual pillar begins at the discontinuity line on the back edge and then moves toward the front side (back-to-front).

To understand this behavior, numerical simulations using the finite element method were performed to map the interfacial stress distribution on the pillars. The stress map (shown in figure 7.16) indicates that on pillars near the force application point, the stress peak is slightly higher on the back edge.

High stress concentrations occur near the pillar edges, where the discontinuities are located, at the interface between the rigid and soft areas. This stress distribution explains why, at the macroscale, the crack propagates from front-to-back among

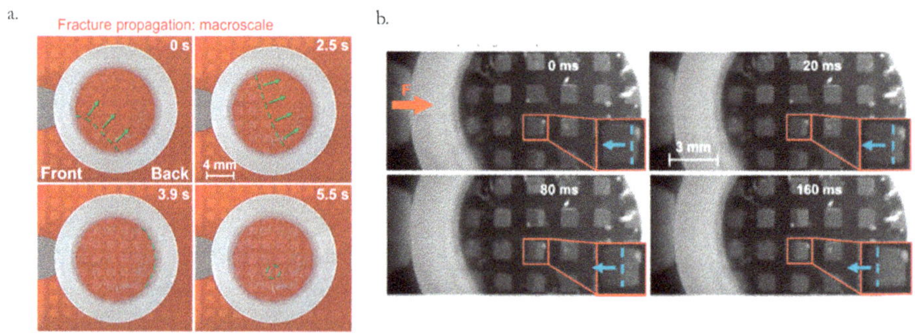

Figure 7.15. (a) Fracture propagation at the macroscale. The green lines and arrows indicate the propagation direction (front-to-back). The green circle shows a soft area just detached (a lighter red color compared to the previous frame). (b) Fracture propagation at the pillar scale: on the individual pillar, the crack is initiated on the discontinuity; but counterintuitively, the crack starts at the back edge of the pillar, and then propagates over the pillar to the front side, with a crack propagation opposite to the force application (back-to-front). (Adapted from [64]. CC BY 4.0.)

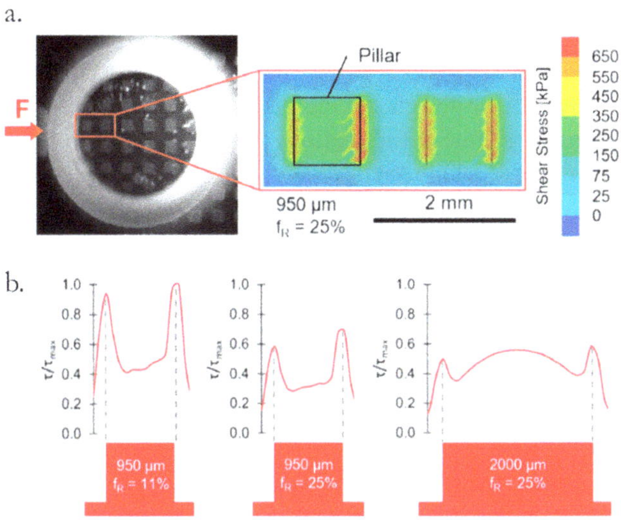

Figure 7.16. (a) Stress distribution and crack propagation direction on discontinuity-enhanced icephobic surfaces. (A) Shear stress distribution at the ice–substrate interface, as obtained by numerical simulations (sample: 25%–950 μm). (B) Stress level comparison for the three pillar geometries, taken from their respective first row of pillars. The values are shown as non-dimensional values τ/τ_{max} (τ_{max} is the maximum stress among the three considered cases). (Adapted from [64]. CC BY 4.0.)

pillars, but on the individual pillar, the propagation occurs from back-to-front (figure 7.14).

From the numerical model, it is noticed that the rigid fraction plays a major role in the stress concentration magnitude, which promotes the crack initiation, however, it cannot fully predict the experimental results, especially for samples with larger discontinuity lines. Therefore, in addition to the rigid fraction, the stress concentration sites play an important role in decreasing the ice adhesion.

7.5 Conclusions

In this chapter, the surface and mechanical properties of rigid and deformable materials have been assessed and their role on the icephobic behavior have been investigated. First, rigid materials were functionalized, obtaining different morphologies at the nano- and micro-scales, depending on the experimental conditions, allowing to enhance the hydrophobic performance of glass slides.

Using silane modification, different wetting and icephobic performance is obtained. Methyl-trichlorosilane-based coatings gives different scale morphology, that promotes hydrophobic and superhydrophobic behavior. In particular, the superhydrophobic coatings avoid liquid water accumulation in drop impact studies. In the ice adhesion tests, rigid coatings present the characteristic sharp detachment, whereas the fluorine-functionalized samples present the sliding of the ice block, resembling a liquid-like layer between the ice and the substrate.

Second, studies on soft substrates have focused on the role of lubrication. The time-dependent behavior of the contact angle, which was previously observed in soft polymers, was confirmed. The mechanical properties of the PDMS allows one to identify the material as a viscoelastic non-ideal rubber, affecting their properties beyond the classical relation for elastic materials. These facts suggest that time- and humidity-dependent mechanical and wetting properties need to be considered when investigating soft material icephobicity.

Additionally, ice adhesion tests on soft materials revealed three distinct ice detachment mechanisms: (i) single detachment, (ii) stick–slip, and (iii) interfacial slippage. The factors influencing these mechanisms are the degree of lubrication and the size of the ice block. It was found that a higher lubricant content in soft materials leads to a reduction in the shear modulus and a decrease in the force required for initial ice detachment. When larger ice blocks are tested, reattachment of the ice block is encouraged, resulting in a stick–slip regime for dry and pristine PDMS. In contrast, a regime of interfacial slippage is observed for lubricated PDMS.

The study on rigid and soft materials confirms that although wetting properties could reduce the ice accretion and may be suitable for anti-icing substrates, once the ice is formed at the substrate, the differences in deformability are more relevant to understand the de-icing mechanism. In particular, rigid materials accumulate an excess of energy at the surface and promote crack initiation, while deformable materials can absorb that energy and show a low ice adhesion force, but ice can eventually slide without really detaching, reaching a slippage limit, depending on the surface viscoselastic properties.

Regarding surface-discontinuities on composite rigid–soft surfaces, it is shown that reducing the rigid fraction improves the icephobic performance, moreover, by increasing the discontinuity line density, the ice adhesion is further reduced. This suggest the possibility to improve the performance by optimizing the design of the discontinuous surfaces.

As a concluding remark, all experimental studies conducted on the different type of surfaces confirm that τ_{ice}, simply defined as F/A, is not an absolute value, as it depends on several experimental conditions. Therefore, it cannot be considered an intrinsic material property and can only be used for a relative comparison, making it crucial to fully describe the testing system and the de-icing mechanism, along with the values of F and τ_{ice}, in order to enable comparisons among different materials.

Acknowledgments

The author expresses gratitude to Roberto Pelusio and Alice Colombo for their dedication and support for some of the experiments presented here.

References

[1] Golovin K, Dhyani A, Thouless M D and Tuteja A 2019 Low-interfacial toughness materials for effective large-scale deicing *Science* **364** 371–5

[2] Huré M, Olivier P and Garcia J 2022 Effect of Cassie–Baxter versus Wenzel states on ice adhesion: a fracture toughness approach *Cold Reg. Sci. Technol.* **194** 103440

[3] Subramanyam S B, Rykaczewski K and Varanasi K K 2013 Ice adhesion on lubricant-impregnated textured surfaces *Langmuir* **29** 13414–8

[4] Chen L, Huang S, Ras R H A and Tian X 2023 Omniphobic liquid-like surfaces *Nat. Rev. Chem.* **7** 123–37
[5] Zhang L, Guo Z, Sarma J and Dai X 2020 Passive removal of highly wetting liquids and ice on quasi-liquid surfaces *ACS Appl. Mater. Interfaces* **12** 20084–95
[6] Kendall K 1971 The adhesion and surface energy of elastic solids *J. Phys. D: Appl. Phys.* **4** 1186
[7] Beemer D L, Wang W and Kota A K 2016 Durable gels with ultra-low adhesion to ice *J. Mater. Chem.* A **4** 18253–8
[8] Wang Z, Zhao Z, Wen G, Zhu Y, Chen J, Jing X, Sun S, Zhang L, Liu X and Chen H 2023 Fracture-promoted ultraslippery ice detachment interface for long-lasting anti-icing *ACS Nano* **17** 13724–33
[9] Wang C, Fuller T, Zhang W and Wynne K J 2014 Thickness dependence of ice removal stress for a polydimethylsiloxane nanocomposite: Sylgard 184 *Langmuir* **30** 12819–26
[10] Ibáñez-Ibáñez P F, Montes Ruiz-Cabello F J, Cabrerizo-Vílchez M A and Rodríguez-Valverde M A 2022 Ice adhesion of PDMS surfaces with balanced elastic and water-repellent properties *J. Colloid Interface Sci.* **608** 792–9
[11] Nosonovsky M and Hejazi V 2012 Why superhydrophobic surfaces are not always icephobic *ACS Nano* **6** 8488–91
[12] He Z, Xiao S, Gao H, He J and Zhang Z 2017 Multiscale crack initiator promoted super-low ice adhesion surfaces *Soft Matter* **13** 6562–8
[13] He Z, Zhuo Y, Zhang Z and He J 2021 Design of icephobic surfaces by lowering ice adhesion strength: a mini review *Coatings* **11** 1343
[14] Zhuo Y, Xiao S, Amirfazli A, He J and Zhang Z 2021 Polysiloxane as icephobic materials—the past, present and the future *Chem. Eng. J.* **405** 127088
[15] Zhao B, Bonaccurso E, Auernhammer G K and Chen L 2021 Elasticity-to-capillarity transition in soft substrate deformation *Nano Lett.* **21** 10361–7
[16] Xu Q, Wilen L A, Jensen K E, Style R W and Dufresne E R 2020 Viscoelastic and poroelastic relaxations of soft solid surfaces *Phys. Rev. Lett.* **125** 238002
[17] Liu Y, Ma L, Wang W, Kota A K and Hu H 2018 An experimental study on soft PDMS materials for aircraft icing mitigation *Appl. Surf. Sci.* **447** 599–609
[18] Glover J D, McLaughlin C E, McFarland M K and Pham J T 2020 Extracting uncrosslinked material from low modulus sylgard 184 and the effect on mechanical properties *J. Polym. Sci.* **58** 343–51
[19] Qi H, Lei X, Gu J, Zhang Y, Gu X, Zhao G and Yu J 2023 Low modulus of polydimethylsiloxane organogel coatings induced low ice adhesion *Prog. Org. Coat.* **177** 107435
[20] Zheng H, Liu G, Nienhaus B B and Buddingh J V 2022 Ice-shedding polymer coatings with high hardness but low ice adhesion *ACS Appl. Mater. Interfaces* **14** 6071–82
[21] Müller A, Wapler M C and Wallrabe U 2019 A quick and accurate method to determine the Poisson's ratio and the coefficient of thermal expansion of PDMS *Soft Matter* **15** 779–84
[22] Johnston I D, McCluskey D K, Tan C K L and Tracey M C 2014 Mechanical characterization of bulk Sylgard 184 for microfluidics and microengineering *J. Micromech. Microeng.* **24** 035017
[23] Petit J and Bonaccurso E 2014 General frost growth mechanism on solid substrates with different stiffness *Langmuir* **30** 1160–8

[24] Xue L, Pham J T, Iturri J and Del Campo A 2016 Stick-slip friction of PDMS surfaces for bioinspired adhesives *Langmuir* **32** 2428–35
[25] Moučka R, Sedlačík M, Osička J and Pata V 2021 Mechanical properties of bulk sylgard 184 and its extension with silicone oil *Sci. Rep.* **11** 19090
[26] Chen L, Bonaccurso E, Deng P and Zhang H 2016 Droplet impact on soft viscoelastic surfaces *Phys. Rev.* E **94** 063117
[27] Yang L, Liu X, Wang J and Zhang P 2023 An experimental study on complete droplet rebound from soft surfaces: critical weber numbers, maximum spreading, and contact time *Langmuir* **40** 2165–73
[28] Regulagadda K, Gerber J, Schutzius T M and Poulikakos D 2022 Microscale investigation on interfacial slippage and detachment of ice from soft materials *Mater. Horiz.* **9** 1222–31
[29] Wong W S Y, Hauer L, Naga A, Kaltbeitzel A, Baumli P, Berger R, D'Acunzi M, Vollmer D and Butt H J 2020 Adaptive wetting of polydimethylsiloxane *Langmuir* **36** 7236–45
[30] Butt H J, Berger R, Steffen W, Vollmer D and Weber S A L 2018 Adaptive wetting—adaptation in wetting *Langmuir* **34** 11292–304
[31] He Z, Zhuo Y, He J and Zhang Z 2018 Design and preparation of sandwich-like polydimethylsiloxane (PDMS) sponges with super-low ice adhesion *Soft Matter* **14** 4846–51
[32] He Z, Zhuo Y, Wang F, He J and Zhang Z 2019 Understanding the role of hollow sub-surface structures in reducing ice adhesion strength *Soft Matter* **15** 2905–10
[33] Zeng C, Shen Y, Tao J, Chen H, Wang Z, Liu S, Lu D and Xie X 2022 Rationally regulating the mechanical performance of porous PDMS coatings for the enhanced icephobicity toward large-scale ice *Langmuir* **38** 937–44
[34] Irajizad P, Al-Bayati A, Eslami B, Shafquat T, Nazari M, Jafari P, Kashyap V, Masoudi A, Araya D and Ghasemi H 2019 Stress-localized durable icephobic surfaces *Mater. Horiz.* **6** 758–66
[35] Wang J, Wu M, Liu J, Xu F, Hussain T, Scotchford C and Hou X 2021 Metallic skeleton promoted two-phase durable icephobic layers *J. Colloid Interface Sci.* **587** 47–55
[36] Jiang X, Lin Y, Xuan X, Zhuo Y, Wu J, He J, Du X, Zhang Z and Li T 2023 Stiffening surface lowers ice adhesion strength by stress concentration sites *Colloids Surf.* A **666** 131334
[37] Chen C, Fan P, Zhu D, Tian Z, Zhao H, Wang L, Peng R and Zhong M 2023 Crack-initiated durable low-adhesion trilayer icephobic surfaces with microcone-array anchored porous sponges and polydimethylsiloxane cover *ACS Appl. Mater. Interfaces* **15** 6025–34
[38] He Z, Zhuo Y, Wang F, He J and Zhang Z 2020 Design and preparation of icephobic PDMS-based coatings by introducing an aqueous lubricating layer and macro-crack initiators at the ice–substrate interface *Prog. Org. Coat.* **147** 105737
[39] Kim J H, Kim M J, Lee B, Chun J M, Patil V and Kim Y S 2020 Durable ice-lubricating surfaces based on polydimethylsiloxane embedded silicone oil infused silica aerogel *Appl. Surf. Sci.* **512** 145728
[40] Ulman A 1996 Formation and structure of self-assembled monolayers *Chem. Rev.* **96** 1533–54
[41] Ospina C 2025 Rigid and soft materials for icephobicity: the role of surface and mechanical properties *PhD Thesis* University of Milano-Bicocca, Milano
[42] Akbari R and Antonini C 2021 Contact angle measurements: from existing methods to an open-source tool *Adv. Colloid Interface Sci.* **294** 102470
[43] Stendardo L, Gastaldo G, Budinger M, Pommier-Budinger V, Tagliaro I, Ibáñez-Ibáñez P F and Antonini C 2023 Reframing ice adhesion mechanisms on a solid surface *Appl. Surf. Sci.* **641** 158462

[44] Tagliaro I, Radice V, Nisticò R and Antonini C 2024 Chitosan electrolyte hydrogel with low ice adhesion properties *Colloids Surf.* A **700** 134695
[45] Gresham I J and Neto C 2023 Advances and challenges in slippery covalently-attached liquid surfaces *Adv. Colloid Interface Sci.* **315** 102906
[46] Antonini C, Villa F, Bernagozzi I, Amirfazli A and Marengo M 2013 Drop rebound after impact: the role of the receding contact angle *Langmuir* **29** 16045–50
[47] Zhao X, Khatir B, Mirshahidi K, Yu K, Kizhakkedathu J N and Golovin K 2021 Macroscopic evidence of the liquidlike nature of nanoscale polydimethylsiloxane brushes *ACS Nano* **15** 13559–67
[48] Arianpour F, Farzaneh M and Jafari R 2016 Hydrophobic and ice-phobic properties of self-assembled monolayers (SAMs) coatings on AA6061 *Prog. Org. Coat.* **93** 41–5
[49] Balordi M, Cammi A, Santucci de Magistris G and Chemelli C 2019 Role of micrometric roughness on anti-ice properties and durability of hierarchical super-hydrophobic aluminum surfaces *Surf. Coat. Technol.* **374** 549–56
[50] Mostofi Sarkari N *et al* 2025 Experimental debate on the overlooked fundamental concepts in surface wetting and topography vs ice adhesion strength relationships *J. Colloid Interface Sci.* **682** 825–48
[51] Ibáñez-Ibáñez P F, Montes Ruiz-Cabello F J, Cabrerizo-Vílchez M A and Rodríguez-Valverde M A 2021 Contact line relaxation of sessile drops on PDMS surfaces: a methodological perspective *J. Colloid Interface Sci.* **589** 166–72
[52] Ibáñez-Ibáñez P F, Montes Ruiz-Cabello F J, Cabrerizo-Vílchez M A and Rodríguez-Valverde M A 2022 Mechanical durability of low ice adhesion polydimethylsiloxane surfaces *ACS Omega* **7** 20741–9
[53] Ospina C, Ibáñez-Ibáñez P F, Tagliaro I, Stendardo L, Tosatti S and Antonini C 2025 Low ice adhesion on soft surfaces: elasticity or lubrication effects? *J. Colloid Interface Sci.* **677** 494–503
[54] Yeong Y H, Milionis A, Loth E and Sokhey J 2018 Self-lubricating icephobic elastomer coating (SLIC) for ultralow ice adhesion with enhanced durability *Cold Reg. Sci. Technol.* **148** 29–37
[55] Krawczyk J, Croce S, McLeish T C B and Chakrabarti B 2016 Elasticity dominated surface segregation of small molecules in polymer mixtures *Phys. Rev. Lett.* **116** 208301
[56] Xie Q, Hao T, Wang C, Kang Z, Shi Z and Zhang J 2021 The mechanical mechanism and influencing factors of ice adhesion strength on ice-phobic coating *J. Mar. Sci. Eng.* **9** 315
[57] Chaudhury M K and Kim K H 2007 Shear-induced adhesive failure of a rigid slab in contact with a thin confined film *Eur. Phys. J.* E **23** 175–83
[58] Viswanathan K, Sundaram N K and Chandrasekar S 2016 Stick-slip at soft adhesive interfaces mediated by slow frictional waves *Soft Matter* **12** 5265–75
[59] Yamaguchi T, Ohmata S and Doi M 2009 Regular to chaotic transition of stick-slip motion in sliding friction of an adhesive gel-sheet *J. Phys.: Condens. Matter* **21** 205105
[60] Golovin K and Tuteja A 2017 A predictive framework for the design and fabrication of icephobic polymers *Sci. Adv.* **3** e1701617
[61] Golovin K, Kobaku S P R, Lee D H, DiLoreto E T, Mabry J M and Tuteja A 2016 Designing durable icephobic surfaces *Sci. Adv.* **2** e1501496
[62] Mohseni M, Dijvejin Z A and Golovin K 2021 Designing scalable elastomeric anti-fouling coatings: shear strain dissipation via interfacial cavitation *J. Colloid Interface Sci.* **589** 556–67

[63] Stendardo L, Gastaldo G, Budinger M, Tagliaro I, Pommier-Budinger V and Antonini C 2024 Why the adhesion strength is not enough to assess ice adhesion on surfaces *Appl. Surf. Sci.* **672** 160740

[64] Ibáñez Ibáñez P F, Stendardo L, Ospina C, Chaudhary R, Tagliaro I and Antonini C 2024 Discontinuity-enhanced icephobic surfaces for low ice adhesion *J. Colloid Interface Sci.* **679** 403–10

Chapter 8

The use of thermally activatable adhesion promoters for permanent and durable immobilization of non-reactive polymeric coatings

Alexandros A Atzemoglou, Niccolò Bartalucci, Anny C Ospina Patiño, Carlo Antonini, Santiago J Garcia, Mark W Tibbitt, Samuele G P Tosatti and Stefan Zürcher

The problem of ice formation and accumulation on surfaces negatively impacts many aspects of daily life. Current strategies focus on the development of icephobic coatings that aim to repel water droplets and reduce the adhesion of ice. To function properly, icephobic coatings often require complex synthesis steps to be immobilized on surfaces, and permanent bonding with a substrate after synthesis is often not feasible. As the adhesion strength of coatings constitutes an important property that determines the performance and durability of adhesive coating systems, we immobilized polymers on different types of substrates with the use of an adhesion promoter to assess their adhesion strength. For that, lap shear tests were used to assess the adhesive bond strength of bound samples, using cyclic olefin copolymers (COCs) or epoxy as adhesives. The use of an adhesion promoter layer, capable of forming covalent bonds with any neighboring C–C bond via nitrene insertion, can increase the adhesion between materials that do not possess inherent chemical affinity, therefore creating an appropriate coating platform for adhesives that would not bind otherwise. Thermal curing of the samples with adhesion promoters containing the reactive 4-azido-3,5-difluorophenoxy (DFPxA) moiety led to increased adhesion strength. Additionally in this work, we immobilized different commonly used polymeric materials as monolayers on an aluminum substrate with the use of an adhesion promoter layer containing the reactive DFPxA moiety. Dynamic contact angle (dCA) and x-ray photoelectron spectroscopy (XPS) were used to assess the successful bonding of the reactive moiety with a wide range of different polymeric materials, demonstrating its universal attachment to neighboring

doi:10.1088/978-0-7503-6009-8ch8

bonds. Moreover, we investigated the adhesion strength of ice on immobilized monolayers of the coated polymeric films, which would not bind otherwise, revealing a correlation between adhesion strength and surface wettability.

8.1 Adhesion promoter compounds in surface engineering

Each material possesses unique characteristic properties that make it suitable for specific applications. However, there are instances where the inherent properties of a material may not fulfill certain requirements [1]. Thus, approaches to combine functionalities originating from various materials to accomplish a desired performance are needed [2, 3].

The surfaces of materials play crucial roles in applications in numerous fields as they define the interaction of a material with the surrounding environment [4]. In the field of surface engineering, polymeric coatings constitute a primary approach to modify surfaces to introduce desired functionalities without compromising the properties of the underlying bulk material [5–7]. Permanent deposition of coatings is therefore important for long-term stability and durability, since otherwise the coatings when exposed to harsh environmental conditions and without strong adhesion may peel off or delaminate, leading to premature failure and loss of protection for the substrate [8–10]. Polymeric coatings are widely used as they offer increased flexibility due to the wide range of properties that are achieved using films with different chemistries, thicknesses, and processing conditions [11]. An inherent property of amorphous polymers is the transition from a hard, glassy state to a softer rubbery state [12, 13]. Many polymers exhibit a glass transition temperature at approximately 100 °C [14–16] and when used as substrates the methods used for surface modification are limited by this property.

Nevertheless, efficient and permanent surface modifications to acquire desired surface functionalities are not always straightforward and can require time-intensive and complex processes that require in-depth analysis of the material surface chemistry [17, 18]. This is because not all polymeric materials that act as coatings have chemical affinity with multiple different substrates. The use of intermediate layers, called primers or adhesion promoters, can facilitate generalizable coating protocols of surfaces with polymeric layers [19]. More specifically, versatile adhesion promoters are bifunctional materials that can strongly bind both to various underlying substrates and covalently to arbitrary polymeric coatings, by using thermally or photolytically activated reactive groups, such as azides [20]. Thus, their use is of great benefit as they can increase adhesion between surfaces that would not bind otherwise. Aryl azides are considered stable reactive compounds, which are easy to synthesize and can form adhesion promoters for effective permanent immobilization of polymers through C–H insertion reactions [21], see figure 8.1.

To overcome these limitations, we designed and synthesized aryl azide containing adhesion promoters for versatile surface modification at temperatures below 100 °C. We focused on demonstrating the versatility of a reactive adhesion promoter towards effective binding with multiple polymers of different wettability. Further, we investigated the ice adhesion strength on the coated polymeric monolayers,

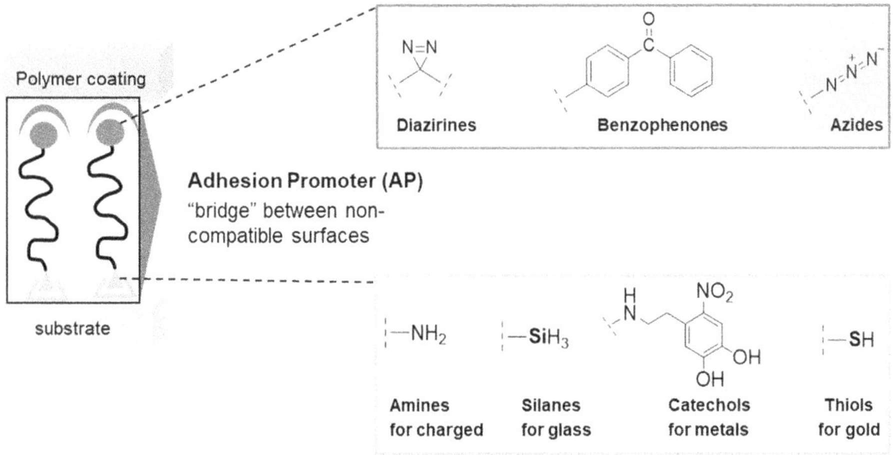

Figure 8.1. Bifunctional adhesion promoters containing aryl azides work as a chemical 'bridge' for surfaces that do not exhibit chemical affinity and would not bind otherwise. (Adapted from [77].)

presuming that thermal and elastic effects deriving from the substrate will be eliminated using a monomolecular polymeric layer tethered to the surface by another adhesion promoting monolayer. Moreover, we investigated the influence of an activated adhesion promoter on making a coating system more robust and durable, a crucial factor to avoid premature failure.

The findings we present could evolve the versatile surface modification of materials either with peculiar shapes, as well as materials that would deform when heated above 100 °C or when UV activation is not feasible.

In our previous work [22], we were able to design and synthesize reactive compounds containing aryl azides, where the activation of the azide can occur thermally already at temperatures below 100 °C. These moieties were capable of undergoing insertion reactions with a polymeric material in close proximity for efficient permanent surface modification, rendering them appropriate candidates for adhesion promoters that are to be used on materials sensitive to light or on polymers having low glass transition temperatures and low melting points. The most reactive synthesized compound was an o,o-difluoro substituted p-phenoxy azide (DFPxA moiety) and the important novelty was that the activation barrier was lowered to ~70 °C. This newly synthesized compound was compared to the commercially available perfluorophenyl azide (PFPA moiety) known to decompose at temperatures higher than 120 °C.

Based on the synthesized aryl azide compounds, we formulated adhesion promoters to assess their efficiency to create a stable chemical 'bridge' between materials that lack reactive end groups, and thus allowing for versatile and permanent surface modification. The differently substituted heat sensitive aryl azide moieties were grafted to different units that allowed for effective surface immobilization. The adhesion promoter containing the DFPxA moiety activated and successfully covalently bonded to polymer films at the lowest temperature, at ~70 °C. Thus, the fact that the compounds can be activated at a lower temperature, especially those containing the DFPxA moiety, render them suitable adhesion promoters for permanent immobilization of coatings on polymeric materials with low glass transition temperature or low melting point, on materials with peculiar shapes or light-blocking materials where light activation is challenging or impossible.

8.2 The effect of surface-tethered monomolecular thin film wettability on ice adhesion

8.2.1 Introduction

The formation and accumulation of ice on exposed surfaces is associated with detrimental impacts on the safety and operation performance of instruments and facilities in various industries, such as aircraft [23, 24], power lines [25], off-shore platforms, wind turbines, and solar panels [26]. For safety critical applications, such as in aircrafts, de-icing methods include energy inefficient electric heating and the time-consuming application of de-icing fluids, which are environmentally problematic [27, 28]. A more appealing and universal approach is the development of economical coatings to which ice minimally adheres or can be easily removed under its own weight or due to natural factors, such as wind. This would have broad applications improving the safety and reliability of low temperature operations [29]. In order for ice to debond by natural forces or vibration, the ice adhesive strength should decrease to less than 100 kPa [30].

Understanding the mechanisms of ice adhesion and the underlying surface properties for minimizing ice–substrate interactions would enable the successful design of coatings known as 'icephobic' coatings [31–34]. Typically they leverage functionalities such as superhydrophobicity to repel water droplets, delay ice nucleation, and reduce ice adhesion significantly [35] and they are meant to be combined with de-icing strategies.

Alongside surface morphology, ice adhesion strength is commonly associated with the surface water wettability/hydrophobicity. However, there have been contradicting findings regarding the effect of surface wettability on ice adhesion strength, with some studies suggesting a decrease in ice adhesion strength on high contact angle surfaces due to their low surface energy [36, 37], whereas others have found little relation between the two parameters [29, 38, 39]. This discrepancy suggests that the relationship between water contact angle and ice adhesion strength is complex and varies depending on additional surface properties.

The design of highly structured icephobic surfaces usually requires complex protocols and modifications of the surface chemistry. Moreover, coating monolayers of polymers exhibiting icephobic properties is not feasible on substrates without chemical affinity since any physisorbed coating would likely wash off during rinsing. Here, we demonstrate the potential of an adhesion promoter carrying a low temperature activated aryl azide, which was functionalized with a nitro-catechol group, to successfully function as an intermediate for the permanent surface modification of aluminum substrates with different polymeric coatings.

Catechol-functionalized polymers covalently link to metallic substrates, via the formation of coordinate bonds with their oxides and hydroxides. However, multifunctional, cross-linked polymeric films are often formed due to the self-oxidation of the catechol moieties [40]. In order to repeatedly obtain a monomeric catechol film, Zürcher *et al* demonstrated that the introduction of electron-withdrawing groups, such as nitro groups, is required [41].

Since the adhesion promoter covalently binds both to the aluminum substrate through the nitro-catechol and to the deposited polymers through the insertion of the nitrenes, increased adhesion and stability can generally be achieved. In that manner, any structural, mechanical, or thermal effects caused by the substrate to the polymeric coating is eliminated. This way the actual chemical and physical composition of the polymer molecules are assumed to be the sole affecting factors of ice adhesion, and thus the measurements. In that respect, understanding the behavior of different classes of polymeric materials towards icing is critical and thus the force necessary to detach ice formed on a surface was measured, using a custom-made test rig to perform the horizontal shear tests.

Dynamic contact angle (dCA) and x-ray photoelectron spectroscopy (XPS) were used to assess the successful bonding of the reactive moiety with a wide range of different polymeric materials, demonstrating its universal attachment to neighboring bonds.

8.2.2 Results and discussion

Different polymeric materials were chosen based on their hydrophilicity to assess the ice adhesion strength once applied as thin, monomolecular films on an aluminum substrate (figure 8.2). Surface qualitative and quantitative characterization was performed with dCA and XPS to evaluate the successful coating mediated by an adhesion promoter consisting of a nitro-catechol and of a DFPxA reactive moiety (figure 8.3(a)). For polymeric substrates a polyethylenimine (PEI) based AP functionalized with DFPxA was synthesized (figure 8.3(b)). The elemental composition of a very thin surface layer of approximately up to 10 nm was determined using XPS [42].

Different polymeric materials of various molecular weights were tested, and the choice was based on the degree of hydrophilicity, since multiple studies suggest decrease of ice adhesion strength with low surface wettability [36, 37]. By testing a wide range of materials, the binding ability towards insertion on non-specific bonds could be demonstrated.

8.2.2.1 Aluminum characterization and deposition of the adhesion promoter

The reference substrate material chosen was aluminum alloy Al6082, a relevant material for aerospace applications, since the parts on airplane wings requiring anti-icing surfaces are usually made of Al6082 [43]. Next, the adhesion promoter used was based on the reactive DFPxA moiety, which was functionalized with a nitro-catechol end group for single site attachment to the Al6082 substrate (figure 8.3(a) and figure 8.4).

The freshly cleaned substrates were characterized by a high surface energy/low contact angle (figure 8.5) and the presence of elemental aluminum (Al), oxidized Al,

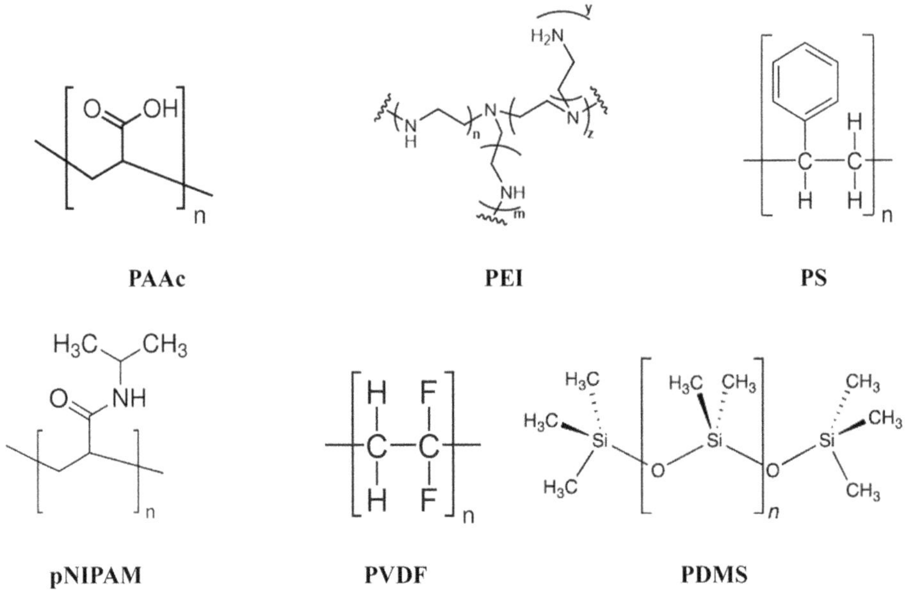

Figure 8.2. The proposed polymeric materials and their respective molecular structures, to test the binding efficacy with the ND-DFPxA adhesion promoter layer. (Adapted from [77].)

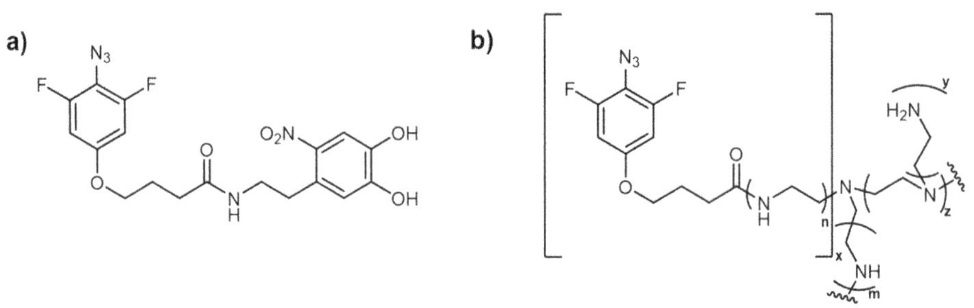

Figure 8.3. 4-azido-3,5-difluorophenoxy moiety functionalized with (a) nitrodopamine to act as an adhesion promoter compound for aluminum substrates (ND-DFPxA) and (b) polyethylenimine for glass fiber and polypropylene substrates (PEI-g-DFPxA). (Adapted from [77].)

oxygen (O), magnesium (Mg), and carbon (C) of adventitious contamination (table 8.1). Moreover, as expected in aluminum substrates due to their complex fabrication [44], a small amount of fluorine contamination was detected, as well as traces of silicon. No other elements, specifically nitrogen, were detected. The successful deposition of the adhesion promoter onto Al substrates was confirmed by an increase in carbon, the presence of nitrogen and fluorine, and the subsequent decrease of aluminum and oxygen as a consequence of attenuation of substrate photoelectrons through the adhesion promoter layer. Upon subtraction of the aluminum, as well as the aluminum and magnesium oxide contributions, the film composition of the adhesion promoter differed from the theoretically calculated

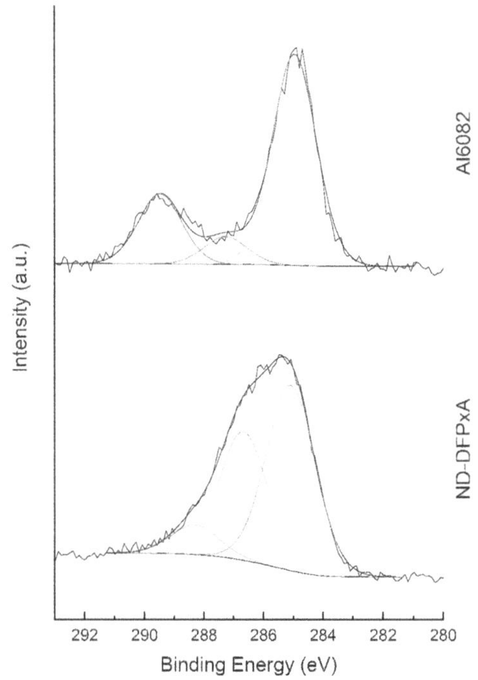

Figure 8.4. Detailed XPS spectra of the C 1s region on the bare Al6082 before and after deposition of the adhesion promoter. Curve-fitted carbon component C–C, in red; C–N, C–O in green; C–F, C=O in blue; and O–C=O in purple. (Adapted from [77].)

stoichiometry. The main deviation was an increased oxygen content, which was attributed to aluminum surface hydroxyl groups. Due to the presence of the DFPxA groups, an increase in advancing contact angle value of 58° with respect to the 14° contact angle of the bare Al substrate was also observed (figure 8.5).

8.2.2.2 Deposition of polymeric materials

The deposition of the polymers was also confirmed by the same two techniques. For substrates with different coatings, the water dynamic contact angles ranged from 13° for the hydrophilic PAAc to 113° for hydrophobic PDMS for advancing contact angle. The increase or decrease in advancing contact angle of the new layer, when compared with the pure adhesion promoter was due to the hydrophilicity/hydrophobicity of the newly bound polymers.

Regarding the XPS analysis, it is worth noting that our system consists of layers of the adhesion promoter and the polymer. When an overlayer is added to the characterized system, this results in attenuation of the intensity signal from the layers below. Thus, a high signal originating from the substrate and the adhesion promoter layer, represents a thin deposited polymer layer. The more attenuated the signal becomes, the thicker the polymeric layers [45]. Hence, the molecular weight of the polymers constitutes a critical parameter, as higher molecular weight could lead to increased density of polymers, and denser polymer would lead to a thicker coating

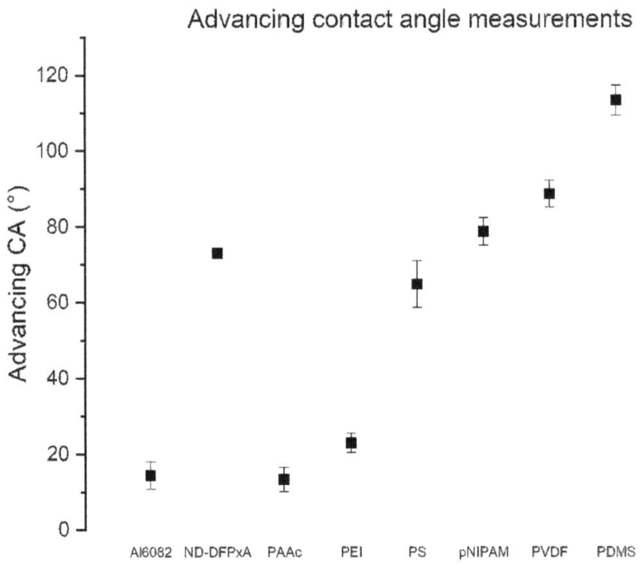

Figure 8.5. Advancing dCA measurements of the freshly cleaned aluminum substrate (Al6082) and of various functionalized surfaces. (Adapted from [77].)

layer that could mask the layers beneath. Thus, depending on the thickness of the coated layer, the composition could be comprised of elements originating from the substrate, the adhesion promoter layer or the top polymeric material. Moreover, according to Serrano *et al* [46], effective attachment of the adhesion promoter differs depending on the polymer used and it appears to be a crucial parameter for the formation of homogeneous coatings.

For PAAc, the substrate signal as well as the signal of the adhesion promoter decreased due to the masking effect of the high molecular weight polyacrylic acid polymer that consists of carbon and oxygen. The composition of the overlayer shows an increase in carbon and a slight increase in oxygen, due to the C=O of the carboxyl bond and it matches with the theoretically calculated one. The measured advancing contact angle was 13°, which agrees with the highly hydrophilic nature of the polymer.

Regarding the chemical composition of PEI, the signals of the substrate and the adhesion promoter were mostly masked by a layer formed from the highly branched polymer that consists of carbon and nitrogen. The main deviation from the theoretical composition was the presence of oxygen in the overlayer, which could be attributed to the aluminum oxides. Hence, a homogeneous PEI film was not successfully achieved.

For pNIPAM, a thin layer was formed from the low molecular weight polymer verified by the high content of C–N and C=O bonds. Again, the presence of fluorine in the composition indicated that the overlayer thickness of pNIPAM was thin and the calculated overlayer composition therefore derived from both layers of adhesion promoter and bound pNIPAM.

For PS, when comparing the chemical composition with the adhesion promoter, one can observe that the presence of an overlayer leads to a decrease of fluorine and films. The samples were characterized by dCA and the apparent normalized atomic concentrations were measured using XPS. In the case of polymer films, the compositions were determined by subtraction of the aluminum and aluminum oxide contributions. Calculated polymer compositions are given in brackets for comparison. (Adapted from [77].)

Chemistry	dCA, Adv/Rec (°)		XPS (normalized At.-%)								
			Mg 1s	Na 1s	Zn 2p3	Al 2p	C 1s	F1s	N 1s	O 1s	Si 2p
Al6082	14 (± 3.6)/14.8 (± 7.2)	At%	9.0	—	—	15.9	19.6	0.5	0.0	53.9	0.4
Adhesion promoter	73 (± 0.8)/17.5 (± 4.4)	At%	1.5	0.0	1.3	12.1	38.5	4.2	6.9	35.4	0.0
		Norm. At%					49.8	5.4	9.0	35.8	0.0
		Theor. At%					(59.4)	(6.2)	(17.2)	(17.2)	(0.0)
PAAc 250 kDa	13 (± 3.2)/5.8 (± 2.7)	At%	0.0	0.0	0.0	12.3	49.5	0.0	0.0	38.2	0.0
		Norm. At%					60.6	0.0	0.0	39.4	0.0
		Theor. At%					(60)	(0.0)	(0.0)	(40)	(0.0)
PEI 25 kDa	23 (± 2.5)/8.8 (± 0.6)	At%	0.0	0.0	0.0	3.2	70.3	0.0	7.5	18.7	0.4
		Norm. At%					73.3	0.0	7.8	18.5	0.4
		Theor. At%					(75)	(0.0)	(25)	(0.0)	(0.0)
PS 280 kDa	65 (± 6.1)/11.9 (± 4.1)	At%	0.5	1.1	2.0	8.1	52.1	2.4	4.8	28.9	0.0
		Norm. At%					61.8	2.9	5.7	29.6	0.0
		Theor. At%					(100.0)	(0.0)	(0.0)	(0.0)	(0.0)
pNIPAM 40 kDa	79 (± 3.7)/16.9 (± 2.3)	At%	1.1	0.0	2.7	14.3	39.7	1.6	4.6	35.9	0.0
		Norm. At%					53.3	2.2	6.2	38.4	0.0
		Theor. At%					(76.2)	(0.0)	(11.9)	(11.9)	(0.0)
PVDF 275 kDa	89 (± 3.6)/35.8 (± 2)	At%	6.7	0.0	1.0	21.3	19.3	4.4	0.0	47.2	0.0
		Norm. At%					35.4	8.2	0.0	56.4	0.0
		Theor. At%					(50.0)	(50.0)	(0.0)	(0.0)	(0.0)
PDMS 330 kDa	113 (± 4)/58.4 (± 3.9)	At%	0.0	0.0	0.0	3.9	51.9	0.9	1.4	25.8	16.3
		Norm. At%					55.2	0.9	1.5	25	17.4
		Theor. At%					(50.0)	(0.0)	(0.0)	(25.0)	(25.0)

an increase in carbon and more specifically in the C–C bond contribution, as expected from the chemical structure of this polymer. The fact that all elements were present in the overlayer suggests that only a rather thin PS film was coated and thus acquisition of signal originating from the substrate, the adhesion promoter and the polymer was possible.

For PVDF, effective coating of the polymer was not achieved, and only partial coating was observed. The overlayer should consist of carbon and fluorine alone. The low fluorine contribution as well as the exceedingly high percentage of Al detected originating from the substrate indicated that PVDF was not successfully attached to form a uniform layer. This is also verified by the dynamic water contact angle measured at 88.5°, which is relatively low compared to the 145° for a uniform PVDF layer [47, 48].

Lastly for PDMS, a reasonably thick layer was deposited masking most of the signal of the substrate. In the composition of the overlayer, small amounts of nitrogen and fluorine were detected due to the adhesion promoter layer as expected from the structure, and the silicon atomic concentration (at%) was increased. At the same time, the contribution of C–C bonds was vastly increased and, in combination with the measured contact angle of 113°, the successful attachment of PDMS was confirmed (figure 8.6).

Successful deposition of the polymers was also confirmed by XPS, except for the case of PVDF and partly PEI. The data revealed that in some cases, the composition originated from all the different layers, suggesting the deposition of quite thin polymeric films. Additionally, interpretation of the content of the different carbon components deriving from the different carbon bonds is essential for definitive determination of successful coating of the polymers (table 8.2). Oxygen concentration showed the greatest deviation across samples which, if higher, is typically attributed to the surface aluminum oxides and hydroxides.

Table 8.2. Normalized C 1s content of different carbon components. The relatively high content of the C=O peak, as observed in most of the chemistries described so far, can be attributed to the presence of the adhesion promoter. (Adapted from [77].)

	Relative peak area (%)			
Chemistry	C–C	C–N, C–O	C–F, C=O	O–C=O
Binding energy	285 ± 0.0 (eV)	286.52 ± 0.09 (eV)	288.75 ± 0.23 (eV)	289.85 ± 0.0 (eV)
Al6082	68.3	9.0	—	22.7
Adhesion promoter	53.9	37.4	8.6	—
PAAc 250 kDa	55.9	17.6	17.6	9.0
PEI 25 kDa	67.5	20.1	12.4	—
PS 280 kDa	65.7	24.4	10.0	—
pNIPAM 40 kDa	56.8	31.2	12.0	—
PVDF 275 kDa	63.2	21.5	15.3	—
PDMS 330 kDa	78.5	14.1	7.3	—

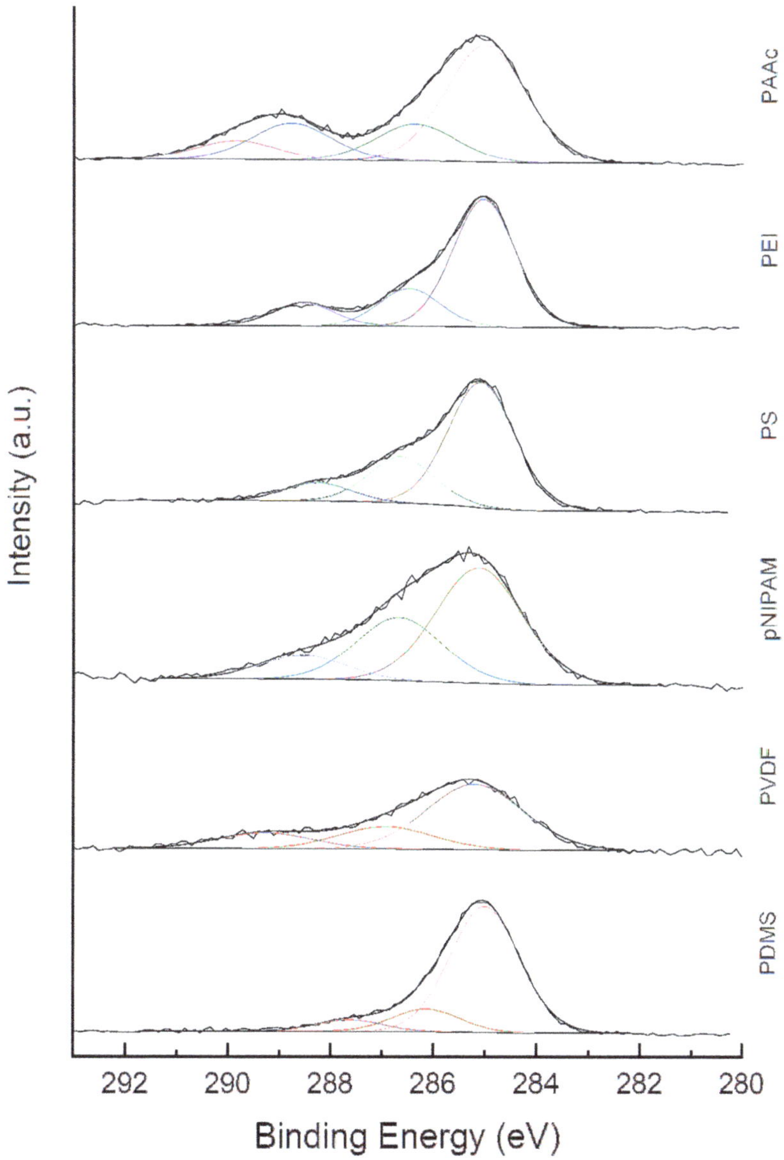

Figure 8.6. Detailed XPS spectra of the C 1s region on the bare Al6082 and various functionalized surfaces. Curve-fitted carbon component C–C, in red; C–N, C–O in green; C–F, C=O in blue; and O–C=O in purple. (Adapted from [77].)

8.2.2.3 Ice adhesion strength correlates with the surface wettability

When ice forms on surfaces with features that allow it to lodge firmly as on hydrophilic surfaces, the water molecules at the interface between the surface and the ice tend to align in a way that maximizes hydrogen bonding interactions, leading to mechanical interlocking [49]. Therefore, hydrophilic surfaces present higher ice adhesion compared to hydrophobic ones, where fewer hydrogen bonds are present [50].

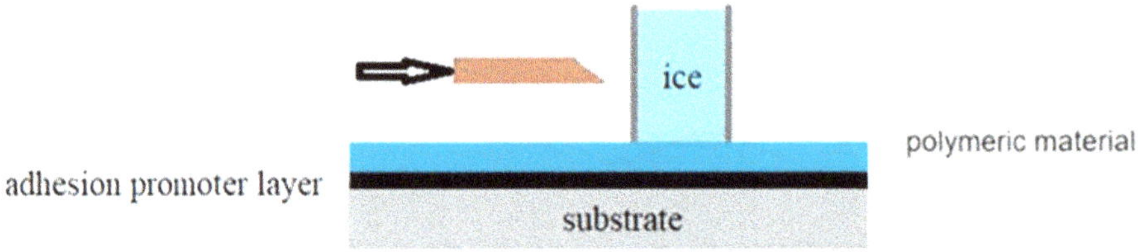

Figure 8.7. Graphical representation of ice adhesion measurements on a standard sample. (Adapted from [77].)

The force–time curves showed a sharp detachment of the ice block for all samples, indicating a behavior of rigid-like materials. Even the PDMS sample showed a sharp detachment, indicating variation between monolayer thick PDMS coating and bulk PDMS, which behave as an elastic material that in combination with the high hydrophobicity yields low strength values of ice adhesion [51]. The force data measured were used to calculate the adhesion strength by dividing the values with the surface area of the formed ice block (figure 8.7).

Two sets of experiments were performed with different diameters of ice block formation, to assess the impact of the ice–substrate interface on the adhesion strength. The first approach was to investigate the detachment of an ice block with diameter of 8 mm [26].

Most of the measurements on the hydrophilic samples—Al6082, PAAc and PEI—exhibited adhesion strength values that are out of range, since the maximum force value (110 N) that the sensor could measure was reached and thus acquisition of the force value in some instances was not achieved. As the surface wettability decreased (increasing contact angles) the adhesion strength of the ice decreased, suggesting a correlation between ice adhesion strength and surface wettability (figure 8.8).

Since the acquisition of force values on most of the hydrophilic samples was not achieved, we also decided to investigate the ice adhesion strength using an ice block with a smaller diameter of 5.4 mm wide.

In this case, force values of the ice detachment were acquired in all measurements, apart from for the PAAc samples. The data demonstrate the same pattern as the previous data set, where high ice adhesion strength was associated with the most hydrophilic samples and decreased upon increasing hydrophobicity of the sample (figure 8.8). Overall, although ice adhesion strength values were acquired, the trend is confirmed principally for the hydrophobic coatings and our data set was mostly dominated by hydrophilic samples. Therefore, more measurements with wider wettability surface ranges are necessary to better understand the key correlation.

Furthermore, all values of ice adhesion strength were much higher than 100 kPa, a fact that could be attributed to the thickness of the deposited polymers, which is in the range of a few nanometers, and they behave like rigid materials showing a sharp detachment of ice. Thus, they present increased stiffness compared to soft elastomeric coatings known to lead to lower ice adhesion strength due to their lower intrinsic surface energy and higher flexibility [28, 52].

Moreover, according to Gao and McCarthy [53], the wettability (or repellency) of a substrate from a thermodynamic viewpoint can be associated with the formation

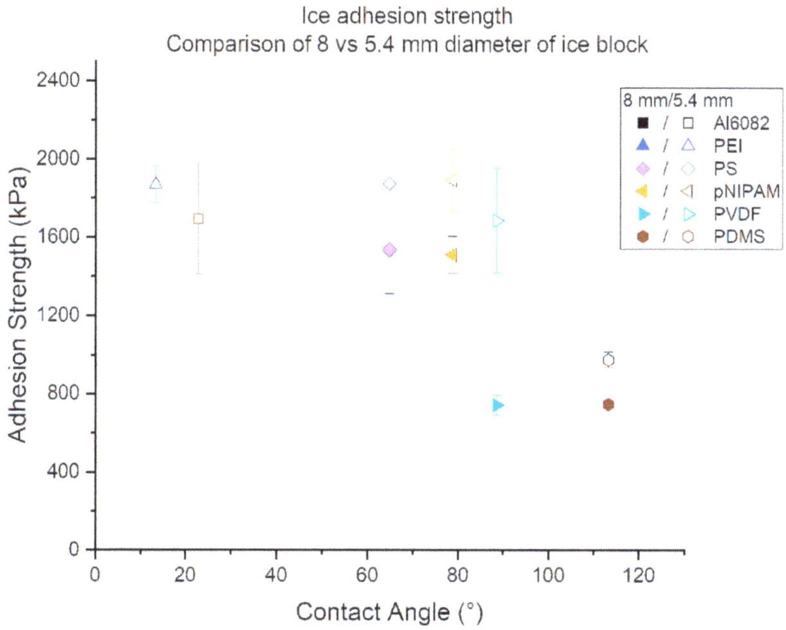

Figure 8.8. Ice adhesion strength values of the different polymeric coated samples. Comparison of ice block diameters of 8 mm versus 5.4 mm. The adhesion strength values of the Al6082, PEI and PAAc samples with an 8 mm ice block and PAAc samples with a 5.4 mm ice block were not determined, since the maximum measurable force value from the sensor was reached. (Adapted from [77].)

and elimination of interfacial areas. Thus, the forces required to separate surfaces connected by a capillary bridge of water are governed by the receding contact angle θ_{rec}. Meuler *et al* [29] observed a strong correlation between the measurements of the average strength of ice adhesion and the practical work of adhesion scaling parameter $[1 + \cos\theta_{rec}]$ for liquid water, suggesting that the 'icephobicity' of nominally smooth surfaces can be predicted simply by measuring the receding contact angle for water droplets on the substrate. They also suggested that the average strength of ice adhesion should follow a linear line which approaches zero, and consequently their data were fitted with the constraint that the linear correlations pass through the origin.

For this specific data set by extrapolating for the best linear fitting of figure 8.9 ($R^2 = 0.92$), the expected receding contact angle of an icephobic polymeric monolayer coating (ice adhesion strength <100 kN) according to our findings, would be a receding angle of 83° or more.

Our data are mostly dominated by hydrophilic samples and only the hydrophobic PDMS point deviates substantially. Moreover, they are located in a completely different region both of ice adhesion strength values, and receding contact angles as well. Thus, direct comparison with previous findings from Meuler *et al* is not possible.

A factor behind this variation could be the difference in the roughness of the substrate material, which was measured to be approximately 0.5 μm, and lower in all cases compared to Meuler *et al.* This difference in the roughness may influence

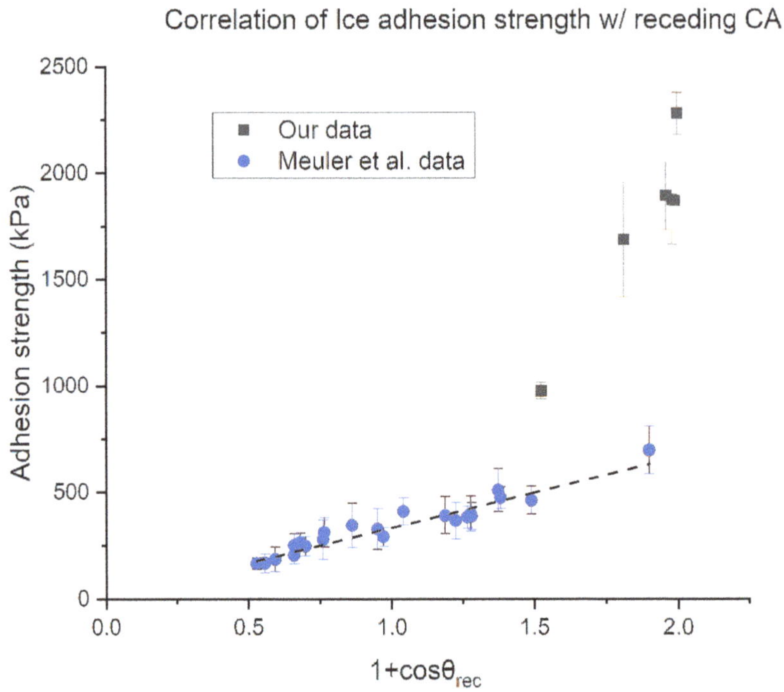

Figure 8.9. Ice adhesion strength values against [1 + cosθ_{rec}]. Our data are compared to data from Meuler *et al* [29]. (Adapted from [77].)

the surface topography, providing sites for water retention and ice nucleation, leading to lower measured receding contact angles and to increased interlocking of ice [54–56]. Also, Mostofi Sarkari *et al* [57] have shown that previous correlations based on the receding contact angle have produced contradictory results. Nevertheless, most of the samples present low receding angles and only a few hydrophobic samples were measured, resulting in an uneven distribution of data points. Thus, this validates that further investigation of the surface topography and the underlying mechanisms is essential.

8.2.3 Conclusion

In this study, we proved that the use of an adhesion promoter layer containing the DFPxA moiety provides a versatile approach for the functionalization of aluminum surfaces with a variety of polymers in a reliable manner. We have used this approach to test the binding capability of the adhesion promoter with multiple polymers that span different wetting properties. The adhesion promoter possesses the ability to coordinatively bind to aluminum surfaces through nitro-catechol end groups and, upon activation, covalently bond to polymers in its close proximity through the reactive nitrene groups, which undergo insertion reaction to any C–H, C–C, or C–N bonds. Validation of the polymer deposition was achieved by means of different surface characterization techniques, such as dynamic contact angle and XPS. Moreover, regarding the behavior of the functionalized polymeric monolayers for ice detachment, they demonstrated a rigid-like material behavior resulting in sharp detachment of the ice block. For that, a horizontal shear (or push) test was used to acquire the force values, and the analysis of the data revealed a correlation between increased ice adhesion strength values and high surface wettability, whereas samples with high contact angles exhibited lower values of adhesion strength, in agreement with previous studies.

8.3 Ultrathin adhesion promoter films for gluing and bonding applications

8.3.1 Introduction

In the field of surface engineering, polymeric coatings constitute a primary approach to modify surfaces to introduce desired functionalities without compromising the properties of the underlying bulk material [58]. Adhesion strength is a crucial property in coatings as it significantly influences the durability and performance of the coating system. Moreover, it is a critical factor in determining the efficacy and longevity of coatings, as it ensures that the coating remains attached even under stress in harsh operating conditions, reducing the need for frequent reapplications or repairs [59, 60]. Understanding the factors that influence adhesion strength, such as substrate and coating composition, or innovative coating techniques are therefore essential for the development of robust and high-performance coating systems [61].

Mechanical characterization of the adhesion strength is performed with the commonly used lap shear tests to evaluate the effect of an adhesion promoter [62, 63]. These tests are essential for quantifying adhesion strength via the acquisition of the shear force necessary to separate the bound materials and for evaluating the performance of different adhesive systems. The outcomes offer valuable insights into the bonding ability of adhesives, facilitating the design of adhesive materials with improved adhesion properties for a variety of applications. The lap shear adhesion strength is influenced by various factors, such as the type of adhesive used, the presence of cross-linkers, and the curing process [64, 65].

In this work, we immobilized polymers on different types of substrates with the use of an adhesion promoter for assessing their adhesion strength. For that, lap shear tests were used to assess the adhesive bond strength of bound samples, using cyclic olefin copolymers (COCs) or epoxy as adhesives. The use of an adhesion promoter layer, capable of forming covalent bonds with any neighboring C–C bond via nitrene insertion, increased the adhesion strength of the adhesives up to four times compared to samples without an adhesion promoter, improving their mechanical durability, therefore creating an appropriate coating platform for adhesives that would not bind otherwise. Thermal curing of the samples results in simultaneous activation of the adhesion promoter to covalently bond with the adhesive, while curing of the adhesive occurred.

8.3.2 Results and discussion

An adhesion promoter layer can increase the adhesion strength of bound systems. To evaluate the adhesion strength of bound specimens containing an adhesion promoter layer, measurements of the force necessary to separate them under shear loading were performed according to the widely used single lap shear test [66] (figure 8.10).

Aluminum alloy 5005 (Al5005), glass fiber (GF) reinforced with epoxy, and polypropylene (PP) specimens were used as substrate materials. Aluminum was chosen as a relevant material for aerospace applications, since the parts on airplane wings requiring anti-icing surfaces are made of this material [67]. Polymeric materials and glass fiber reinforced polymers were chosen as relevant materials for multiple industrial applications [68, 69]. Regarding the adhesion promoters used, the DFPxA and PFPA reactive moieties were functionalized by additional end groups for them to attach to the different substrates. More specifically, for attachment to the Al5005 substrate, DFPxA and PFPA reactive moieties were functionalized with a nitrodopamine catechol (ND-DFPxA and ND-PFPA) known to strongly bind to aluminum (figure 8.2(a)). In the case of attachment to the PP and GF substrates, the reactive moieties were grafted to the branched PEI containing amine groups. The PEI-g-DFPxA and PEI-g-PFPA adhesion promoters bind electrostatically to glass and plasma treated polymers and/or covalently to any nearby carbon atom from the

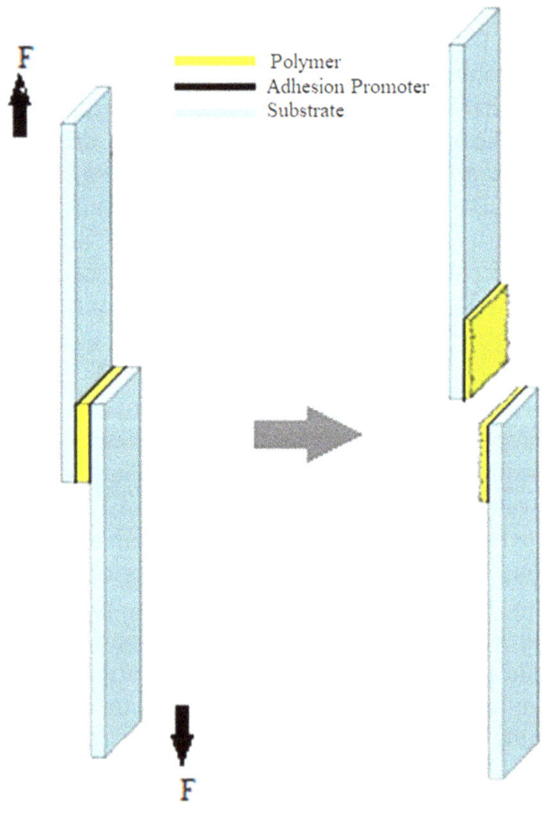

Figure 8.10. The prepared bound samples were tested by the single lap shear test, where the force necessary to separate the two specimens is measured. (Adapted from [77].)

substrate, the coating or within the adhesion promoting layer, after activation of the azides (figure 8.2(b)). Finally, the top coatings/adhesives used were COCs, a material where no adhesive strength is expected due to its inertness and lack of functionality, and a low temperature one-component epoxy as a commercial high strength adhesive. These materials were chosen to physico-chemically represent industry-relevant functionalities, such as high transparency, high thermal stability, hydrophobicity, chemical resistance, a barrier to oxygen and water vapor, and durability [70–72].

To prepare the samples, two identical pieces coated with an adhesion promoter layer containing a reactive aryl azide and the desired adhesive (COC or epoxy) were prepared separately (figure 8.11). Solvent bonding at an elevated temperature of the two pieces was performed by addition of two droplets of the coating material. The overlapping area of the two pieces was 12.5 mm in length and 25 mm in width and they were finally gripped with a screw clamp. The resulting specimens were cured at 80 °C in an oven for 2 h to activate the azides, leading to the formation of covalent bonds with the polymeric topcoat. Control samples that were not coated with an adhesion promoter layer were prepared and measured for comparison.

With the designed experiments, the increase of adhesion strength of the newly synthesized adhesion promoters containing the DFPxA reactive moiety with an activation temperature ~80 °C was quantified and compared with samples containing adhesion promoters with the PFPA moiety. The latter are well established adhesion promoters known to become activated when exposed to temperatures higher than 120 °C. Consequently, PFPA-containing adhesion promoters were not expected to be activated at 80 °C, the temperature to which the samples were exposed, in contrast to the recently developed DFPxA containing adhesion promoters. At the same time, control samples containing no adhesion promoter but only COC or epoxy were prepared and cured using the same conditions.

When COC was used as an adhesive polymer, the adhesion strength was highest for the samples containing DFPxA adhesion promoters, which suggests that they were successfully attached both to the substrate and the top-coating of COC after activation. The measured strength values exhibited an almost 100% increase in strength when used for the Al5005 and PP substrates, and in the case of glass fiber we observed a four times increase in adhesion. For PP and Al5005, the data suggest that

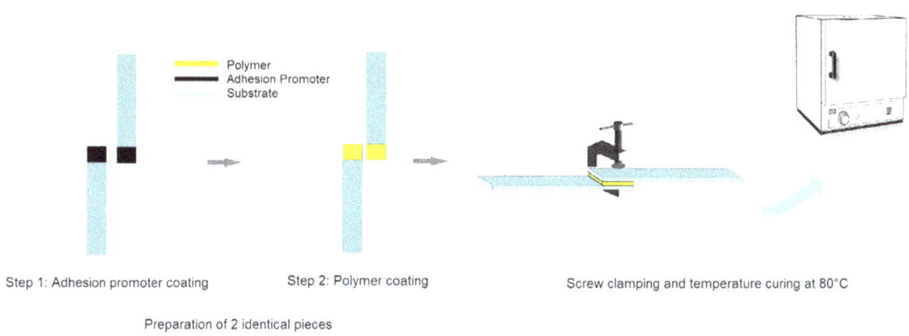

Figure 8.11. Preparation of bound samples for lap shear tests. (Adapted from [77].)

even the use of the PFPA-containing adhesion promoters, expected not to be activated at 80 °C, slightly increased the adhesion strength of the samples, presumably since they provides a more suitable platform for the coating of COC on the substrate (figure 8.12).

In the next experiment, a low temperature curing 'one-component' epoxy was used. It provides unlimited working time at room temperature, but most importantly it can be cured simultaneously with the adhesion promoter's activation. Using such an adhesive in between the two bound specimens, the values of adhesion strength were significantly higher compared to COC, since the cured epoxy is a very strong material. The epoxy adhered well to bare Al5005, and the use of a non-activated adhesion promoter (ND-PFPA) did not confer additional substantial benefit. For the activated ND-DFPxA adhesion promoter, the adhesion strength of the system was increased slightly, suggesting that formation of covalent bonds with epoxy increased its strength, nevertheless adhesion was already suitable.

The adhesion using the glass fiber reinforced epoxy substrate was quite different. Samples without adhesion promoter resulted in the highest adhesion strength of the whole series, since the substrate already contains an epoxy material and the affinity between the substrate and epoxy adhesive was extensive. The application of an adhesion promoting layer led to a decrease of the adhesion strength, since the epoxy was not directly coated on the glass fiber substrate and the affinity is lower. Nonetheless, the use of the activated PEI-g-DFPxA doubled the adhesion strength compared to PEI-g-PFPA, suggesting the formation of covalent bonds with the epoxy glue.

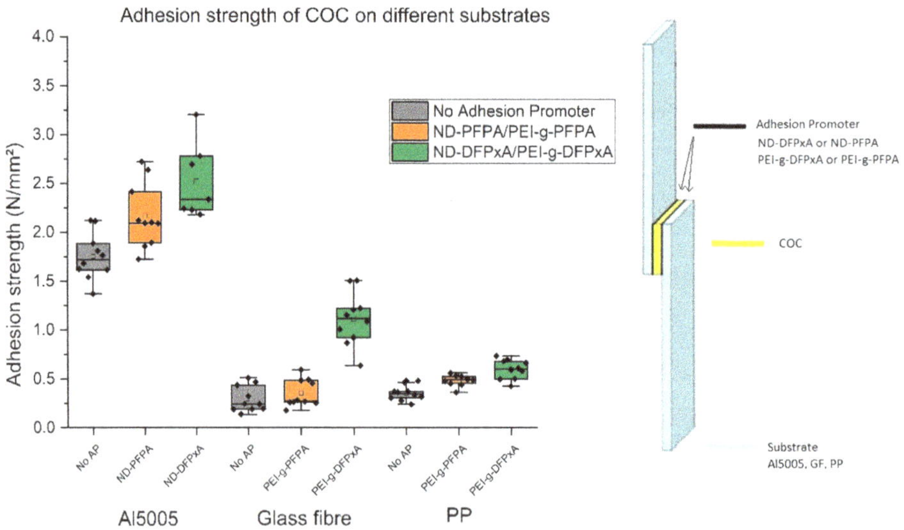

Figure 8.12. Comparison of the adhesion strength between all the substrates and conditions when COC was used as an adhesive. The gray boxplots correspond to the samples without any adhesion promoter layer, the orange to the samples with adhesion promoters containing the PFPA moiety which were expected not to be activated at the curing temperature of 80 °C and lastly the green boxplots correspond to the samples with adhesion promoters containing the DFPxA moiety which were expected to be activated at 80 °C and covalently attach to the COC. (Adapted from [77].)

Lastly, the outcome of the tests with PP specimens is very straightforward. It is well known that epoxy adhesives do not work well with non-polar polymeric materials since their affinity is low [73, 74], as validated from the very low adhesion strength of the control samples without adhesion promoter. The use of both adhesion promoters created an appropriate coating platform for epoxy, and acquisition of the force values for the assessment of the adhesion strength was not possible, since failure of the PP substrate itself occurs. It is evident, however, that the use of an adhesion promoter increased vastly the capability of an epoxy to be coated on polypropylene, materials that do not have affinity with each other (figure 8.13). That the epoxy can react with amino groups of the adhesion promoter was expected, nevertheless, this is an unexpected finding since it was not anticipated that the non-activated PEI-g-PFPA would bind so strongly on the PP substrate.

After the outcome of the above performed experiments, we demonstrated that we could immobilize numerous polymeric coatings with the use of a thermally activated, nanoscale thick adhesion promoter layer, increasing their adhesion strength. As a next step, we considered bonding two specimens together without deformation or degradation, just by using an adhesion promoter layer. For that reason, polymethyl methacrylate (PMMA) specimens were chosen as a polymer demonstrating a low glass transition temperature, of approximately 105 °C, that is non UV-C transparent as well [75]. Thus, the only possible way to bond such materials is using a low temperature activated adhesion promoter such as PEI-g-DFPxA.

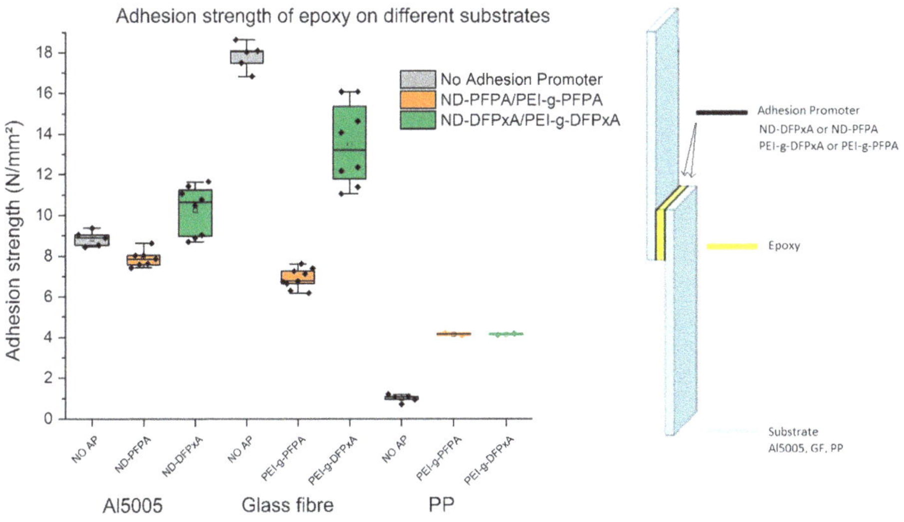

Figure 8.13. Comparison of adhesion strength between all substrates and conditions when low temperature curing epoxy was used as the adhesive. The gray boxplots correspond to the samples without any adhesion promoter layer, the orange to the samples with adhesion promoters containing the PFPA moiety which were expected not to be activated at the curing temperature of 80 °C and lastly the green boxplots correspond to the samples with adhesion promoters containing the DFPxA moiety which were expected to be activated at 80 °C and covalently attach to the epoxy. The adhesion strengths of PP in the cases of PEI-g-PFPA and PEI-g-DFPxA at approximately 4.2 N mm^{-2} do not correspond to failure of the epoxy rather than failure of the PP substrate material. (Adapted from [77].)

The bonding of two PMMA specimens at a temperature higher than its glass transition led to deformation and rearrangement of the PMMA macromolecular chains which entangled upon transition back to brittle behavior [76], represented as high measured adhesion force. This finding validated the need for a system that is activated at temperatures below 100 °C, and thus the temperature chosen to test the effect of the adhesion promoters was 90 °C. It is clear that the adhesion force after curing is the highest when the activated PEI-g-DFPxA was used for bonding (figure 8.14).

These preliminary results show that this idea is promising, and that PEI-g-DFPxA can be used to bond materials at temperatures below 100 °C. Nevertheless, there are some factors that need to be considered, such as the difficulty presented

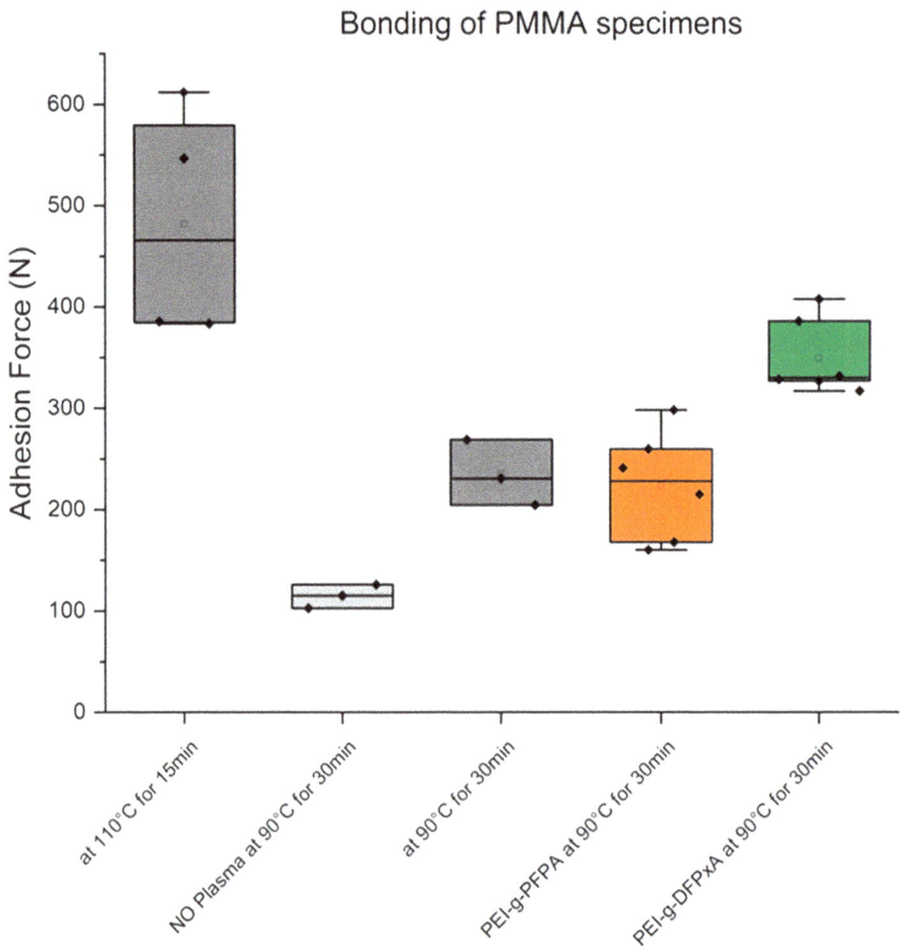

Figure 8.14. Force adhesion measurements of thermally bonded PMMA sheets with and without the use of adhesion promoters. PMMA exhibits a glass transition temperature at approximately 105 °C, thus it shows high adhesion force due to deformation and rearrangement of the PMMA chains. At 90 °C no deformation occurs and the activated PEI-g-DFPxA results in the higher adhesion force at this temperature. (Adapted from [77].)

when trying to bond substrates which are not perfectly flat, rather having a microscale structure, with a nanometer thick adhesion promoter layer. Investigation of the optimal parameters and conditions is still ongoing.

8.3.3 Conclusion

Here, we demonstrated that an engineered adhesion promoter can create an appropriate coating platform for the application of coatings and increase the adhesion strength of the system after activation and formation of covalent bonds with the adhesive. The adhesion strength of samples containing an adhesive (such as COC and epoxy), using different activated or not activated adhesion promoters, on three different substrates (Al5005, glass fiber and PP) was investigated by performing lap shear measurements. It was demonstrated that intrinsically purely adhered materials can be covalently coated, and improved durability of coatings under mechanical stresses can be achieved with the use of an activated adhesion promoter. Thermal activation of the samples in the experimental set-up resulted in simultaneous activation of the azides to covalently bind with the adhesive, while curing of the adhesive occurring. The adhesion promoters containing the reactive DFPxA moiety, which is demonstrated to be activated already at 70 °C, can improve the adhesion between Al or glass fiber substrates and a non-activated polymer such as COC. Moreover, they constitute promising materials for increasing adhesion between non-polar polymeric substrates such as PP and epoxy at low temperatures.

8.4 Results overview

DFPxA-based adhesion promoters that are activatable at the lowest temperature were investigated for their binding capability with different polymer materials but also for their effect on the adhesion strength of a coating system. Our findings indicate that the thermally activated adhesion promoter can non-specifically insert to polymers that do not carry functional end groups via the formation of covalent bonds, leading to their permanent immobilization on different substrates. Thus, we successfully surface-tethered monolayers of different polymers with diverse surface wettability, and by measuring the adhesion strength of a formed ice block on top, we correlated the high surface wettability with decreased ice adhesion strength values. Moreover, it was demonstrated that bound samples containing an activated adhesion promoter and an adhesive polymer can lead to an increase of adhesion strength by up to four times when cured at low temperatures of ~80 °C, compared to non-activated samples.

References

[1] Li F *et al* 2018 Advanced composite 2D energy materials by simultaneous anodic and cathodic exfoliation *Adv. Energy Mater.* **8** 1–8

[2] Zhao X, Cai Y, Wang T, Shi Y and Jiang G 2008 Preparation of alkanethiolate-functionalized core/shell Fe_3O_4@Au nanoparticles and its interaction with several typical target molecules *Anal. Chem.* **80** 9091–6

[3] Quan Z *et al* 2015 Additive manufacturing of multi-directional preforms for composites: opportunities and challenges *Mater. Today* **18** 503–12
[4] Granozzi G 2023 Impact of surface science in current science and technology: some basic considerations *Surfaces* **6** 533–5
[5] Alaei Shahmirzadi M A and Kargari A 2018 *Nanocomposite Membranes* (Amsterdam: Elsevier)
[6] Li C J 2010 Thermal spraying of light alloys *Surface Engineering of Light Alloys: Aluminium, Magnesium and Titanium Alloys* (Woodhead: Cambridge) pp 184–241
[7] Vieira T, Castanho J and Louro C 2006 Hard coatings based on metal nitrides, metal carbides and nanocomposite materials. PVD process and properties *Materials Surface Processing by Directed Energy Techniques* (Amsterdam: Elsevier) pp 537–72
[8] Camperi J, Pichon V and Delaunay N 2020 Separation methods hyphenated to mass spectrometry for the characterization of the protein glycosylation at the intact level *J. Pharm. Biomed. Anal.* **178** 112921
[9] Konai M M, Bhattacharjee B, Ghosh S and Haldar J 2018 Recent progress in polymer research to tackle infections and antimicrobial resistance *Biomacromolecules* **19** 1888–917
[10] Zhou G and Groth T 2018 Host responses to biomaterials and anti-inflammatory design—a brief review *Macromol. Biosci.* **18** 1–15
[11] Michelmore A 2016 *Thin Film Growth on Biomaterial Surfaces* (Amsterdam: Elsevier)
[12] Hale A, Macosko C W and Bair H E 1991 Glass transition temperature as a function of conversion in thermosetting polymers *Macromolecules* **24** 2610–21
[13] Lindholm E A, Stolov A A, Dyer R S, Slyman B and Burgess D 2009 Reliability of optical fibers in a cryogenic environment *Fiber Optic Sens. Appl. VI* **7316** 73160Z
[14] Eßbach C, Fischer D and Nickel D 2021 Challenges in electroplating of additive manufactured ABS plastics *J. Manuf. Process.* **68** 1378–86
[15] Sastri V R 2010 Commodity thermoplastics *Plastics in Medical Devices* (Amsterdam: Elsevier) pp 73–119
[16] Samavedi S, Poindexter L K, Van Dyke M and Goldstein A S 2014 *Synthetic Biomaterials for Regenerative Medicine Applications* (Amsterdam: Elsevier)
[17] Hasan A and Pandey L M 2015 Review: polymers, surface-modified polymers, and self assembled monolayers as surface-modifying agents for biomaterials *Polym. Plast. Technol. Eng.* **54** 1358–78
[18] Ratner B D 1995 Surface modification of polymers: chemical, biological and surface analytical challenges *Biosens. Bioelectron.* **10** 797–804
[19] Brannon H 2014 The development of novel adhesion promoters for waterborne coatings and polypropylene car bumpers *PhD Thesis* University of Birmingham
[20] Kost J, Bleiziffer A, Rusitov D and Rühe J 2021 Thermally induced cross-linking of polymers via C,H insertion cross-linking (CHic) under mild conditions *J. Am. Chem. Soc.* **143** 10108–19
[21] Abu Bakar R, Li Y, Hewitson O P, Roth P J and Keddie J L 2022 Azide photochemistry in acrylic copolymers for ultraviolet cross-linkable pressure-sensitive adhesives: optimization, debonding-on-demand, and chemical modification *ACS Appl. Mater. Interfaces* **14** 30216–27
[22] Atzemoglou A A, Bartalucci N, Donat F, Tibbitt M W, Tosatti S G P and Zürcher S 2025 Development of low temperature activatable aryl-azide adhesion promoters for versatile surface modifiers *ACS Appl. Eng. Mater.* **3** 867–82

[23] Civil Aviation Authority 2000 *Aircraft Icing Handbook* (New Zealand: Safety Education and Publishing Unit) pp 1–20

[24] Lynch F T and Khodadoust A 2001 Effects of ice accretions on aircraft aerodynamics *Prog. Aerosp. Sci.* **37** 669–767

[25] Laforte J L, Allaire M A and Laflamme J 1998 State-of-the-art on power line de-icing *Atmos. Res.* **46** 143–58

[26] Stendardo L *et al* 2023 Reframing ice adhesion mechanisms on a solid surface *Appl. Surf. Sci.* **641** 158462

[27] PRO-ACT 1998 Air force aircraft and airfield deicing/anti-icing the role of deicing and anti-icing in the air force types of deicing and anti-icing agents *Fact Sheet* TI-16621 PRO-ACT 1–7

[28] Wang C, Fuller T, Zhang W and Wynne K J 2014 Thickness dependence of ice removal stress for a polydimethylsiloxane nanocomposite: Sylgard 184 *Langmuir* **30** 12819–26

[29] Meuler A J, Smith J D, Varanasi K K, Mabry J M, McKinley G H and Cohen R E 2010 Relationships between water wettability and ice adhesion *ACS Appl. Mater. Interfaces* **2** 3100–10

[30] Yang S, Xia Q, Zhu L, Xue J, Wang Q and Chen Q M 2011 Research on the icephobic properties of fluoropolymer-based materials *Appl. Surf. Sci.* **257** 4956–62

[31] Golovin K, Kobaku S P R, Lee D H, DiLoreto E T, Mabry J M and Tuteja A 2016 Designing durable icephobic surfaces *Sci. Adv.* **2** e1501496

[32] Jellinek H H G 1962 Ice adhesion *Can. J. Phys.* **40** 1294–309

[33] Marengo M 2022 *The Surface Wettability Effect on Phase Change* (Cham: Springer)

[34] Shen Y, Wu X, Tao J, Zhu C, Lai Y and Chen Z 2019 Icephobic materials: fundamentals, performance evaluation, and applications *Prog. Mater Sci.* **103** 509–57

[35] Huang J, Lou C, Xu D, Lu X, Xin Z and Zhou C 2019 Cardanol-based polybenzoxazine superhydrophobic coating with improved corrosion resistance on mild steel *Prog. Org. Coat.* **136** 105191

[36] Yeong Y H, Milionis A, Loth E, Sokhey J and Lambourne A 2015 Atmospheric ice adhesion on water-repellent coatings: wetting and surface topology effects *Langmuir* **31** 13107–16

[37] Subramanyam S B, Rykaczewski K and Varanasi K K 2013 Ice adhesion on lubricant-impregnated textured surfaces *Langmuir* **29** 13414–8

[38] Kulinich S A and Farzaneh M 2009 How wetting hysteresis influences ice adhesion strength on superhydrophobic surfaces *Langmuir* **25** 8854–6

[39] He Z, Vågenes E T, Delabahan C, He J and Zhang Z 2017 Room temperature characteristics of polymer-based low ice adhesion surfaces *Sci. Rep.* **7** 1–7

[40] Sever M J, Weisser J T, Monahan J, Srinivasan S and Wilker J J 2004 Metal-mediated cross-linking in the generation of a marine-mussel adhesive *Angew. Chem. Int. Ed.* **43** 448–50

[41] Malisova B, Tosatti S, Textor M, Gademann K and Zürcher S 2010 Poly(ethylene glycol) adlayers immobilized to metal oxide substrates through catechol derivatives: influence of assembly conditions on formation and stability *Langmuir* **26** 4018–26

[42] Krishna D N G and Philip J 2022 Review on surface-characterization applications of x-ray photoelectron spectroscopy (XPS): recent developments and challenges *Appl. Surf. Sci. Adv.* **12** 100332

[43] Singh P, Singh R K and Das A K 2024 Optimization of heat treatment cycle for cast-Al6082 alloy to enhance the mechanical properties *Eng. Res. Express* **6** 015036

[44] Greenwood N N and Earnshaw A 1984 *Chemistry of Elements* 2nd edn (Butterworth Heinemann) ch 7 pp 216–67

[45] Tosatti S, Michel R, Textor M and Spencer N D 2002 Self-assembled monolayers of dodecyl and hydroxy-dodecyl phosphates on both smooth and rough titanium and titanium oxide surfaces *Langmuir* **18** 3537–48

[46] Serrano Â *et al* 2013 Nonfouling response of hydrophilic uncharged polymers *Adv. Funct. Mater.* **23** 5706–18

[47] Kamaz M, Sengupta A, Gutierrez A, Chiao Y H and Wickramasinghe R 2019 Surface modification of PVDF membranes for treating produced waters by direct contact membrane distillation *Int. J. Environ. Res. Public Health* **16** 685

[48] Gontarek-Castro E, Rybarczyk M K, Castro-Muñoz R, Morales-Jiménez M, Barragán-Huerta B and Lieder M 2021 Characterization of PVDF/graphene nanocomposite membranes for water desalination with enhanced antifungal activity *Water* **13** 1279

[49] Ling E J Y, Uong V, Renault-Crispo J S, Kietzig A M and Servio P 2016 Reducing ice adhesion on nonsmooth metallic surfaces: wettability and topography effects *ACS Appl. Mater. Interfaces* **8** 8789–800

[50] Emelyanenko K A, Emelyanenko A M and Boinovich L B 2020 Water and ice adhesion to solid surfaces: common and specific, the impact of temperature and surface wettability *Coatings* **10** 1–23

[51] Ibáñez-Ibáñez P F, Montes Ruiz-Cabello F J, Cabrerizo-Vílchez M A and Rodríguez-Valverde M A 2022 Ice adhesion of PDMS surfaces with balanced elastic and water-repellent properties *J. Colloid Interface Sci.* **608** 792–9

[52] Fu Q *et al* 2014 Development of sol gel icephobic coatings: effect of surface roughness and surface free energy *Thin Films 2014. The 7th Int. Conf. on Technological Advances of Thin Films and Surface Coatings*

[53] Gao L and Mccarthy T J 2008 Teflon is hydrophilic. Comments on definitions of hydrophobic, shear versus tensile hydrophobicity, and wettability characterization *Langmuir* **24** 546–50

[54] Wang X and Zhang Q 2020 Role of surface roughness in the wettability, surface energy and flotation kinetics of calcite *Powder Technol.* **371** 55–63

[55] Wang J, Wang S and Hou X 2024 Study on the dependence between surface topography and icephobic behavior of Ni–Cu–P ternary coatings *Adv. Eng. Mater.* **26** 1–10

[56] Cui W, Jiang Y, Mielonen K and Pakkanen T A 2019 The verification of icephobic performance on biomimetic superhydrophobic surfaces and the effect of wettability and surface energy *Appl. Surf. Sci.* **466** 503–14

[57] Mostofi Sarkari N *et al* 2025 Experimental debate on the overlooked fundamental concepts in surface wetting and topography vs ice adhesion strength relationships *J. Colloid Interface Sci.* **682** 825–48

[58] Pruchniewski M *et al* 2023 Nanostructured graphene oxide enriched with metallic nanoparticles as a biointerface to enhance cell adhesion through mechanosensory modifications *Nanoscale* **15** 18639–59

[59] Vengadaesvaran B, Rau S R, Ramesh K, Puteh R and Arof A K 2010 Preparation and characterisation of phenyl silicone-acrylic polyol coatings *Pigm. Resin Technol.* **39** 283–7

[60] Díaz Téllez J P, Harirchian-Saei S, Li Y and Menon C 2013 Adhesion enhancement of biomimetic dry adhesives by nanoparticle *in situ* synthesis *Smart Mater. Struct.* **22** 105031

[61] Nedela O, Slepicka P and Švorcík V 2017 Surface modification of polymer substrates for biomedical applications *Materials* **10** 1115

[62] Gomez-Lopez A, Grignard B, Calvo I, Detrembleur C and Sardon H 2022 Accelerating the curing of hybrid poly(hydroxy urethane)-epoxy adhesives by the thiol-epoxy chemistry *ACS Appl. Polym. Mater.* **4** 8786–94

[63] Zong H, Fang C, Lin Z, Yan Q and Lin X 2020 A novel polyether polyol contains repeating cyclohexane units and its application on reactive polyurethane adhesive *Polym. Adv. Technol.* **31** 2535–44

[64] Samanta S, Banerjee S L, Bhattacharya K and Singha N K 2021 Graphene quantum dots-ornamented waterborne epoxy-based fluorescent adhesive via reversible addition-fragmentation chain transfer-mediated miniemulsion polymerization: a potential material for art conservation *ACS Appl. Mater. Interfaces* **13** 36307–19

[65] Zhou J *et al* 2015 Adhesion properties of catechol-based biodegradable amino acid-based poly(ester urea) copolymers inspired from mussel proteins *Biomacromolecules* **16** 266–74

[66] Abdolah Zadeh M, Van Der Zwaag S and Garcia S J 2016 Adhesion and long-term barrier restoration of intrinsic self-healing hybrid sol–gel coatings *ACS Appl. Mater. Interfaces* **8** 4126–36

[67] Xia A *et al* 2022 Fabrication of an anti-icing aluminum alloy surface by combining wet etching and laser machining *Appl. Sci.* **12** 2119

[68] Wang R *et al* 2021 Mass-synthesized solution-processable polyimide gate dielectrics for electrically stable operating OFETs and integrated circuits *Polymers* **13** 3715

[69] Han M J and Khang D Y 2015 Glass and plastics platforms for foldable electronics and displays *Adv. Mater.* **27** 4969–74

[70] Zhao W and Nomura K 2016 Copolymerizations of norbornene and tetracyclododecene with α-olefins by half-titanocene catalysts: efficient synthesis of highly transparent, thermal resistance polymers *Macromolecules* **49** 59–70

[71] Hong M, Yang G F, Long Y Y, Yu S and Li Y S 2013 Preparation of novel cyclic olefin copolymer with high glass transition temperature *J. Polym. Sci.* A **51** 3144–52

[72] Dallaev R, Pisarenko T, Papež N, Sadovský P and Holcman V 2023 A brief overview on epoxies in electronics: properties, applications, and modifications *Polymers* **15** 3964

[73] Xie Y, Hill C A S, Xiao Z, Militz H and Mai C 2010 Silane coupling agents used for natural fiber/polymer composites: a review *Composites* A **41** 806–19

[74] Singh K, Nanda T and Mehta R 2017 Compatibilization of polypropylene fibers in epoxy based GFRP/clay nanocomposites for improved impact strength *Composites* A **98** 207–17

[75] Garcia-Gonzalez D, Rusinek A, Bendarma A, Bernier R, Klosak M and Bahi S 2020 Material and structural behaviour of PMMA from low temperatures to over the glass transition: quasi-static and dynamic loading *Polym. Test.* **81** 0–18

[76] Yan Y, Sun Y, Su J, Li B and Zhou P 2023 Crazing initiation and growth in polymethyl methacrylate under effects of alcohol and stress *Polymers* **15** 1375

[77] Atzemoglou A A 2025 Development of low temperature activated aryl azide adhesion promoters as versatile surface modifiers *PhD Thesis* ETH Zürich https://doi.org/10.3929/ethz-b-000701943

Chapter 9

Icephobic gradient polymer coatings deposited via iCVD

Gabriel Hernandez Rodriguez and Anna Maria Coclite

The development of coatings using initiated chemical vapor deposition (iCVD) represents a promising strategy to meet the growing demand for icephobic technologies that can be integrated into existing anti-icing/de-icing systems. The iCVD technique allows for the deposition of thin films with precise characteristics, and its application for creating ice-resistant surfaces is an area of significant interest. In this chapter, we focus on the design of gradient polymer coatings via iCVD for anti-icing applications. Gradient coatings are characterized by a change in chemical composition throughout their entire thickness. This variation allows for the optimization of coating properties, creating a transition between different compositions that can effectively respond to environmental conditions. We chose to combine two types of compositions: one that ensures high stability, to withstand mechanical and environmental stresses, and one that promotes icephobicity, i.e. the ability to repel ice. This combination allows for coatings that not only reduce ice accumulation but also maintain high durability over time.

The icephobicity of the gradient polymer coatings was evaluated through various experiments designed to demonstrate how the icephobic properties could be enhanced by tailoring the gradient polymer design. Supercooled droplet impact experiments were conducted to test the coatings under various environmental conditions. These experiments simulated real-world scenarios to assess the endurance and practicality of the gradient polymer films as anti-icing materials. The results demonstrated the coatings' ability to resist ice formation even in challenging conditions. Finally, the gradient polymer coatings were applied to aluminum surfaces and combined with an electromechanical de-icing system. This integration aimed to enhance the performance of existing anti-icing technologies, showing the potential for these gradient coatings to complement and improve conventional de-icing systems.

doi:10.1088/978-0-7503-6009-8ch9

9.1 Introduction

In this chapter, a new concept for ice prevention, control, and mitigation based on gradient polymer coatings developed using initiated chemical vapor deposition (iCVD) is presented. Gradient polymer coatings have a chemical composition that changes gradually within their thickness from the interface with the substrate to the interface with the environment. This molecular architecture, unfeasible by any other known polymer deposition technique, was easily performed in one step via iCVD without the need for a pre- or post-substrate treatment. Furthermore, the manufacturing of ice protection systems need to be suitable to successfully functionalize large areas, several shapes, and different materials. iCVD offers all of this. It is a vacuum-based polymerization technique which allows one to build polymer thin films with great tuneability and without the use of solvents. Nevertheless, its use for ice protection systems remains unexplored.

The gradient polymer coating structure that will be the object of this chapter, started with a poly-siloxane composition and ended with a perfuorinated polymer composition. The poly-siloxane being very cross-linked ensures stability to the coating and enhances the adhesion of the coating to the sbstrate. The choice of ending the structure with a perfluorinated composition aimed to obtain hydrophobicity and icephobicity. In addition, the used perfluoropolymer shows strong crystalline aggregation which can be tuned by changing the iCVD condition, with direct consequences on the wettability. The surface's wettability exhibited an apparent hydrophobic nature, yet with high values of water contact angle hysteresis and roll-off angles. In addition, it was observed that when the polymeric structure was adjusted, specifically by increasing the thickness of the upper section of the gradient structure, a systematic change in the crystallographic orientation at the surface plane occurred, leading to a discontinuity in the surface energy.

It was hypothesized that the surface energy discontinuity created by the random orientations of the surface crystallites affected the organization of the water molecules at localized points at the surface, hence originating the icephobic properties. The analysis of the icephobic properties supported this hypothesis, as it was observed that icephobic properties were stronger with the increment of the top section thickness. The gradient polymer coatings exhibited a broad spectrum of icephobic performance by reducing the ice adhesion, depressing the freezing point of water, delaying the drop freezing, inhibiting ice nucleation and frost propagation, and reducing the freezing probability of supercooled drops upon impact. In contrast to the common approach of designing high-repellency surfaces to minimize the contact time between water and the surface and prevent ice nucleation and frost build-up, we demonstrate that strong pinning of micro-condensed droplets can actually delay ice formation and provide better control over its spread. This method proved to be more effective than trying to completely prevent frost and ice formation.

Industrial applications require robust materials that can withstand harsh environments; therefore, durability is crucial for the implementation of such materials. Unlike conventional homogeneous coatings, gradient polymer coatings exhibited extraordinary durability and adhesive properties. A thorough study was carried out on the interaction of supercooled droplets upon impact with gradient polymer coatings. Unlike typical static tests of icephobic properties, these experiments reflect the dynamic conditions of real-world environments, where multiple factors affect ice formation. The results confirmed the icephobic effectiveness of the gradient coatings, considering the dynamic interactions between supercooled droplets and surfaces, thereby ensuring the reliability of ice prevention strategies in practical applications.

Furthermore, a hybrid de-icing/anti-icing system was formed by combining the coating with an electromechanical de-icing system based on piezoelectric actuators. Multiple de-icing experiments were performed in an icing wind tunnel to mimic the icing in outdoor conditions. The results demonstrated how the nanometric nature of the coating did not interfere with the vibrations generated by piezoelectric actuators, resulting in a promising system to mitigate ice formation. The action–response times were under a second, and the coatings exhibited exceptional resistance to mechanical stresses even after multiple icing and de-icing cycles. The ice detachment mechanism across larger areas was typical of coated surfaces, showing fewer cohesive fractures. Ultimately, the entire ice block detached with less energy compared to experiments with an uncoated plate.

In the following paragraphs, each of the aspect related with this research will be detailed, starting from an overview of the iCVD process and its application to obtain icephobic coatings. Then the gradient polymer coatings will be fully described in terms of their structure and properties, highlighting in particular their durability and their icephobic performance. Finally, the hybrid de-icing/anti-icing system will be presented.

9.2 The iCVD process

9.2.1 iCVD fundamentals

Initiated chemical vapor deposition is a technique by which solid thin films are simultaneously polymerized and deposited from vapor precursors. This technique was developed at MIT by Professor Karen K Gleason in the early 2000s as a variation on the classical chemical vapor deposition approach [1]. The fundamental difference is the presence of an initiator in the reaction scheme [2]. The initiator compound, usually a peroxide, is thermally decomposed, and free radicals are formed which then polymerize the monomers following the mechanism of free-radical polymerization. The process proceeds in the complete absence of solvents. This allows one to achieve of ultra-high purities, avoids purification or drying processes, enhances conformality in complex geometries and particularly in sharp edges (figure 9.1(a)), and overcomes the monomer's miscibility issue when using more than one. iCVD is known for requiring a small and controlled amount of precursor per deposition and generating almost zero waste [3]. The deposition process typically occurs at temperatures $\leqslant$40 °C, which enables one to coat delicate substrates sensitive to high temperatures. The construction of several polymeric architectures such as homopolymers, random copolymers, gradient polymers, stacked structures, or even microstructures can be achieved easily. In the last few

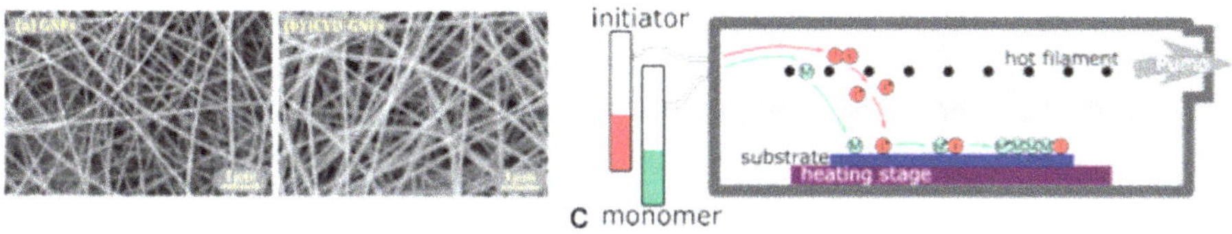

Figure 9.1. (a) and (b) Scanning electron microscope (SEM) images of a conformal coating deposited by iCVD gelatin nanofibers (GNFs) as bare (a) and coated by iCVD (b). (c) Schematic representation of the steps involved in the iCVD deposition process. ((a) and (b) adapted with permission from [5]. Copyright (2020) Elsevier. (c) Reproduced from [6]. CC BY 4.0.)

years, it has been demonstrated that solid compounds can also be vaporized and deposited using iCVD [4].

Figure 9.1(b) schematically represents the iCVD mechanism. In steps 1 and 2, the monomer(s) and initiator are fed into the reactor in the vapor phase. When the monomer is in the liquid phase and its vapor pressure is low, constant heating will ensure a continuous vapor delivery. When the monomer is in the solid phase, the sublimation process might be an alternative to deliver the monomer in vapor phase. In steps 3 and 4, the monomer(s) pass unaffected through the heated filaments (200 °C–300 °C) and adsorb on the cooled substrate (10 °C–40 °C), while the initiator is thermally decomposed forming free radicals. In step 5, the radicals will react and polymerize the adsorbed monomers. The reaction occurs via free-radical polymerization in which through the active center created by the free radicals, the growth and propagation of the polymeric chain occur by a fast-sequential addition of monomer units. The growth end occurs intentionally by stopping the monomer flow, leading to the combination of two growing chains or disproportionation reactions [7].

The iCVD system fundamentally consists of a reactor chamber, vacuum system, monomer(s) and initiator storage, and pressure/temperature controllers. Several iCVD reactor configurations can be found in the literature [2, 7], however, the filament array is a common denominator in their design and is usually positioned a couple of centimeters above the substrate surface. To maintain the substrate temperature as constant, a recirculating system is connected to the substrate stage. iCVD usually operates in a continuous mode, however, batch depositions are also possible. It has been used to successfully coat semiconductor components, metals, polymers, fabrics, paper, powder, porous structures, three-dimensional structures, and encapsulate liquids [8].

iCVD is an adsorption-driven process, the thickness and conformality achieved during the process are directly linked to the monomer(s) adsorption. The adsorption is predicted through the saturation ratio, S, defined as the ratio between the monomer's partial pressure, P_m, and the monomer's saturation pressure, P_{sat}:

$$S = P_m/P_{sat}.$$

This value estimates the concentration of the adsorbed monomer at a given temperature and the deposition rate, and determines the film's conformality. Experimental evidence suggests that uniform coverage is achieved for most of the monomers in a range of $0.3 < S < 0.8$, although successful depositions have been reported outside that range. The equilibrium between the liquid and vapor phase inside the reactor, i.e. $S = 1$, leads to the condensation regime, where monomer droplets form on the substrate surface. When the deposition is controlled within this regime, the droplets can polymerize forming solid microstructures. When $S < 0.1$, the kinetics change from being adsorption-limited to reaction-controlled, i.e. in this regime the polymerization is limited by the amount of monomer(s) present on the surface.

9.2.2 iCVD and anti-icing technologies—a brief overview of the state-of-the-art

The iCVD process has been applied in numerous fields, such as drug delivery systems, medical device coatings, dielectric layers, encapsulation of electronic components, battery coatings, capacitors, water-repellent fabrics, antibacterial textiles, air filtration systems, water purification membranes, optical coatings, anti-fouling coatings, biomedical sensors, food-packing barrier films, anti-fog coatings, solar cell encapsulation, functional adhesives, etc [2]. Nonetheless, up to today only two contributions are available concerning anti-icing coatings. In 2018, Sojoudi *et al* [9] used iCVD to form a bilayer coating on silicon and steel from poly-divinylbenzene (pDVB) and poly-perfluorodecyl acrylate (pPFDA) on top. The icephobic properties of the coating were assessed by the reduction of ice adhesion and durability with sand erosion tests. The coating successfully reduced the ice adhesion on steel with 180 ± 85 KPa (equivalent to 18 N assuming a toughness-dominated detachment), however, after only four de-icing cycles, the ice adhesion increased to 235 ± 68 KPa (equivalent to 23 N). This represents a significant loss of icephobic properties. Interestingly, after the sand erosion test the wettability was barely affected, from 157° ± 4.5° to 150.3° ± 3°. This is an example that wettability does not always correlate with icephobicity [10].

In 2023 Huang *et al* [11] used iCVD to create in one step a nanotextured surface using ethyleneglycol diacrylate (EGDA) and perfluorodecyl acrylate (PFDA). The icephobicity exhibited by this surface was observed in the icing and frosting delay. By depositing PFDA at $S \geqslant 1.0$, the monomer condensed in nanodroplets which then were polymerized following a 'vapor–liquid–solid' mechanism. The condensed nanodroplets served as nucleation centers for the polymerization to take place at the liquid–solid interface. The polymer chain growth due to the continuous adsorption of vapor monomers at the vapor–liquid interface led to the formation of a nanocone array structure. Through a systematic study, it was found that taller arrays led to enhanced icephobicity and wettability. For the best sample the icing delay was 540 s, no information about ice adhesion reduction was reported.

9.3 The gradient polymer coatings

Gradient polymer coatings are characterized by a gradual transition from species A to species B, forming a vertical structure with homopolymer-like sections at both ends and a copolymer region in between, all without a distinct interface. Some examples of gradient polymer structures are present in the iCVD literature, e.g. [12, 13]. The composition of gradient polymers can be precisely customized throughout the entire thickness of the coating, allowing for distinct properties in each section of the structure. To date, the development of these structures has been limited to vapor deposition techniques.

Schröder *et al* [14] demonstrated the mechanical and chemical benefits of a gradient polymer structure made of poly 1,3,5-trivinyl-1,3,5-trimethylcyclotrisiloxane (PV_3D_3) in the bottom section and poly(tetrafluoroethylene) (PTFE) in the top section. In addition, they proposed that gradient polymers could be useful for anti-icing coatings but without evaluating the icephobic properties of the material.

In this work, a gradient polymer with a bottom section made of poly 1,3,5,7-tetravinyl-1,3,5,7-tetramethylcyclotetrasiloxane (pV4D4) and a top section of poly 1H,1H,2H,2H-perfluorodecyl acrylate (pPFDA) was evaluated. The decision to use PFDA as a fluorinated monomer was based on the strong hydrophobicity and oleophobicity exhibited by its polymer, depending on the crystalline aggregation and the growth regimes [15–17]. A fine tuning of the iCVD deposition conditions enabled the control on the orientation of the crystallites. In figure 9.2 the lamellar crystalline aggregation of the pPFDA is shown, together with the x-ray diffraction patterns of several polymers deposited by iCVD at different initiator/monomer ratios. The diffraction peak observed in pPFDA at the lowest scattering angle corresponds to 3.2 nm, which represents the size of the bilayer lamella structure formed when the pendant fluorinated groups of two polymer chains align 'face-to-face' after polymerization. This structure is identified as the Smectic B phase. The second-order peak corresponds to a single pendant group. The presence of these peaks in the specular direction indicates that the lamella is oriented 'out-of-plane', meaning it is positioned perpendicularly to the surface of the sample.

The structure of the gradient polymer coatings developed in this work is schematized in figure 9.3(a). It can be divided into three distinct sections, each with different properties: the bottom section, which is in direct contact with the

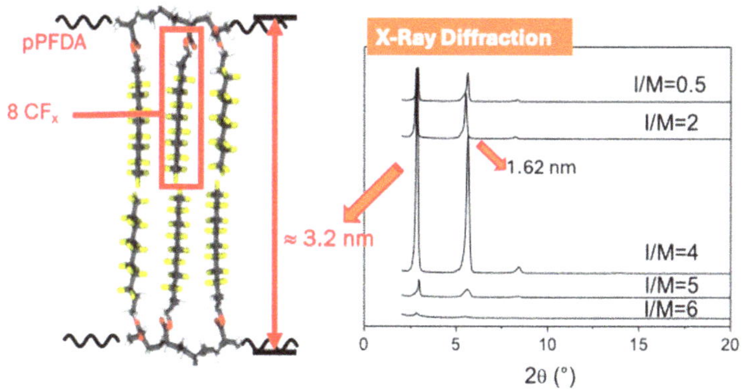

Figure 9.2. Chemical structure of the pPFDA, showing its lamellar crystalline aggregation. On the right, x-ray diffraction patterns of the pPFDA polymers deposited at different initiator over monomer (*I*/*M*) ratio. (Adapted with permission from [15]. Copyright (2012) Elsevier.)

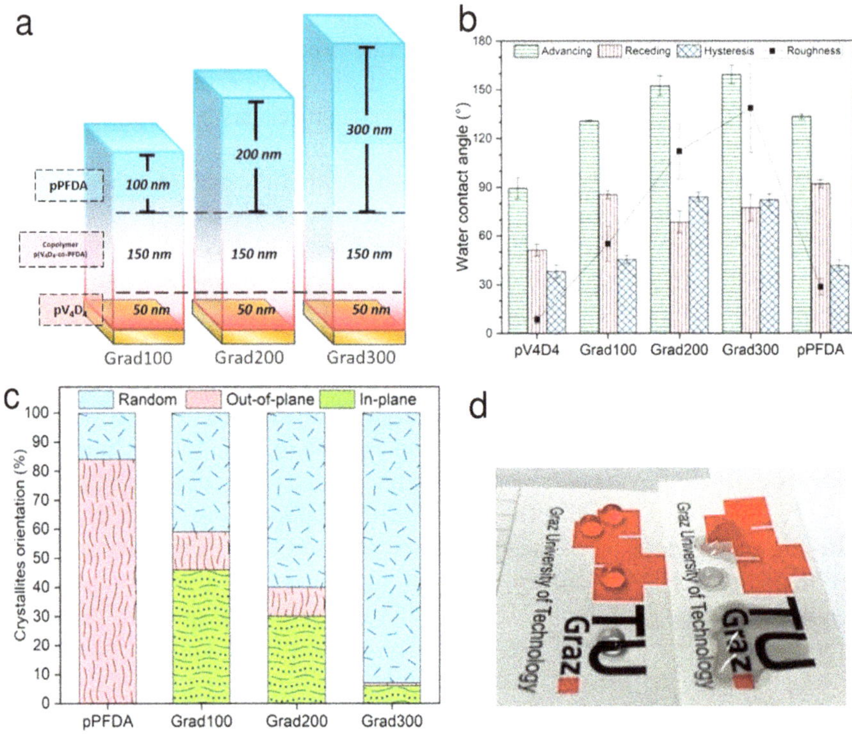

Figure 9.3. (a) Schematic representation of the structure of a gradient polymer where no interphases are distinguished. (b) Comparison of the advancing, receding, WCA hysteresis, and roughness of the different sections that constitute the gradient polymer and the different top section thicknesses. (c) Crystallite orientation fraction of the pPFDA and gradient polymers. (d) Photograph of drops of water placed on a coated and uncoated piece of paper. The coated paper on the left did not get wet from the drops, while the uncoated paper on the right shows halos around the drops, typical of wet paper. ((a)–(c) adapted from [18]. CC BY 4.0.)

substrate, is a V_4D_4 homopolymer. The middle section forms when the monomer PFDA is gradually introduced into the reactor, leading to the gradual creation of a copolymer p(V_4D_4-co-PFDA). This copolymer starts with a high concentration of V_4D_4 and ends with a higher concentration of PFDA units. The top section forms above the copolymer, after the flow of V_4D_4 is completely stopped, causing the PFDA monomer units to polymerize into a PFDA homopolymer at the surface. The copolymer section allows for a smooth transition between the two homopolymers with different chemical and physical properties, thus preventing a sharp interface. Gradient polymers with top PFDA sections of 100, 200, and 300 nm (labeled as Grad100, Grad200, and Grad300, respectively) were deposited to investigate their impact on the coating's surface properties.

The infrared spectra confirmed that the synthesis was successful, as the key characteristic signals of the homopolymers pPFDA and pV_4D_4 were present in the gradient polymers. This indicates that the functional groups were effectively retained after polymerization and deposition [18]. X-ray photoelectron spectroscopy (XPS) measurements showed no significant difference in the elemental composition between the various gradient polymers and pure pPFDA, suggesting that the top sections of the gradient polymers have a chemical composition as pure as that of the homopolymer pPFDA.

As shown in figure 9.3(c), the gradient polymer surfaces exhibited hydrophobic behavior, with advancing contact angles greater than 120°, which increased as the thickness of the top section grew. At the same time, the contact angle hysteresis was notably high, indicating a 'sticky' surface where water droplets are not easily removed. To assess the drop mobility, water contact angles were measured on a tilted plane to determine the roll-off angle—the angle at which a droplet begins to slide off the surface. These measurements confirmed low drop mobility: for all the gradient polymers, the roll-off angle was greater than 90°, and the droplets remained stationary even when the surface was tilted by up to 180°. Only for 20 μl droplets did gravity overcome the capillary adhesion forces, causing the droplets to slide. Figure 9.3(c) also displays the water contact angles of the homopolymers: pV_4D_4 does not exhibit a strong hydrophobic behavior, as anticipated, while pPFDA has an advancing contact angle similar to that of Grad100. The variation in wettability between the three gradual coatings and pPFDA—despite all sharing the same surface chemistry—could be attributed to differences in surface morphology and crystalline aggregation.

Atomic force microscopy (AFM) imaging revealed that the surfaces exhibited features with varying shapes, sizes, and distributions. pPFDA displayed nanoscale roughness (RMS = 8 ± 1 nm), with small and widely scattered aggregates characteristic of island growth, as shown also in the literature [17]. In contrast, the gradient polymers showed spherical aggregates that became more densely packed as the top section increased, leading to a noticeable rise in roughness. The surface roughness measured by AFM is shown in figure 9.3(b). The differences in topography can be attributed to the underlying layer on which the pPFDA section grows: the homopolymer grew on a silicon substrate, a smooth and clean surface, while in the gradient polymer, the top section grew on a copolymer surface made up of islands with irregular shapes and greater roughness than silicon and pV_4D_4. These surface inhomogeneities created nucleation points, from which the PFDA growth formed the spherical aggregates observed in the gradient polymers.

The structural analysis of the gradient polymers was done through grazing incidence x-ray diffraction measurements [18]. As discussed in figure 9.2, the presence of Bragg peaks in the specular direction suggests an out-of-plane orientation of the pPFDA crystallites. However, it was observed that in the gradient polymers, the orientation of the crystallites was not always out-of-plane, and in particular, it varied with the thickness of the top pPFDA section. The GIXD analysis allowed for the estimation of the crystallite fraction at different orientations and the distribution of crystallite orientations for both pPFDA and the gradient polymers is shown in figure 9.3(c). The pure pPFDA composition is estimated to be predominantly made up of crystallites oriented 'out-of-plane' (84%), with the remaining part being randomly oriented. As the gradient polymers were formed, there was a significant shift towards randomly oriented structures, along with the appearance of 'in-plane' crystallites, which occurred at the expense of the 'out-of-plane' oriented crystallites. The 'in-plane' orientation of the crystallites refers to lamellae that are parallel to the substrate. In this case, the fluorinated CF_2 and CF_3 groups are directly exposed at the surface. This orientation inherently produces an extremely low-energy surface, in contrast to the 'out-of-plane' orientation where the CH_2 chains are exposed at the surface.

In a randomly oriented configuration, both 'in-plane' and 'out-of-plane' orientations coexist. In other words, in the gradient polymers, regions with the carbon backbone exposed at the surface are present alongside regions where the fluorinated groups are exposed, all within the same plane. These two types of surface exposure have very different surface energies, creating a discontinuity in surface energy that becomes more pronounced as the thickness of the top section increases. We believe this discontinuity, along with the changes in surface roughness, is responsible for the difference in wettability observed between pure pPFDA and the gradient polymers. The discontinuity results in distinct surface properties due to the varying molecular interactions between the coating and water molecules.

Finally, figure 9.1(d) shows water droplets placed on both uncoated and coated paper. First, the paper remains intact after the iCVD process, highlighting the gentle nature of the deposition method, which does not damage the substrate. Second, the coated paper does not absorb the water, thanks to the hydrophobic coating applied to its surface, unlike the bare paper which gets wet.

9.4 Icephobic performance of gradient polymers

Parts of this section have been reproduced from [18]. CC BY 4.0.

The icephobic performance of the gradient polymers was evaluated through drop freezing delay experiments, and evaluation of the ice adhesion force, condensation and frost formation [18].

Drop freezing delay tests involve placing five water droplets on a surface under specific humidity and temperature conditions, then recording how long it takes for each droplet to freeze. The gradient polymer surfaces demonstrated remarkable ability in delaying the freezing process. During these experiments, as the relative humidity increased, the formation of frost varied across the coatings, due to the naturally higher water content in the surrounding environment (see figure 9.4(a)). Specifically, at low humidity levels ($<10\%$), frost formed in a dendritic pattern along the surface. At medium humidity ($\sim 50\%$), the frost grew similarly close to the surface but appeared noticeably thicker. At higher humidity levels ($>70\%$), a large volume of frost developed, with dendrites growing mostly in an out-of-plane direction. Additionally, a consistent increase in freezing delay was observed from Grad100 to Grad300.

To assess ice adhesion, the force required to push an ice mold onto the surface of the coatings was measured. The experiments revealed a distinct fracture propagation, suggesting that ice detachment follows a toughness-dominated mechanism. This is particularly notable because, at such small ice mold diameters (8–12 mm), detachment is typically governed by stress and results in immediate failure.

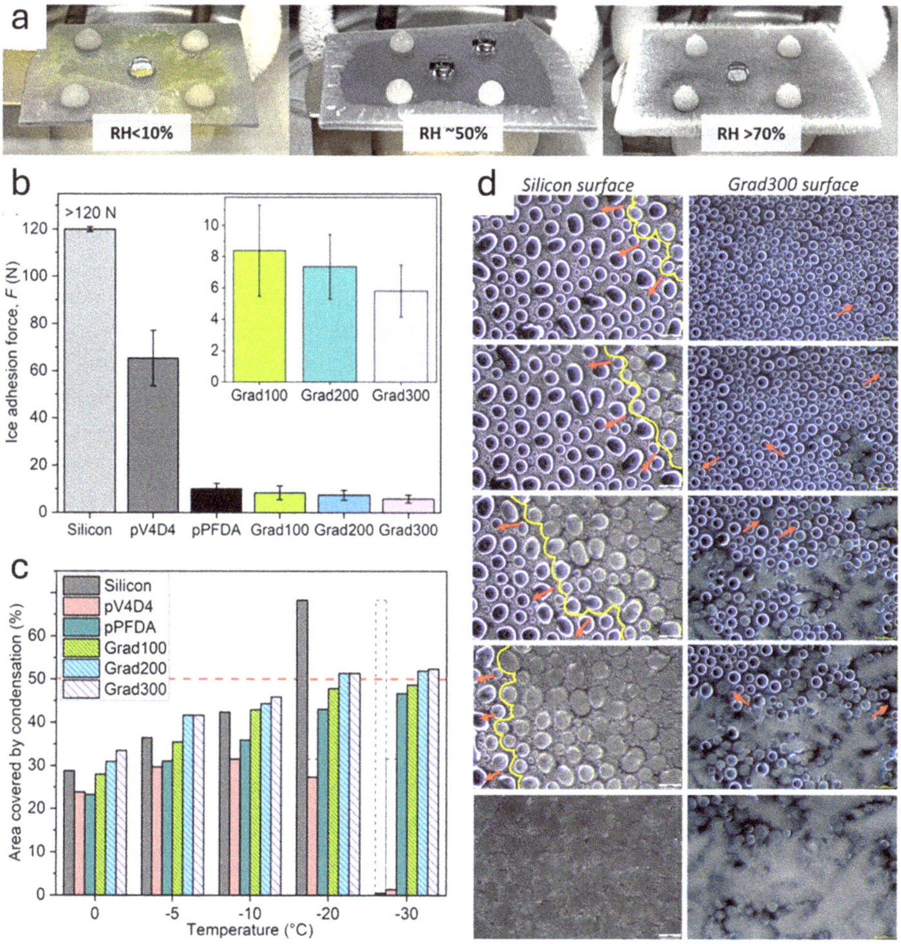

Figure 9.4. (a) Photographs of the Grad300 surface with five water drops during the frost formation experiment and exposed to different relative humidities. (b) Ice adhesion force for various coatings, with the inset illustrating the ice adhesion strength for the gradient coatings. As the thickness of the top section increased, the ice adhesion force decreased. (c) Plot of the area covered by condensation at different temperatures. The reduction in coverage at −30 °C for the silicon and pV_4D_4 samples was due to freezing. The dashed lines indicate the maximum area covered reached. (d) Optical micrographs showing the differences in the freezing mechanisms and the resulting frost. The frost front is indicated by a yellow line and the nucleation points by red arrows. (Adapted from [18]. CC BY 4.0.)

Therefore, gradient polymers are categorized as low-interfacial toughness (LIT) materials, meaning that the force needed to detach the ice is not only low but also independent of the interfacial area [19]. Figure 9.4(b) compares the ice adhesion force of gradient polymers with silicon, pV_4D_4, and pPFDA. The gradient polymers exhibited the lowest ice adhesion, with the trend showing a clear improvement in icephobicity as the top section thickness increased. Grad300, in particular, demonstrated an adhesion reduction factor (ARF) of at least 20 times lower than the silicon substrate. A key point of interest is the comparison between the ice adhesion force of pPFDA and Grad300. Both films had the same total thickness (500 nm), but their adhesion forces were quite different: 10 N for pPFDA and 6 N for Grad300. Despite no chemical differences between the two samples and a significantly higher contact angle hysteresis on Grad300 (as discussed earlier in figure 9.3(b)), the variation in icephobicity cannot be attributed to hydrophobicity. The only material property that could account for the reduced ice adhesion between these two surfaces is the difference in crystal orientation. Specifically, the random crystal orientation in Grad300 seems to play a key role in reducing ice adhesion.

As the temperature drops, water from the surrounding environment condenses onto the surface. The amount and distribution of this condensed water play a crucial role in determining when and how the surface freezes. Therefore, condensation was observed and measured before freezing occurred. Figure 9.4(c) illustrates the condensation at different temperatures, shown as the percentage of surface area covered by condensed water. As expected, condensation increased with decreasing temperature, covering a larger area. A clear pattern emerged among the gradient polymers: as the top section thickness increased, so did the amount of condensation. Interestingly, below −20 °C, more than half of the gradient polymer surface was covered by condensation without any signs of freezing (consistent with the freezing delay results), in contrast to silicon and pV_4D_4, where most of the condensed water froze. The liquid state was identified by the reflection of the microscope light on the droplets, while the frozen state was characterized by a loss of reflection and an increase in opacity.

As the temperature remained low, the condensed water from the air eventually froze, a process commonly referred to as ‘condensation frosting’. The temperature was kept at −20 °C, and the development of the condensation frosting was monitored and quantitatively assessed by analysing microscopy images. Several distinct mechanisms of condensation frosting were identified, which are key to understanding why gradient polymers show promise as icephobic materials. Notably, the propagation of frost was delayed on the gradient polymer surfaces. The propagation of condensation frosting followed different mechanisms on each surface. On silicon, the process was directional, continuous, and rapid. Nucleation began at the edges and quickly spread across the entire observable area through the formation of bridges between neighboring droplets, a phenomenon known as ‘interdrop ice propagation’, as studied by Boreyko *et al* [20] This mechanism occurs due to a simultaneous evaporation and rapid deposition process from the liquid droplets surrounding a frozen one. The propagation rate on silicon was 0.28 $mm^2\ s^{-1}$, meaning the entire observable area was covered in under 10 s. Although the frost front was not sharply defined, its progress can be tracked, as shown in figure 9.4 (d). The resulting condensation frosting consisted of a tightly packed layer of frozen droplets with indistinct boundaries, topped by a dense dendritic layer. On gradient polymer surfaces, frost formation started sporadically in various droplets across the observed area and progressed slowly (figure 9.4(d)). The droplets froze individually, at a very slow pace, with no immediate freezing of neighboring droplets; however, rapid out-of-plane dendritic growth was observed from the tips of the droplets. The dendrite growth occurred faster than the lateral frost propagation, likely due to direct desublimation of water vapor from the surrounding air onto the dendrites [21]

Heterogeneous nucleation prevalently governed the process, therefore no chain reaction nor a propagation front were identified. The size and position of the droplets remained consistent, and when droplets merged with nearby ones, larger dry areas were formed. The low-energy surface and strong pinning effect helped maintain a reduced contact area with the substrate, limiting the mobility of the droplets. This was important during nucleation and propagation, as the larger, stable dry zones increased the thermodynamic barrier for propagation, thereby delaying the condensation frosting process. The frost formed at a rate of 0.006 mm^2 s^{-1} for Grad300, which is 47 times slower than on silicon. The condensation frosting consisted of a discontinuous ice layer made up of scattered frozen droplets with delicate dendrites on top, and surprisingly, some remaining dry patches, creating an overall airy frost structure (figure 9.4(d)).

The consistent trend observed in the gradient polymers throughout the icephobic tests showed that as the top section thickness increased, icephobicity improved. While other factors, such as chemistry, topography, wettability, and thermal conductivity, can influence the water-coating interaction, we propose that the primary factor responsible for the icephobic properties is the surface energy discontinuity caused by variations in crystallographic orientations due to the top section thickness. It has been shown that surfaces with contrasting energy levels affect the organization of water molecules, leading to a delay in ice nucleation and frost propagation [22]. On a molecular level, the random orientation of the fluorinated groups disrupts the organization of water molecules, creating weaker bonds with the surface, which is reflected macroscopically in lower ice adhesion.

9.5 Durability

The cross-hatch adhesion test was successfully employed to determine the adhesion of the iCVD coatings to the substrates. The samples were analysed following the ASTM D3359 standards and examined once the adhesive tape was removed from the surface of the coating. Figure 9.5(a) shows the surface of pPFDA and Grad100 before and after the test. The surfaces prior to adhering to the tape showed a clean and homogeneous coating surface in which the scratched lattice pattern is clearly distinguished. After adhering and removing the tape, no recognizable differences were observed in the gradient coatings (Grad100, Grad200, and Grad300): the edges of the cuts were completely smooth and none of the squares of the lattice detached. This corresponds to the strongest level of adhesion observable with this test. In contrast with the gradient polymer coatings, the homopolymer pPFDA showed evident damage: the coating flaked along the edges of the cuts partly or wholly in large ribbons. As expected for the homopolymer pPFDA, after the test was repeated five times most of the coating was detached. The large difference in the results of the pPFDA and the gradient polymer samples, considering that the surface in all samples is chemically identical and that the substrate is the same, demonstrates that gradient structures successfully provide adhesive properties.

Ice detachment tests were also used as an indirect way to evaluate the durability of the coatings and their resistance to scratches. The gradient polymer coatings

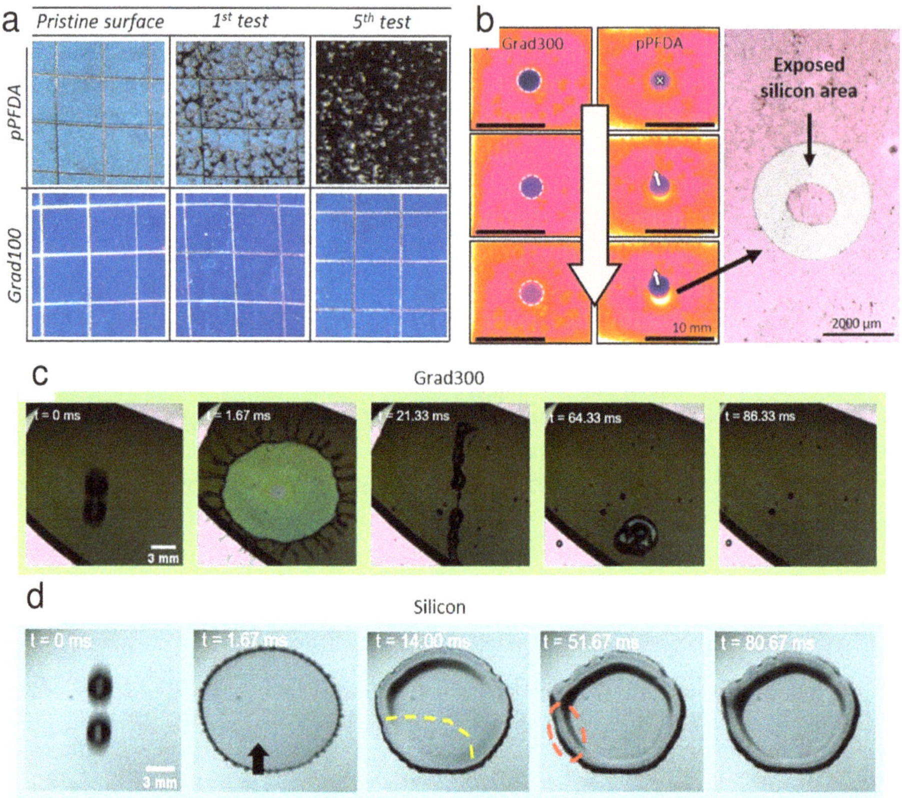

Figure 9.5. (a) Optical microscopy images of the pPFDA and Grad100 surfaces before and after the cross-hatch adhesion test. No damage was visible on the Grad100 surfaces after the tests. (b) Snapshots taken with the infrared camera during the de-icing of the frozen drop in the Grad300 and pPFDA substrate. In Grad300 the drop stays in place, while in pPFDA the drop shifts from its original position, tearing the coating and exposing the silicon substrate. A microscopy image clearly shows the damaged area where the pPFDA coating has been torn. Snapshots of the supercooled drop impacts on (c) Grad300 and (d) silicon with an external airflow of 20 m s^{-1}. The arrow indicates the nucleus, the yellow dashed line indicates the thin ice freezing front and the red dashed line indicates the dendrites. ((a) and (b) adapted from [18]. CC BY 4.0. (c) and (d) adapted from [23]. CC BY 4.0.)

demonstrated long-lasting performance with no signs of damage, while pV_4D_4 and pPFDA coatings were significantly damaged after several repetitions of the experiments. While drops on pPFDA surfaces rarely froze, when they did, the local stresses within the drop—caused by volume changes—were strong enough to tear the coating and expose the substrate. As a result, the drop would shift from its original position (figure 9.5(b)). This clearly shows that pPFDA coatings are not suitable for icephobic applications, as their durability is too low to withstand even a single freezing event. In contrast, gradient polymer coatings showed no signs of damage, even after multiple freezing events.

The impact and freezing of supercooled water drops were studied on gradient polymer coatings deposited on silicon via iCVD. This tests can also be considered an indirect measure of durability against water impact and erosion, since the surfaces of the coatings are exposed to accelerated drop impact for more than 90 times.

During these types of experiments, the drop impact process is recorded with a high-speed camera and the profile of the impacting drop is observed and correlated to the surface properties. Upon impact with a surface, the drop spreads by flowing in a thin lamella [23]. The spreading diameter of a droplet is influenced by the dynamics of the rim, which are governed by capillary and viscous forces, as well as the inertia of the liquid in the thin lamella. On hydrophobic substrates, the droplet spreads and then the rim recedes. On hydrophilic surfaces, however, the rim is pinned near the point of maximum spreading. In supercooled liquids, the formation and dissolution of ice embryos occur continuously. At times, an ice embryo grows large enough to reach a critical radius. This critical size marks the onset of heterogeneous nucleation, where the ice embryo becomes stable and continues to expand. Once it surpasses this critical size, it triggers the nucleation process [24]. A thin front of ice begins to form, signaling the start of the liquid's solidification. As this process continues, dendritic structures begin to grow. Once the dendrites have spread, the droplet becomes a solid–liquid mixture. If the surface stays cold, the liquid between the partially frozen layers will keep solidifying. This slower phase is referred to as secondary solidification [25].

Figures 9.5(c) and (d) show the key impact scenarios on both the silicon substrate and Grad300 at an airflow velocity of 20 $m\,s^{-1}$. On the silicon substrate, a clear solidification process was observed upon drop impact: all the drops froze during the receding phase. In contrast, on the gradient coatings, every impact caused the drop to rebound, followed by a secondary impact. When the drops were accelerated by the higher airflow velocity, the kinetic energy caused larger lamella spreading across all surfaces. On the gradient polymer surfaces, the drop fragmented into smaller droplets that stayed pinned after the initial impact. After the secondary impact, the main droplets were blown off by the airflow, leaving the tiny secondary droplets still adhered to the surface. When the drop impacted, the freezing probability of supercooled drops upon the impact with gradient polymer coatings systematically decreased as the thickness of the top section of the gradient polymer increased, and it was at least 70% smaller than freezing probability upon impact on silicon.

Interestingly, after at least 90 experiments, the coatings showed no signs of delamination, erosion damage, or any loss in functionality, as evidenced by the consistent full rebound observed in all tests. This further demonstrated the stability and durability of the gradient polymer structure.

9.6 Hybrid de-icing/anti-icing systems

Since Grad300 exhibited the best icephobic properties, it was applied to aluminum plates and paired with an electromechanical de-icing system to evaluate its effectiveness and performance in hybrid de-icing/anti-icing systems. The aluminum plates were coated using iCVD, and then five piezoelectric actuators were attached to their back. More details on this can be found in [26].

Ice accretion was tested on both coated and non-coated aluminum plates placed in an icing wind tunnel (IWT). Before conducting the ice accretion tests, the wettability of the plates was assessed at room temperature by spraying water through the IWT nozzles. Due to the strong pinning effect and hydrophobic nature of the coating, the drops retained a uniform size and distribution, with clear dry zones between them, as discussed in section 9.3. Once a certain drop volume was reached, the pinning effect was overcome, causing the drops to roll off the surface, effectively drying the area. In the non-coated region, the drops spread across the surface, forming a nearly continuous water layer. Despite the noticeable differences in wettability, the formation of various types of ice (glaze, rime, and mixed) in the IWT was unaffected, resulting in uniform ice blocks.

The coated and uncoated aluminum plates, both equipped with the electromechanical de-icing system, were placed in the IWT, where ice accumulated until the surface was covered by a uniform layer of glaze ice, approximately 3 mm thick. The electromechanical system was then activated, exciting the plate at its first extensional mode with a frequency around 13 kHz. This was expected to generate high stresses at specific points in the ice block, leading to crack formation and eventual detachment from the aluminum surface. Non-coated plates were used as controls. In tests with the non-coated plates, only cracks formed, but complete detachment of the ice did not occur.

The half-coated and fully coated plates were tested under identical conditions, with the results shown in figure 9.6. Figure 9.6(b) presents the experiment where a continuous glaze ice block uniformly covered both the coated and non-coated sections of the plate. The piezoelectric actuators were excited for 10 s, using the first extensional resonant mode. After just 0.1 s, several cracks appeared on the left side of the plate and spread across the entire coated area. At 1.5 s, a large piece of ice detached from the coated section, while the cracks continued to propagate toward the non-coated area. It took 4.8 s to completely remove the ice from the coated area. In contrast, the ice remained firmly attached to the non-coated area, despite the formation of cohesive cracks in the ice block during excitation. The test on the half-coated plate highlighted the positive effect of the Grad300 coating on the de-icing process.

The same experiment was repeated with two separate blocks of mixed ice in each area, as shown in figure 9.6(c). The piezoelectric actuators were excited, and after just 0.1 s of excitation, cracks formed sequentially, propagated, and caused ice detachment almost simultaneously on the top-left side of the coated area. This critical event triggered the full detachment of the remaining ice block. As a result, in under one second (0.2 s), the coated area was completely free of ice. In contrast, the ice on the non-coated section remained firmly attached to the substrate throughout the entire experiment, with no signs of cracking or adhesive propagation. The same de-icing performance was observed with glaze ice blocks. The observed de-icing process suggested that the detachment was driven by both cohesive and adhesive fractures. The ice fracture began as a cohesive fracture and then propagated adhesively until the ice detached. The exclusive detachment seen on the coated area highlighted the crucial role of the coating in the detachment process, effectively supporting the electromechanical system.

Figure 9.4(d) illustrates the experiment conducted on the fully coated plate. In this case, after 0.1 s of excitation, the vibrations caused a burst of ice crystals to form

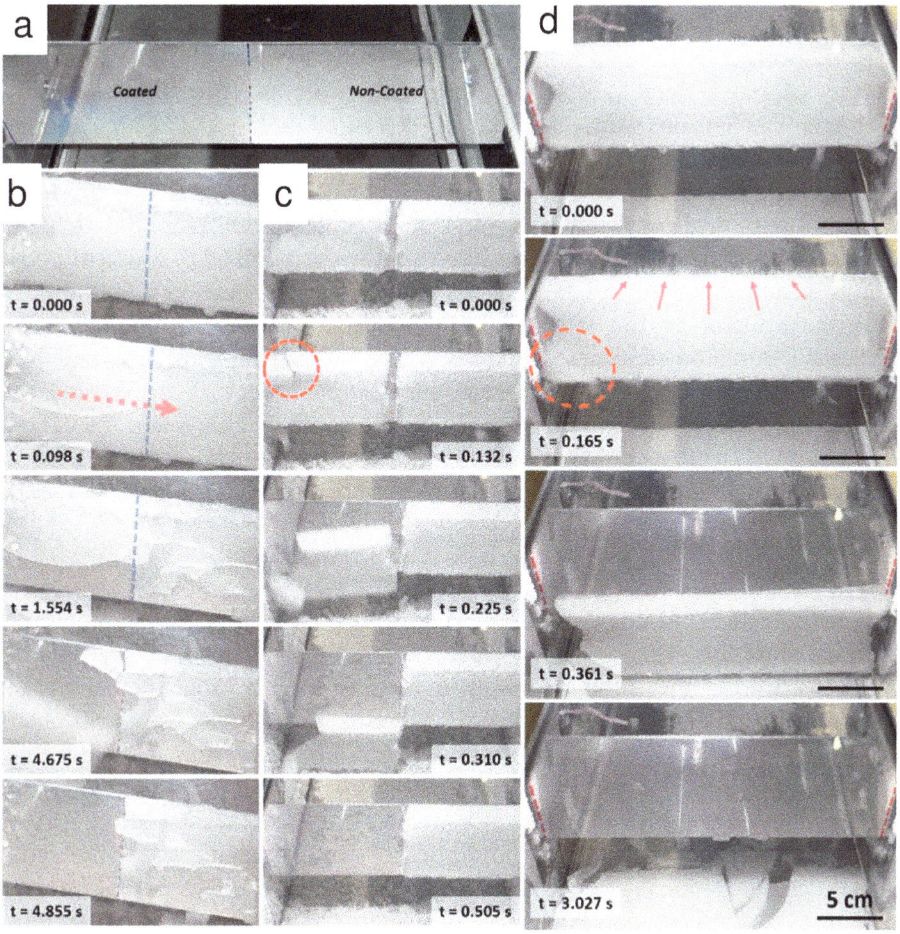

Figure 9.6. (a) A half-coated plate inside the IWT before ice begins to form a block. (b) Sequential images of the half-coated plate with a continuous glaze ice block during excitation. When the plate was excited, multiple cracks appeared on the coated area, propagating towards the non-coated section (dashed arrow). After 4 s, the ice had completely detached from the coated area, while it remained adhered to the non-coated section. (c) Sequential images of the half-coated plate during excitation, showing two separate blocks of mixed glaze and rime ice. When the plate was excited, a crack formed and propagated, detaching a small piece of ice (dashed circle), followed by the detachment of the remaining block in one piece in under 1 s. (d) Sequential images of the fully coated plate during excitation with a continuous glaze ice block. When the plate was excited, a crack formed and propagated, detaching a small piece of ice (dashed circle), followed by the detachment of the rest of the block in one piece in less than 1 s. (Reproduced with permission from [26]. Copyright (2024) Wiley.)

along the top of the ice block. At the same time, a small crack began to form and propagate at the bottom-left corner of the ice block, leading to the complete adhesive detachment of the ice block in one piece. Within just 0.3 s of excitation, the surface of the plate was completely free of ice.

The gradient polymer coating demonstrated remarkable durability, withstanding multiple ice detachment cycles. Ice was accumulated on each plate at least 10 times, and no signs of scratches, delamination, degradation, or loss of effectiveness were observed. Both cohesive and adhesive fractures occurred on the same time scale, with the ice block detaching in less than half a second. No differences in the detachment mechanisms were observed between small ice blocks (70 cm^2) and larger ones (140 cm^2), indicating that the critical length for detachment is smaller than the characteristic sizes of the ice blocks tested in this study. As a result, the fracture mechanism remained consistent regardless of the area. These findings align with the toughness-dominated detachment mechanism previously linked to gradient polymer coatings, which was observed even with smaller areas (see section 9.4).

A crucial factor in the effective performance of this system is the coating thickness. The nanometer-scale thickness of the coating plays a beneficial role in vibrational systems, as the stresses generated by the actuators are directly transmitted to the ice without being dampened by the coating. This is not the case with softer coatings, which could absorb some of the vibrations, reducing de-icing efficiency. Additionally, no signs of wear were observed after multiple excitations.

9.7 Conclusions

Gradient polymer coatings deposited using iCVD have shown great potential as icephobic materials, demonstrating effectiveness across various aspects. iCVD enables the production of these coatings in a single step, without the need for pre- or post-treatments, making it an appealing option for industrial applications. We show that the icephobic properties of this materials stem from a surface energy discontinuity caused by the random orientation of the fluorinated groups. By adjusting the architecture of gradient polymers, we developed a straightforward and effective method to introduce randomness, thereby enhancing icephobicity by reducing ice adhesion. We provide strong evidence that these icephobic properties originate at the atomic level, where even small changes can lead to significant macroscopic effects.

Fluorinated compounds are highly sought after in the field of icephobic coatings, but their use is often limited due to issues with solubility, compatibility, and stability. However, when incorporated into gradient polymer structures, their stability and resistance to scratches and delamination are significantly enhanced.

An investigation in a cold wind tunnel was carried out to study the effect of supercooled water droplets on gradient polymer-coated substrates at various wind speeds. In the absence of airflow, the freezing probability consistently decreased as the top section thickness of the gradient polymer coatings increased. When the droplets were accelerated by airflow, the freezing probability remained low, averaging only 14%, showing that gradient polymer coatings significantly reduced the likelihood of freezing, even at high impact velocities.

For the first time, iCVD was utilized to create a hybrid ice protection system by combining gradient polymer coatings with piezoelectric actuators. The system, tested in the IWT, exhibited highly effective de-icing performance, regardless of the iced area. This system offers a practical solution for ice mitigation due to its simplicity, low excitation requirements, and rapid response. Additionally, the coating showed exceptional resistance to scratches and delamination after multiple detachment cycles.

In conclusion, gradient polymers produced via iCVD not only exhibited exceptional icephobic properties but also proved to be strong candidates for integration with resonant electromechanical de-icing systems. This combination resulted in hybrid systems that deliver high performance and durability under real-world outdoor conditions.

References

[1] Pryce Lewis H G, Caulfield J A and Gleason K K 2001 Perfluorooctane sulfonyl fluoride as an initiator in hot-filament chemical vapor deposition of fluorocarbon thin films *Langmuir* **17** 7652–5

[2] Gleason K K 2024 Designing organic and hybrid surfaces and devices with initiated chemical vapor deposition (iCVD) *Adv. Mater.* **36** 2306665

[3] Reeja-Jayan B *et al* 2014 A route towards sustainability through engineered polymeric interfaces *Adv. Mater. Interfaces* **1** 1–30

[4] Boscher N D, Wang M, Perrotta A, Heinze K, Creatore M and Gleason K K 2016 Metal–organic covalent network chemical vapor deposition for gas separation *Adv. Mater.* **28** 7479–85

[5] Mansurnezhad R *et al* 2020 Fabrication, characterization and cytocompatibility assessment of gelatin nanofibers coated with a polymer thin film by initiated chemical vapor deposition *Mater. Sci. Eng.* C **110** 110623

[6] Unger K 2021 Smart hydrogels via initiated chemical vapor deposition: humidity, light and pH responsive polymer films for biosensors, drug delivery systems and cell cultivation platforms *Dissertation* Graz University of Technology

[7] Lau K K S and Gleason K K 2006 Initiated chemical vapor deposition (iCVD) of poly(alkyl acrylates): an experimental study *Macromolecules* **39** 3688–94

[8] Bradley L C and Gupta M 2012 Encapsulation of ionic liquids within polymer shells via vapor phase deposition *Langmuir* **28** 10276–80

[9] Sojoudi H, McKinley G H and Gleason K K 2015 Linker-free grafting of fluorinated polymeric cross-linked network bilayers for durable reduction of ice adhesion *Mater. Horiz.* **2** 91–9

[10] Nosonovsky M and Hejazi V 2012 Why superhydrophobic surfaces are not always icephobic *ACS Nano* **6** 8488–91

[11] Huang X *et al* 2023 Chemical vapor deposition of transparent superhydrophobic anti-icing coatings with tailored polymer nanoarray architecture *Chem. Eng. J.* **454** 139981

[12] Unger K, Salzmann P, Masciullo C, Cecchini M, Koller G and Coclite A M 2017 Novel light-responsive biocompatible hydrogels produced by initiated chemical vapor deposition *ACS Appl. Mater. Interfaces* **9** 17408–16

[13] Montero L, Baxamusa S H, Borros S and Gleason K K 2009 Thin hydrogel films with nanoconfined surface reactivity by photoinitiated chemical vapor deposition *Chem. Mater.* **21** 399–403

[14] Schröder S, Polonskyi O, Strunskus T and Faupel F 2020 Nanoscale gradient copolymer films via single-step deposition from the vapor phase *Mater. Today* **37** 35–42

[15] Coclite A M, Shi Y and Gleason K K 2012 Controlling the degree of crystallinity and preferred crystallographic orientation in poly-perfluorodecylacrylate thin films by initiated chemical vapor deposition *Adv. Funct. Mater.* **22** 2167–76

[16] Coclite A M, Shi Y and Gleason K K 2012 Grafted crystalline poly-perfluoroacrylate structures for superhydrophobic and oleophobic functional coatings *Adv. Mater.* **24** 4534–9
[17] Perrotta A, Christian P, Jones A O F, Muralter F and Coclite A M 2018 Growth regimes of poly(perfluorodecyl acrylate) thin films by initiated chemical vapor deposition *Macromolecules* **51** 5694–703
[18] Hernández Rodríguez G, Fratschko M, Stendardo L, Antonini C, Resel R and Coclite A M 2024 Icephobic gradient polymer coatings deposited via iCVD: a novel approach for icing control and mitigation *ACS Appl. Mater. Interfaces* **16** 11901–13
[19] Mohseni M, Recla L, Mora J, Gallego P G, Agüero A and Golovin K 2021 Quasicrystalline coatings exhibit durable low interfacial toughness with ice *ACS Appl. Mater. Interfaces* **13** 36517–26
[20] Boreyko J B and Collier C P 2013 Delayed frost growth on jumping-drop superhydrophobic surfaces *ACS Nano* **7** 1618–27
[21] Sommers A D, Gebhart C W and Hermes C J L 2018 The role of surface wettability on natural convection frosting: frost growth data and a new correlation for hydrophilic and hydrophobic surfaces *Int. J. Heat Mass. Transf.* **122** 78–88
[22] Esmeryan K D, Castano C E, Mohammadi R, Lazarov Y and Radeva E I 2018 Delayed condensation and frost formation on superhydrophobic carbon soot coatings by controlling the presence of hydrophilic active sites *J. Phys. D: Appl. Phys.* **51** 055302
[23] Rodríguez G H, Ding M, Roisman I V, Hussong J and Coclite A M 2025 Study of supercooled water drop impact on icephobic gradient polymer coatings *Adv. Mater. Interfaces* **12** 2400723
[24] Schremb M, Roisman I V and Tropea C 2017 Transient effects in ice nucleation of a water drop impacting onto a cold substrate *Phys. Rev.* E **95** 022805
[25] Zhang X, Liu X, Wu X and Min J 2020 Impacting-freezing dynamics of a supercooled water droplet on a cold surface: rebound and adhesion *Int. J. Heat Mass Transf.* **158** 119997
[26] Hernández Rodríguez G *et al* 2024 Icephobic gradient polymer coatings coupled with electromechanical de-icing systems: a promising ice repellent hybrid system *Adv. Eng. Mater.* 2401532

Chapter 10

Unveiling durability of icephobic coatings by standardized environmental tests for industrial applications

Theodoros Dimitriadis and Anna-Maria Coclite

The environmental durability of icephobic coatings is critical for industrial applications where surfaces are exposed to harsh weather conditions. This work focuses on using standardized test methods (artificial and natural weathering) to evaluate the long-term performance of icephobic coatings under near-operational conditions, especially superhydrophobic and hydrophobic solutions. The study provides a strong foundation for measuring the effectiveness and durability of icephobic coatings in a variety of industrial contexts by following stringent standards (ISO, ASTM, military standards). The evaluation includes testing the resistance to a wide range of environmental influences such as direct water-jet impact, salt-fog, temperature fluctuations, natural weather conditions, and a military-standardized altitude–temperature–humidity test, along with de-icing tests to simulate near-operational environmental conditions. The findings indicate significant performance disparities between coatings, emphasizing the importance of consistent testing techniques in order to follow trustworthy durability standards. Hence, it provides practical insights into selecting and deploying icephobic coatings in industries that require long-term resilience, laying the groundwork for including standardized durability requirements as a criterion for icephobic coatings.

10.1 Introduction

Extreme weather conditions and especially ice are major challenges in a variety of industries, including the aerospace, automotive and energy fields. The build-up of ice on exposed surfaces can lead to mechanical failure, reduced efficiency and compromised safety. To address this issue, many de-icing and anti-icing procedures

doi:10.1088/978-0-7503-6009-8ch10

have been developed, which can be characterized primarily as active or passive approaches [1]. However, the demand for effective icephobic coatings has increased as these industries look for passive protection measures that can prevent or delay ice adhesion, accumulation and frost formation, reducing energy costs and increasing operational safety [2]. While much progress has been made in the development of icephobic coatings, e.g. textured surfaces, lubricants, superhydrophobic, hydrophobic, either smooth or rough coatings, and many other combinations and novelties [3–9], there is still a fundamental gap in testing their durability in near-operational environmental conditions similar to those encountered in service [10–13]. This discrepancy has led to concerns about the long-term performance and reliability of icephobic coatings when used in real-world applications where environmental exposure exceeds controlled laboratory conditions.

Traditional testing for icephobic coatings has mostly focused on determining ice adhesion strength, delay of ice formation, etc, which is usually measured under controlled icing conditions in a laboratory environment [14–18]. By aiming to inform the design criteria for effective icephobic materials, whilst this approach provides useful information on initial performance, it does not consider the wide range of environmental conditions to which these coatings are exposed in practice. Real-world conditions present a more complex array of challenges due to the diverse forms of precipitation and environmental variables such as temperature, humidity, corrosion, etc. The combined effect of these factors can accelerate the ageing process [19–21], leading to a gradual loss of anti-icing performance over time. Therefore, evaluating durability as an isolated property, e.g. only de-icing cycles, is insufficient; it should be evaluated together with the ability of the coating to withstand operational stressors that mimic real-world environmental conditions.

Standardized durability testing protocols are widely used in the coating industry, particularly for protective and functional coatings in several industrial sectors. Organisations such as the International Organisation for Standardisation (ISO), the American Society for Testing and Materials (ASTM), military standards (MIL-STD) and Institute of Printed Circuits (IPC) and many others have established strict criteria for evaluating the resistance of coatings to various climatic and mechanical stresses. By providing established benchmarks, these standards allow researchers and industry experts to evaluate the performance of coatings, predict their durability and verify their reliability in a range of applications.

The paradox, in the field of icephobic coatings, is that standardized environmental durability tests are generally neglected. Many studies [14, 15, 18, 22–30] focus only on measuring icephobic properties without considering the full range of environmental conditions to which these coatings are exposed. Others are conducting environmental tests, such as temperature variations, but without following any standardized method [3, 10, 11], which lack of repeatability and consistency. This gap makes it difficult to compare the results of different studies and limits the practical applicability of coatings in harsh environments, which often involve complex conditions with multiple parameters beyond just icing.

To bridge this gap, this work uses standardized environmental test methods to evaluate the durability of various fluorinated icephobic coatings designed for industrial applications: a superhydrophobic coating (SHS), a contrast wettability hydrophobic-textured hydrophilic coating, and a gradient hydrophobic coating. These coatings have been developed using three distinct deposition methods: spraying and curing at room temperature, application as a powder and baked at 400 °C, and using initiated chemical vapour deposition (iCVD). Fluorinated coatings and their application procedures were chosen based on their demonstrated performance in industrial applications such as friction reduction, water repellence, icing prevention, and electrical insulation.

Standardized processes, such as ISO, ASTM, and MIL-STD, were used to assess their durability under a variety of near-operational harsh environments. These tests provided insights into the degradation of icephobic coatings after exposure to water-jet impact, salt-fog, temperature variations, humidity with cold sub-cycles, combined environmental stressors, and year-long natural weathering. These methodologies recreate near-operational weather conditions, providing a solid foundation for evaluating the long-term performance of icephobic coatings and developing design criteria for industrial use. Because there is no standardized test for ice adhesion, a shear push test lap was utilized before and after environmental exposure to determine durability and effectiveness.

In addition, this study presents advanced surface functionalization techniques, including iCVD, for the development of a vertically discontinuity-enhanced coating. Discontinuities are characterized by controlled gradients [31] or abrupt changes [32] in material or surface properties that disrupt homogeneity and enhance specific interfacial phenomena in any dimensional axis (x, y or z). These transitions, such as variations in wettability, are designed to exploit changes in material behaviour to improve functionality, such as increasing pumpless liquid flow or increasing durability [32]. Controlled gradients in the direction of one axis [31, 33] are utilized to achieve hydrophobic [31, 34] and icephobic properties [14, 31] without relying on the superhydrophobic structures traditionally considered as passive icephobic candidates.

The iCVD process, based on free radical polymerization, is particularly suitable for the production of ultrathin, uniform coatings with high adhesion strength on complex substrates. It offers precise control over surface chemistry and functionality, making it perfect for producing coatings that can be applicable in a plethora of applications. For industrial applications, the iCVD process is particularly desirable as it is a one-step, all-dry vapour and solvent-free process [35–37]. This work presents one of the first studies to examine iCVD-based icephobic coatings through a full series of standardized durability tests that mimic environmental stresses in near-operating conditions.

Evaluating the durability of icephobic coatings under real-world conditions shows how material selection, surface roughness and coating thickness affect environmental resistance. This chapter takes a significant step toward bridging the gap between laboratory performance and real-world applications by employing industry-standard methods. The results highlight the importance of standardized environmental durability assessments in the field of icephobic coatings, which will provide a consistent framework for the selection and deployment of durable, long-lasting solutions in a variety of operational situations. This methodology advocates for research practices that better reflect near-operational conditions, establish a standard for future studies and improve the practicality of icephobic coatings in industrial applications.

10.2 Durability assessment of icephobic coatings

Durability, a term rooted in the Latin 'durare' meaning to 'last', is fundamentally the ability of a coating to withstand various stresses, environmental exposure, and operational wear over time, retaining its protective and functional qualities [10, 19, 38–40]. Durability is not simply a matter of longevity; it is closely tied to a coating's ability to endure while fulfilling its intended purpose, particularly under near-operational conditions that address real-world applications. This includes resisting physical, chemical, and environmental stressors that, in combination, can lead to gradual degradation of performance. As with other durable materials, coatings are expected to perform under a wide range of weathering conditions defined by the application environment, usage demands, and expected service life [19].

Durability of coatings refers to resistance to a variety of criteria, which are roughly classed as environmental, mechanical, and chemical elements [10, 19, 38, 41]. Environmental stressors, such as variable temperatures and UV radiation can cause thermal degradation and oxidation, gradually degrading the coating structure. In contrast, chemical factors such as hydrolysis and corrosion can cause molecular breakdown, reducing the coating's protective qualities. Mechanical forces such as erosion, cracking, and abrasion all contribute to degradation, finally resulting in adhesion loss from the substrate or loss of important qualities such as hydrophobicity. These breakdown mechanisms vary in severity, but together they endanger the reliability and lifetime of coatings, especially in industrial applications where high-performance coatings are required [19].

Despite the significant amount of novel icephobic materials in the literature, the majority of these research works tailor the environmental durability of an icephobic coating to only reflect its primary objectives; delaying ice, reducing ice accretion and adhesion, and supercooled water repellence. Limited studies tried to incorporate more tests to mimic real-world environmental conditions, but without using standardized methods or benchmarked locations according to industrial needs. Hence, such evaluations do not provide reliable and consistent information about the longevity of the icephobic coatings, nor, most importantly, repeatable experimental conditions.

In near-operational conditions, the functional lifespan of a coating is influenced by both its intrinsic properties and external stresses, including exposure frequency and any maintenance interventions. Durability requirements vary significantly across applications, making standardized testing essential to simulate real-world conditions and predict the ability of the coating to retain its functional qualities over time [19, 38]. For critical applications, accelerated degradation testing can reveal potential failure modes, offering insights into expected performance.

This work applies targeted, standardized tests to assess coatings capable of withstanding near-operational conditions, providing a robust basis for evaluating durability. We examine these testing methodologies, exploring factors influencing coating degradation and methods to ensure that coatings meet engineering requirements.

10.2.1 Superhydrophobic surfaces

Driven by nature, superhydrophobic surfaces exhibit unusual wetting properties useful in water management, ice repellence, and self-cleaning [40, 42]. Nevertheless, unlike natural surfaces with self-healing characteristics, artificial surfaces are frequently constrained by durability problems, reducing their applicability for industrial application. Superhydrophobic surfaces are based on complex micro- and nanoscale features to achieve distinct water interactions, but this design is highly sensitive to environmental and mechanical degradation. Upon damage on their nano-/micro-features they may lose repellence, transitioning from the Cassie–Baxter to the Wenzel state, leading to reduced contact angles and diminished functionality [40, 43]. UV exposure, temperature fluctuations, and contaminants can accelerate this degradation, especially for applications that will directly face weather conditions [44, 45].

In the spirit of scaling up superhydrophobic coatings, different approaches based on wet chemistry and dry etching have been proposed, followed by further engineering the surfaces. Despite numerous techniques presented to date, commercial products have not been developed due to durability issues in real-world applications and production upscaling concerns regarding the use of environmentally harmful chemicals or processes (as per REACH, RoHS list, and other regulations) [32]. In the last decade, many researchers [46–50] explored the scalability of laser techniques, as laser systems offer a one-step fabrication process. However, the substrate and the combination of superhydrophobic patterns remain critical factors. In such cases, durability remains a challenge as many of these superhydrophobic coatings have never been tested under standardized tests including exposure to severe outdoor weather conditions near operational conditions.

10.2.2 Icephobic surfaces

Developing long-lasting icephobic coatings requires careful consideration of both chemical stability and mechanical robustness in harsh environmental conditions. Durability must be prioritized as a critical design criterion, ensuring coatings can maintain their functionality over extended periods. Smooth hydrophobic coatings benefit from standardized testing techniques, but superhydrophobic and others such as the liquid-infused coatings have distinct issues due to their reliance on fine micro- and nanoscale features or the encapsulated liquid, respectively [10]. The paradox is that researchers are debating whether current common standards for evaluating these complex surfaces should be applied or not. Consequently, many studies employ customized testing techniques, which often hinder cross-comparison and broader applicability. Furthermore, research tends to emphasize metrics such as ice adhesion strength and icing delay [51] while frequently overlooking long-term durability considerations.

Since icephobic coatings are constantly exposed to environmental stressors, they must meet durability standards comparable to those of other coatings. In addition to withstanding mechanical ice removal procedures and natural ice shedding [1], effective solutions must resist hydrolysis, UV degradation, corrosion, and physical damage [10, 11, 19, 38]. Reliable long-term performance depends on addressing these aspects, especially in real-world scenarios when several stressors are present.

According to studies [52–56], superhydrophobic surfaces do not by itself ensure better icephobicity in extended or harsh settings, even when they are initially excellent at repelling water. The micro- and nanoscale roughness necessary to preserve superhydrophobicity deteriorates with time due to abrasion and repeated icing/de-icing cycles, increasing ice adhesion and decreasing anti-icing efficacy. Research by Kulinich *et al* [52, 57] and others [53–56, 58] highlights the difficulty of preserving superhydrophobic qualities over long periods of time and demonstrates how weathering accelerates this loss. However, since they lack fragile surface features, smooth hydrophobic coatings frequently show superior durability under similar conditions.

Despite their potential, rough superhydrophobic surfaces are particularly susceptible to mechanical wear. Repeated cycles of ice formation and removal can damage the intricate micro- and nanoscale textures that are essential to their functionality, resulting in reduced water repellence and increased ice adhesion over time [52, 53, 55]. Techniques such as laser texturing have been explored to improve the robustness of these surfaces and create structures that better withstand prolonged abrasion wear. For example, laser textured stainless steel surfaces have shown significantly improved resistance to sand abrasion and freeze–thaw cycles [1], demonstrating a way to improve the longevity of superhydrophobic coatings.

In contrast, smooth hydrophobic surfaces lack the vulnerable texturing of their rough counterparts, making them less susceptible to mechanical degradation. While they may accumulate ice more quickly, they frequently have lower ice adhesion than untreated metals, facilitating de-icing processes [10]. These coatings' endurance is mostly determined by the chemical stability of their hydrophobic layers, which must withstand repeated exposure to corrosive conditions and mechanical de-icing techniques [59–61]. Advances in chemically stable hydrophobic materials may extend the operating lifetime of these surfaces, making them viable candidates for applications requiring mechanical simplicity and durability.

The development of long-lasting icephobic coatings remains a significant issue, particularly when it comes to ensuring consistent performance under a variety of climatic circumstances. Several icephobic surfaces have distinct advantages and limitations in anti-icing applications, underlining the importance of a consistent approach to evaluating their durability. Hence, standardized test procedures are required to enable direct comparisons between different surface types and to guarantee that durability measures correspond to real-world operating needs. The current study employs such methodologies by applying accelerated laboratory testing and natural weathering assessments to smooth, superhydrophobic, and discontinuity-enhanced coatings. By establishing the longevity of these surfaces under various environmental conditions, this study aims to provide uniform benchmarks for future research and promote the development of more reliable icephobic solutions.

10.2.3 Standardized testing methods

Standardized testing techniques are essential for objectively evaluating coating resilience, as they provide a consistent foundation for comparing durability across materials and applications [19, 21, 62]. While coatings have long had well-established testing techniques, potential icephobic coatings lack globally accepted standards [1, 10, 12, 13]. This gap frequently leads to the use of unique testing procedures, which complicates cross-study comparisons and may obscure insights about long-term performance under near-operational situations.

To successfully assess durability, comprehensive, standardized testing must be conducted that takes into consideration a wide range of environmental conditions. These testing methods are often divided into two categories: artificially accelerated tests in controlled laboratory conditions and natural weathering experiments in outdoor locations. Accelerated tests imitate harsh conditions such as temperature changes, moisture cycles, mechanical abrasion, and pressure variations, reducing years of potential environmental stress to shorter timeframes [19, 21, 63]. These tests commonly use ASTM, ISO, and MIL-STD protocols, which provide uniformity and reliability during the first stages of coating development [19, 62].

However, natural weathering tests subject coatings to near-operational conditions, capturing intricate interactions that are sometimes difficult to recreate in the lab. Because of their harsh climates, well-known testing areas such as Florida and Arizona are popular. Florida's subtropical environment, which includes high UV radiation, persistent humidity, air corrosion, and frequent rain, puts coatings to the test in terms of UV deterioration and moisture infiltration. In contrast, Arizona's arid desert environments, characterized by high Sun radiation and significant temperature changes, reveal failure mechanisms associated with thermal ageing and desiccation. According to research [19, 21], real-world exposure tests supplement laboratory assessments by revealing durability issues that may not be apparent in accelerated testing alone.

The inherent diversity of real-world weather conditions highlights the significance of combining artificial and natural standardized tests. Standardized laboratory tests provide reliable and repeatable data on specific performance aspects, whereas standardized natural tests provide a more comprehensive view of coating behaviour under dynamic and compounded environmental conditions such as temperature cycles, corrosion, and moisture exposure. This research ensures that potential icephobic coatings are evaluated with the same reliability and comparability as other industrial coatings required by applying the same rigorous standards used for other industrial coatings, such as those in the automotive, marine, and aviation sectors.

Establishing standard testing procedures boosts confidence in coating long-term functionality, allows for accurate cross-study comparisons, and promotes their widespread use in industrial applications. This chapter emphasizes the need for standardized durability testing, which has previously been applied to a variety of industrial coatings, in evaluating and improving the performance of superhydrophobic and icephobic surfaces. By adhering to established guidelines, this work helps to ensure the reliable integration of icephobic coatings into challenging near-operational conditions.

10.2.3.1 Natural and artificial environmental tests

Several artificial and natural environmental tests can be performed in environmental test chambers and lab-sized custom-builds, as well as in Miami, FL (USA). This is a summary of the selected artificial and natural weathering standardized (and not) tests applied to several test samples coated with potential icephobic coatings in order to simulate near-operational conditions.

Artificial accelerating tests:

- *Salt-fog (S-F) corrosion test*: Samples were exposed to a corrosion chamber for 500 h as per ASTM B117 standard. Wettability was measured both before and after testing, evaluating hydrophobic performance. For contrast wettability coatings, the water transport capacity was assessed post-exposure to determine their functional degradation [32].
- *Thermal cycling*: Testing followed EN 60068-2-14:2023 standards, consisting of 32 cycles between −40 °C and +85 °C, with 2 h isothermal holds at both temperature extremes. Measurements of contact angle and roll-off/slide angle were conducted before and after the cycles to assess the coatings' resistance to thermal stress.
- *Temperature–humidity (referred to as humidity) cycling*: BS EN IEC 60068-2-38:2021 standards were employed to simulate humidity and temperature variations. Samples underwent ten cycles, with phases of steady humidity (93% ± 5%) at 25 °C and cold sub-cycles at −10 °C. This test simulated operational environments with fluctuating conditions.
- *Altitude–temperature–humidity (referred to as altitude) test*: As per MIL-STD-810 Method 520.3 (Procedure III), coatings were exposed to simulated aerospace conditions. Mission profiles included rapid altitude changes and temperature shifts between 43 °C (hot atmosphere) at 0 km and −65 °C (cold atmosphere) at 10 km, alongside varying humidity conditions (<10% to 100%). This test aimed to evaluate coating resilience under aerospace-relevant stressors.
- *Water-jet impact (WJ) test* [32]: A custom-built water-jet system simulated direct water impact. Using a nozzle diameter of 3 mm, water exited at 12 m s^{-1}. Samples were subjected to this high-impact water flow for up to 24 h. Tilted at 45°, coatings were evaluated for erosion and hydrophobic performance post-testing.

- *Ice adhesion strength test*: A critical metric for evaluating anti-icing performance was measured using a horizontal shear test [64]. Ice is generated within a cylindrical mould of 8 and/or 3 mm diameter by adding 7 drops (0.35 ml) of deionized water. The 3 mm mould was used to ensure that it fits on the hard anodized microtextured track. The freezing process takes 15 min with the surface temperature set at −15 °C, and then the water is poured into the mould to freeze. This is an important step since water should not have the time to penetrate the air cushions formed due to the surface texture of the superhydrophobic coating. Measurements are taken immediately after freezing, with the dynamometer moving forward at a speed of 0.01 mm s^{-1} to produce a load (N) versus time (s) graph. Measurements were conducted both before and after weathering tests to assess changes in adhesion strength due to environmental degradation. Microscopic imaging complemented these evaluations, providing visual evidence of coating integrity and microstructural changes.

Natural weathering tests:

- *Outdoor exposure in Miami, Florida, USA*: Year-long exposure followed ASTM G7 standards, targeting Florida's unique combination of high UV radiation, persistent humidity, and frequent rainfall. These conditions provided an internationally recognized benchmark for accelerated weathering. Samples were mounted on an aluminium frame at tilt angles between 5° and 45°. The lower angles minimized water accumulation, while varying tilts maximized sunlight exposure. Monthly assessments of wettability and surface roughness captured degradation trends over time.

10.3 Performance evaluation under standardized tests

10.3.1 Materials and methods

The most well known lightweight engineering substrate is aluminium, which has been utilized in this chapter. The commercially available aluminium alloy substrate used (6026-T9), is non-toxic since it does not contain lead (Pb) and features excellent corrosion resistance [65, 66] and can be easily anodized as well machined. Furthermore, a type III hard anodized coating, which produces a dense oxide layer of 40–50 μm thickness [32] was applied. The anodization process creates nanopores (20–40 nm diameter) that were sealed to enhance corrosion resistance [67, 68]. This choice of substrate aligns with the industrial requirements for lightweight materials that combine mechanical durability with excellent environmental resilience.

Three distinct fluorinated coatings were applied, each selected for their unique properties and application methods. A superhydrophobic nanoparticle emulsion polymer was prepared via cold spraying, followed by curing at room temperature (23 °C, relative humidity ~50%). This coating achieved a water contact angle (WCA) of ~150° and a roll-off angle (RoA) of ~3°, indicative of its superior hydrophobic properties. The second coating, a hydrophobic perfluoroalkoxy polymer (PFA), was applied using an electrostatic spray technique and baked at

400 °C. This process yielded a smooth, uniform surface with a WCA of ~110° and an RoA of ~15°. Furthermore, the PFA coating was selectively laser-ablated and micro-structured by a femtosecond laser system [32]. These samples were supplied by FT Technologies UK Ltd, and further technical specifications are confidential. Lastly, a vertically discontinuity-enhanced coating, a gradient hydrophobic polymer, was synthesized via iCVD, which facilitated the precise formation of three sections. The bottom section consisted of 50 nm of poly(2,4,6,8-tetraethenyl-2,4,6,8-tetramethylcyclotetrasiloxane) (pV4D4), followed by a middle copolymer section of 150 nm (p(V4D4-co-PFDA)), and capped by a 300 nm section of perfluorodecyl acrylate (PFDA) [14]. This configuration provided tailored hydrophobic properties with WCAs od approximately 135° and an RoA of ~20°, as well as excellent adhesion to the substrate according to previous studies.

The coatings were evaluated for their mechanical and chemical robustness using a combination of advanced characterization techniques. Wettability was assessed through static WCA and RoA measurements using the sessile drop method, with droplets of up to 30 μl deposited in controlled atmospheric conditions. Surface roughness, a key parameter in evaluating coating degradation, was measured using a Keyence VHX-7000 microscope. Measurements were taken in both parallel and perpendicular orientations to the surface, with a total of 12 scans for each sample. The scanned line length was 17.5 mm, and an evaluation area of 12.5 mm was used. Filters were set to L-filter (λ_c) of 2.5 mm and S-filter (λ_s) of 8 μm, ensuring precision in capturing the surface roughness profiles before and after weathering tests.

Chemical changes were examined using attenuated total reflectance Fourier transform infrared spectroscopy (ATR-FTIR), which provided insights into the degradation of coating materials over a spectral range of 500–4000 cm^{-1}.

Establishing a well-defined process for assessing the durability of icephobic coatings is critical, especially given this chapter's emphasis on rough surfaces. While material topography has a considerable influence on design criteria [69–73], this study focuses on durability assessments using surface roughness measures after natural weathering testing (figure 10.1). These tests provide important information about the long-term durability of the coatings and the subsequent impact on shear ice adhesion performance [7, 73]. The structured evaluation methodology emphasizes the need to include mechanical and chemical stability criteria in the testing process. By employing wettability as an initial assessment of coating performance, the study creates a baseline for testing durability under standardized environmental circumstances. Following natural weathering, advanced characterization methods are used, such as roughness profiling and ATR-FTIR analysis, to ensure a comprehensive understanding of coating stability.

The final phase of the evaluation process involves shear ice adhesion testing, which directly correlates the durability of the coatings with their icephobic capabilities. This sequential approach underscores how wettability and surface roughness are not merely design features but integral to achieving long-term reliability in engineering applications. Furthermore, these properties enhance self-cleaning capabilities and durability, even under extreme environmental conditions, while maintaining effective icephobicity.

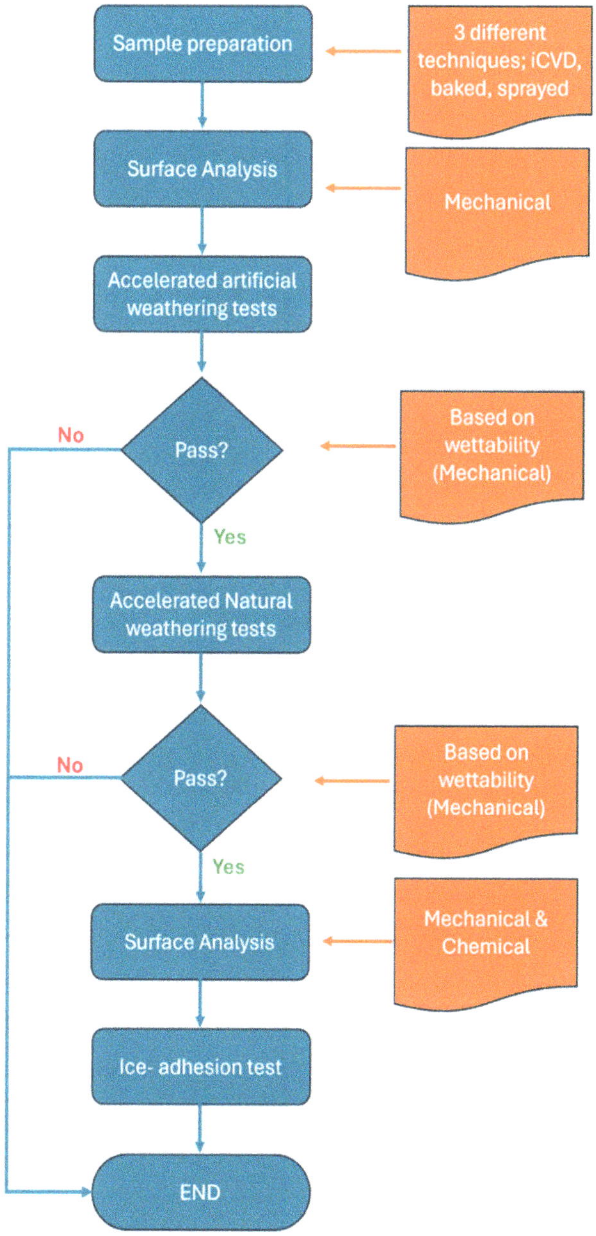

Figure 10.1. Flow chart of the process for evaluating the durability of icephobic coatings. First, the samples are prepared, and then the surface characterization takes place. This step considers only the mechanical stability of the coatings by means of wettability since this is the primary objective-function of the coatings. Then the evaluation of three coatings under standardized environmental tests is carried out. At the end of this process the wettability of the coatings is measured again. If the coatings showed a good mechanical stability in terms of wettability, then they proceed to the natural weathering tests. Next, the surface characterization of the mechanical stability this time include wettability and surface roughness measurements, as well as chemical stability measurements by utilizing an ATR-FTIR. Finally, the samples that have been exposed under natural weather conditions will go through shear ice adhesion tests and a final examination. This time virtual inspection under a microscope will take place. (Adapted from [84].)

10.3.2 Discussion and insights

10.3.2.1 Artificial environmental tests

To assess durability in the first place, a convenient way is by measuring the water contact angle and roll-off angle of the coatings, before and after the environmental tests.

Figure 10.2 presents the wettability results of both pre- and post-assessment for each environmental test conducted. For the superhydrophobic coatings, the water contact angle consistently remains above 130°, signifying retention of substantial hydrophobicity. However, the roll-off angle, a critical factor influencing the self-cleaning capabilities of these materials, increases after each test when compared to the reference sample. While the tests were performed on different samples, the observed increase in RoA highlights the impact of environmental stresses on the material performance. This metric is crucial as exceeding an RoA of 10° transitions the classification of the coatings from superhydrophobic to merely hydrophobic.

In contrast, the novel Gradient 300 coating, characterized by a continuous, smooth transition in properties across the thickness, exhibits significant deterioration in wettability properties, indicative of compromised durability. This is the first time that this specific coating is being evaluated by standardized tests, and especially the B117 salt-fog test. Post-testing, the water contact angle consistently declines by approximately 10°, and the roll-off/slide angle, which is pertinent to assess as a slide angle in this context, exhibits minor increases (except after the salt-fog and the water-jet tests) due to the bottom organosilicon layer which accommodates the thermal energy, following altitude, thermal, and humidity tests. Organosilicon layers, particularly those derived from monomers such as V4D4, are known for their high thermal stability and ability to maintain structural integrity under thermal stress due to their crosslinked siloxane network, as highlighted in previous studies [34, 74, 75]. Notably, after exposure to a corrosive environment and high-impact water-jet, there is a marked escalation in roll-off/slide angles, suggesting pronounced erosion of surface properties.

The hydrophobic PFA, belonging in the horizontal discontinuity—contrast wettability—surface, demonstrates the most robust and stable performance across these accelerated tests, maintaining consistent wettability characteristics throughout

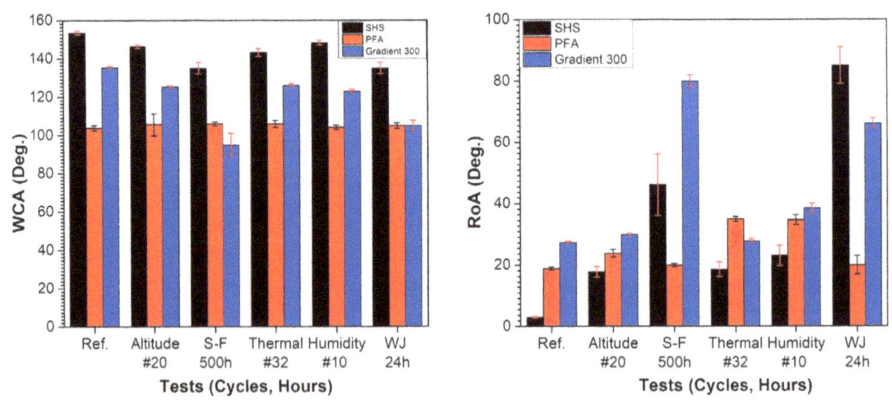

Figure 10.2. (Left) Water contact angle and (right) roll-off/slide angle measurements before (reference) and after several artificial environmental tests. (Adapted from [84].)

the various challenging conditions imposed. The stability of PFA underscores its potential suitability for applications requiring reliable performance under diverse environmental stressors, especially when water transportation throughout an open surface is vital.

10.3.2.2 Natural weathering tests

Following to the flow chart (figure 10.1), only the coatings that were robust in maintaining their wettability during the artificial environmental tests were proceeded towards the natural weathering tests in a verified aggressive location such as Miami, Florida, USA [19, 32, 76], i.e. in this case only the contrast wettable surfaces and the superhydrophobic coatings.

Therefore, by exposing the coatings to such a rigorous environment for an extended duration (12 months) and subsequently assessing their capacity to mechanically withstand ice removal, valuable insights can be gained into the coatings' ability to withstand natural degradation and the impact of such degradation on their anti-icing properties.

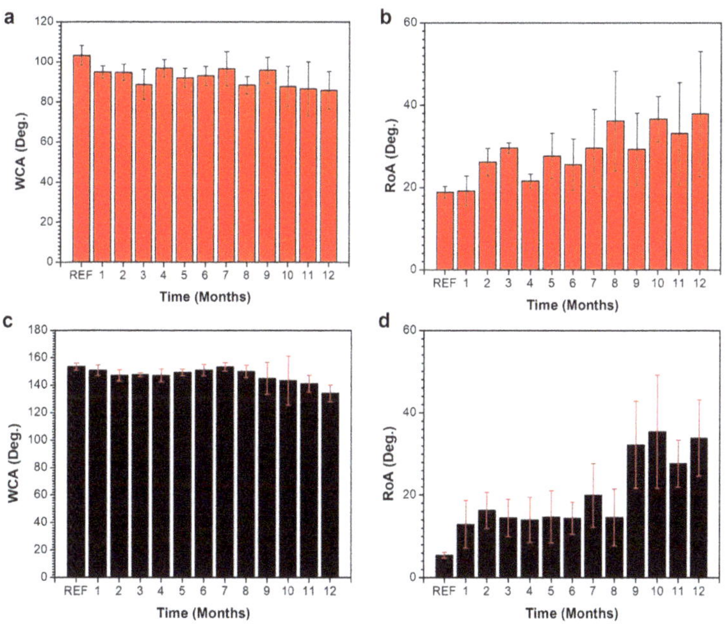

Figure 10.3. Mechanical stability evaluation for contrast wettability surfaces and superhydrophobic coatings. The evaluation of the water contact angle and roll-off/slide angles for both coatings provides insight into their mechanical stability. The WCA measures the surface ability to repel water, while the RoA assesses the ease with which water droplets roll-off the surface. Together, these parameters reflect the integrity of the surface microstructure, as any mechanical degradation, such as abrasion or surface wear, leads to a decrease in WCA and an increase in RoA. The contrast wettable surface ((a), (b)) demonstrates less mechanical degradation over time, maintaining higher WCA and lower RoA compared to the superhydrophobic coating ((c), (d)). This indicates that the contrast wettability surface retains its functional properties better under environmental stressors, showcasing superior mechanical stability. ((a),(b) Adapted from [32]. CC-BY 4.0. (c), (d) Adapted from [84].)

Both types of coatings exhibit a decline of the wettability over time (figure 10.3). The degradation of the coatings implies that two factors should be investigated further, these are the chemical and the mechanical stability. The former is shown in figure (figures 10.4 and 10.5), which compares the freshly prepared samples (untreated) against the first month (M1) and the last month (M12) exposed in Florida, respectively. Both fluorinated polymers show two sharp, intense and characteristic absorption bands at 1141 and 1195 cm^{-1} (PFA), and at 1145 and 1201 cm^{-1} (superhydrophobic), these can be attributed to the symmetric and antisymmetric stretching vibrations of CF_2 bonds. Interestingly, the superhydrophobic coating shows a slight superficial hydroxylation, highlighted by the presence of a wide broad band between 3200 and 3500 cm^{-1} with a maximum peak value at 3322 cm^{-1} (ascribable to O–H stretching), two more defined absorption bands at 2925 and 2854 cm^{-1} (symmetric and antisymmetric stretching), as well as two more defined absorption bands at 1731 cm^{-1} (internal double bonds –CF=CF–) and 1594 cm^{-1} (methylene

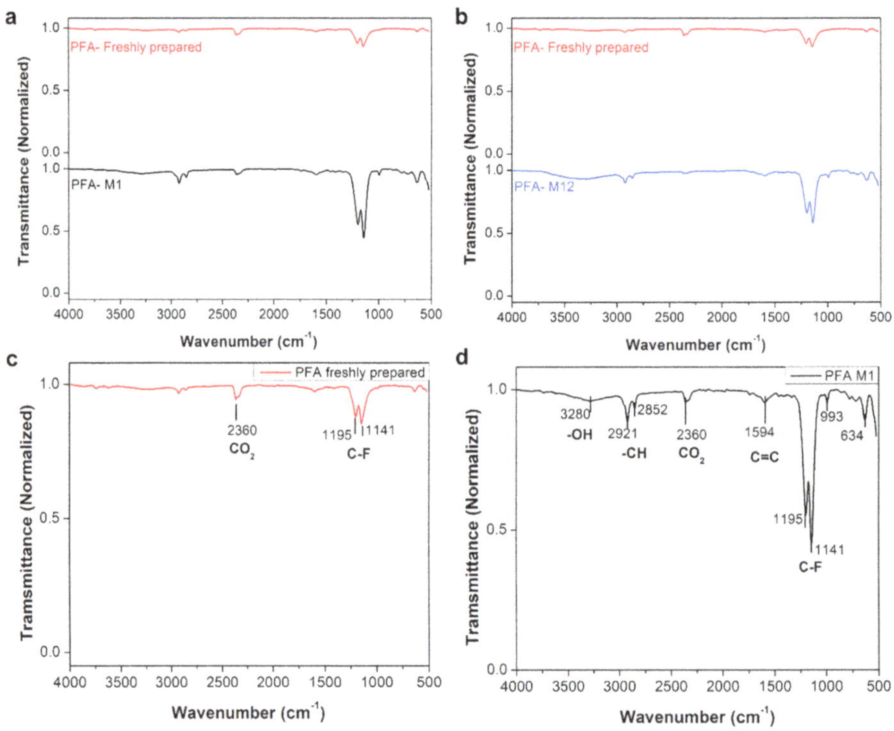

Figure 10.4. Normalized ATR-FTIR spectra of PFA. (a), (b) Spectra of a freshly prepared PFA coating compared to those exposed to natural weathering conditions after one month (M1) and twelve months (M12). (c) Spectra of a freshly prepared PFA coating, highlighting the characteristic peaks of fluorinated components (C–F), and (d) spectra of PFA coatings after exposure to Florida's natural weathering conditions for one month (M1) and twelve months (M12), revealing chemical deterioration over time. Key peaks such as –OH (3280 cm^{-1}), –CH (2921, 2852 cm^{-1}), CO_2 (2360 cm^{-1}), and C=C (1594 cm^{-1}) emerged, indicating oxidative and structural changes. These results demonstrate the progressive environmental degradation of the PFA coatings. (Adapted from [84].)

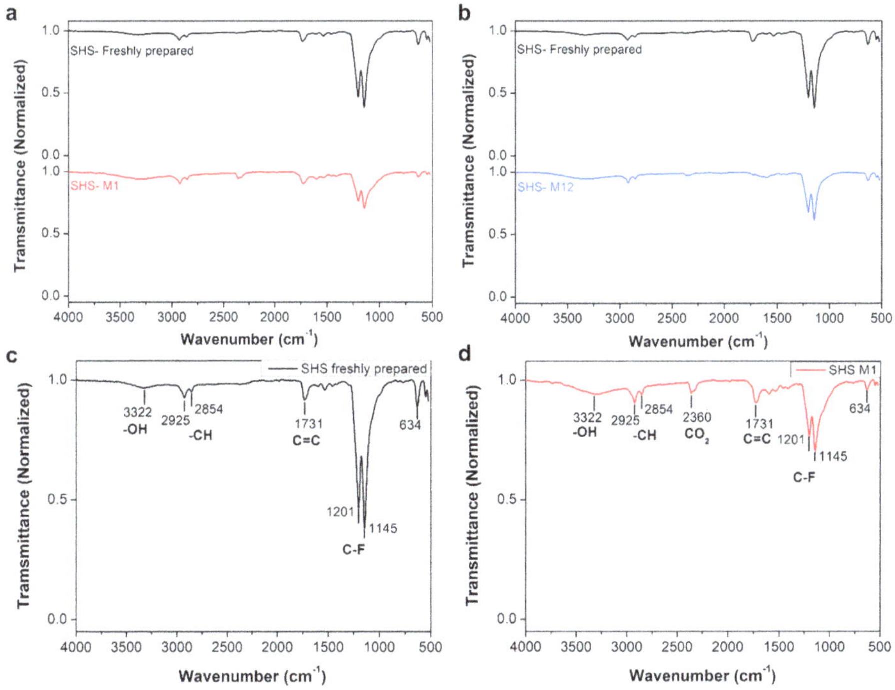

Figure 10.5. Normalized ATR-FTIR spectra of superhydrophobic surfaces. (a), (b) Spectra of freshly prepared superhydrophobic coatings compared to those exposed to natural weathering in Florida for one month (M1) and twelve months (M12), illustrating progressive chemical deterioration. (c) Spectra of freshly prepared superhydrophobic coatings, highlighting characteristic peaks of fluorinated components (C–F). (d) Spectra of superhydrophobic coatings after exposure to natural weathering for M1, revealing the emergence of key peaks such as –OH (3322 cm^{-1}), –CH (2925, 2854 cm^{-1}), CO_2 (2360 cm^{-1}), and C=C (1731 cm^{-1}), indicating oxidative and structural changes. These results emphasize the environmental degradation of the coatings over time. (Adapted from [84].)

CH_2 scissoring) [77–80]. Similar peaks are evident for both fluorinated-based coatings after the first month. The contrast wettable surfaces show also a weak peak in 963 cm^{-1} (=CH_2 out-of-plane blending, terminal methylene) [78] and both coatings reveal another weak peak at 634 cm^{-1} (CF_2 rocking mode) [80].

Furthermore, the mechanical stability is also influenced by the substrate and surface roughness. The surface roughness was assessed in accordance with the technical standards STN EN ISO 21920, herein, a Keyence VHX-7000 microscope was employed to measure both parallel and perpendicular to the surface area at least 12 times each [81].

In both cases the surface roughness changes during the long-term exposure under natural weather conditions. The superhydrophobic coating becomes smoother during the first month $R_a = 3.65 \pm 0.5$ μm and ion the last month the $R_a = 2.7 \pm 0.4$ μm, mainly due to the nanoparticles' smoothness and delamination. Controversially, the surface roughness of the PFA increases in the first month

$R_a = 2.6 \pm 0.3$ μm up to the last month where $R_a = 4.2 \pm 0.4$ μm, due to pinholes, cracks and colour loss. However, the contrast wettability surface distinctly maintains its hydrophobic properties throughout the testing period. In contrast, the superhydrophobic coating demonstrates a significant reduction in its extreme hydrophobicity, beginning as early as the first month of exposure (figure 10.3), indicated by the increased RoA. This differential performance underscores the varying resilience of coating formulations under prolonged exposure to harsh environmental conditions and highlights the importance of tailored material selection for near-operational environments.

The deposition method of the coating plays also a vital role. The hard anodized aluminium is relatively rough ($R_a = 0.8 \pm 0.1$ μm), this also depends on the anodization time, however, the hard anodization followed in this work was type III [82]. The superhydrophobic coating shows a higher surface roughness ($R_a = 4.2 \pm 0.5$ μm), due to both the emulsion nanoparticles and its sprayed application method. However, the PFA powder coating surface roughness is smoother, making the polymer more durable [83]. Here, the baked process which was followed leads to a more uniform and smoother surface ($R_a = 0.21 \pm 0.03$ μm).

10.3.2.3 Ice-adhesion test

Ice adhesion tests are critical for evaluating the effectiveness of icephobic materials, providing insights into the mechanisms of ice prevention and removal on coated surfaces. Various methodologies exist for conducting these tests, among which the shear test is particularly noteworthy. Although it is an unstandardized test, it enables precise measurement of the force required to detach the ice from a surface, thereby quantitatively assessing the anti-icing properties of the coating. In this chapter, the shear ice adhesion test has been utilized to assess the durability of all coatings both prior to and following the environmental durability tests. The ice adhesion strength of each coating is quantitatively reported, providing a direct measure of the material's ability to resist ice formation and adhere to the surface under test conditions.

Furthermore, microscopic imaging has been employed as an additional evaluative tool to identify any microstructural changes or damage to the coatings that may occur as a result of ice adhesion and subsequent removal. These images offer invaluable visual evidence of the physical state of the coatings after undergoing the ice adhesion test, highlighting potential durability issues not immediately apparent from quantitative measurements alone. This comprehensive approach ensures a thorough understanding of each coating's performance under icy conditions, aiding in the optimization of formulations for enhanced anti-icing efficacy and durability in practical applications.

By employing this unstandardized method, prior research works could systematically compare the initial and post-durability ice adhesion strengths of the coatings, thereby quantifying the effects of environmental ageing processes on their performance. This analysis not only highlights the resilience of the coatings but also aids in identifying formulations that maintain their icephobic properties under prolonged exposure to adverse conditions.

10.3.2.3.1 Ice-adhesion test after artificial tests

The method of coating application and the final surface roughness are critical factors influencing the icephobic properties. For instance, the findings in figure 10.6 indicate that the process of spraying nanoparticles to achieve superhydrophobicity inadvertently creates surface textures that enhance mechanical interlocking with ice. This phenomenon diminishes the ability of the material to shed ice effectively, contrasting sharply with the performance of the hydrophobic coatings. The high-temperature baked PFA, despite its roughness, does not achieve the smoothness of the Gradient 300, which is prepared using initiated chemical vapour deposition (iCVD). The iCVD method facilitates the deposition of ultrathin, uniform coatings, enhancing their icephobicity by minimizing physical imperfections that could promote ice adhesion.

Thickness is another pivotal factor affecting the durability of coatings. Figure 10.6 indicates that in tests that do not physically impact the coatings, such as controlled environmental ageing, both PFA and Gradient 300 exhibit comparable durability post-ice adhesion testing. However, the scenario alters significantly under conditions that involve aggressive physical impacts. The ultrathin Gradient 300 does not withstand high-stress tests such as water-jet and salt-fog spray as effectively as the thicker PFA. These tests are critical for assessing the robustness of icephobic coatings, especially for evaluating their ability to maintain integrity and functionality after exposure to harsh environmental conditions. However, the transparent

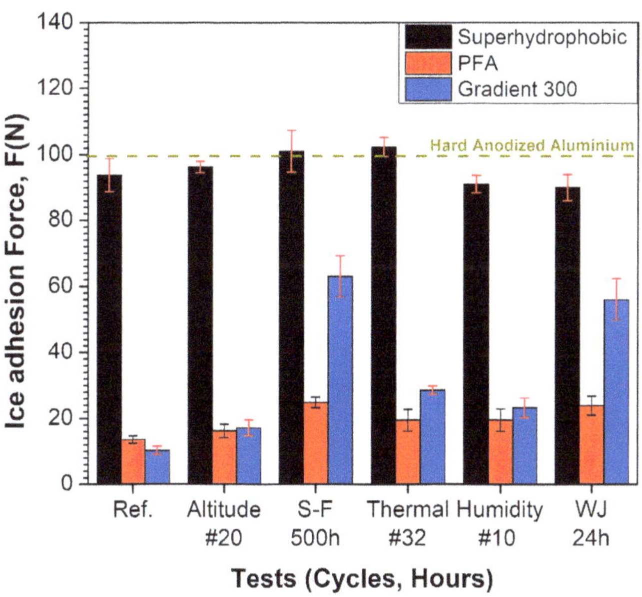

Figure 10.6. The graph shows the ice adhesion strength after the durability test for the three coatings. The substrate, hard anodized aluminium, was measured and placed as an upper limit (dark yellow dashed line). Remarkably the force needed to shed the ice from superhydrophobic coating after exposing to the artificial accelerated tests was the same as the bare substrate. Both the Gradient 300 and PFA were steady for the most of the tests, and both revealed an increase, slight for the PFA, of the input force after the salt-fog and water-jet tests. (Adapted from [84].)

nature of the Gradient 300 makes it difficult to detect possible delamination or degradation during these tests.

In contrast, the PFA coating displays remarkable stability in ice removal over a wide range of durability tests. Its thickness not only adds to its mechanical strength, but it also assures consistent ice shedding performance, which is essential for effective long-term use in icing-prone areas. These comparative studies highlight the relationship between material properties and anti-icing performance, emphasizing the importance of selecting appropriate materials and processes tailored to near-operational needs in order to improve the performance of icephobic coatings.

10.3.2.3.2 Ice-adhesion test after natural weathering

After the natural weathering test which was completed in Florida, the two coatings underwent ice adhesion testing to shed light on the influence of harsh environmental conditions on their anti-icing properties.

Although both the superhydrophobic and PFA coatings are derived from fluorinated polymers, they exhibit unexpectedly high ice adhesion strengths, even when freshly applied (figure 10.7). The enhanced ice adhesion observed in the superhydrophobic coating can be attributed to the water penetration among the nanoparticles, which effectively reduces the air gaps crucial for maintaining its superhydrophobic properties. The experimental configuration involves placing the coating on a Peltier stage set at −15 °C, with relative humidity maintained approximately 0% by continuous nitrogen infusion into the test chamber. When water at room temperature is introduced into the silicon mould and comes into contact with the cold coating surface, it transitions into a supercooled liquid state

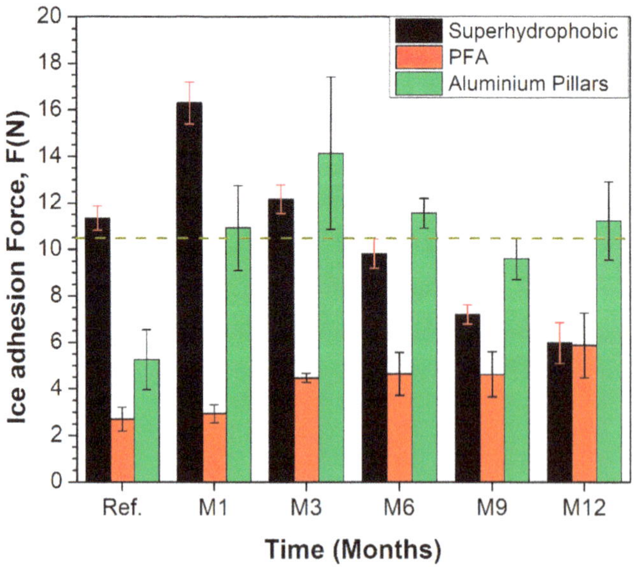

Figure 10.7. The graph shows the ice adhesion strength after the durability test for the superhydrophobic, hydrophobic (PFA) and micro-structured hard anodized aluminium-based pillars. After 12 months both SHS and hydrophobic coatings need approximately the same force to remove the ice. (Adapted from [84].)

without freezing entirely. This scenario likely facilitates the penetration of supercooled water into the nanoparticle-induced air gaps, where it acts as an interlocking mechanism [15, 57, 71]. Remarkably, after a 12 month exposure to the aggressive weather conditions in Florida, the ice adhesion strength of the superhydrophobic coating has reduced to nearly half of its initial value. In contrast, the PFA coating shows a three-fold increase in ice adhesion strength. This divergence likely stems from changes in the surface topography induced by long-term environmental exposure. The superhydrophobic coating, applied via a cold spray method that creates a rough nanoparticle-based surface, contrasts with the high-temperature cured PFA, which forms a uniform layer.

During ice removal tests, there is a noticeable detachment of the superhydrophobic coating material along with the ice figure 10.8(c), highlighting a potential

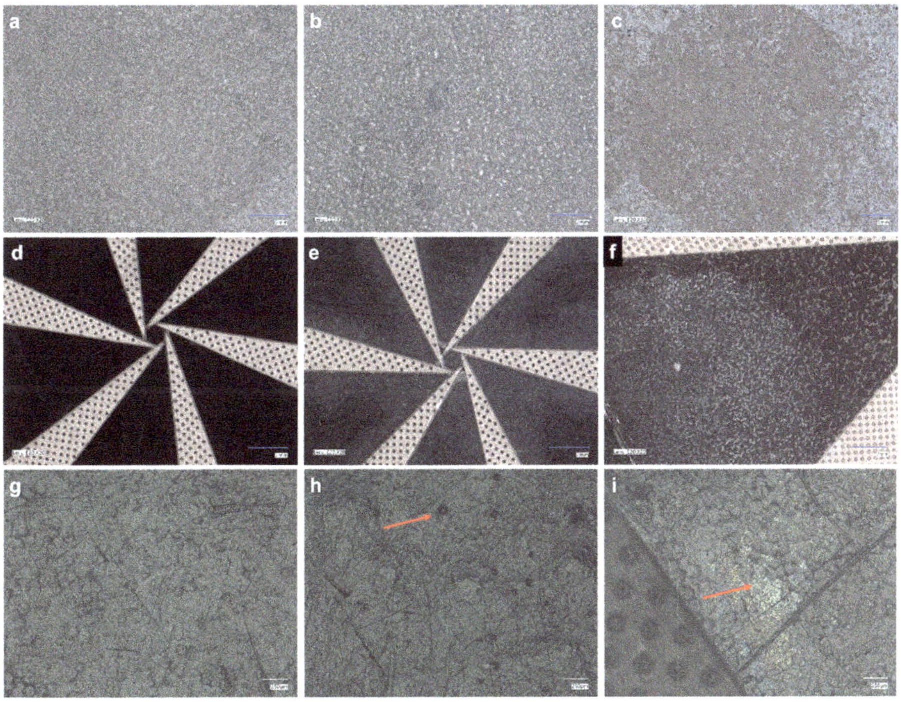

Figure 10.8. Images of the two coatings under environmental exposure over time. (a)–(c) The superhydrophobic coating at three stages: fresh, after a month, and after 12 months of exposure in Florida. After a month, whitish marks appear, potentially due to oxidation, as confirmed by ATR-FTIR. By 12 months, these marks vanish, but changes in gloss and surface degradation outside the ice adhesion circle indicate environmental wear, such as corrosion, erosion, and UV damage. The circular mark created by the ice detachment mould further supports this observation. (d)–(f) The contrast wettability coating with PFA islands: fresh, after a month, and after 12 months of exposure in Florida. Over time, carbonized material on the pillars shows partial degradation, correlating with increasing ice adhesion force. By 12 months, a single PFA island exhibits delamination of the topcoat, though this is an isolated incident among the 12 samples tested. Wettability and ice adhesion measurements were avoided on the degraded island. The scale bar is 2 mm for all images. (g)–(i) A closer view (scale bar 250 μm) of the PFA coating, in which degradation effects are visible. (Adapted from [84].)

compromise in coating integrity under ice adhesion forces. This phenomenon further emphasizes the need for a detailed examination of the mechanical durability of icephobic surfaces under practical use conditions. Meanwhile, the PFA coating, though smoother, shows signs of degradation from environmental exposure, particularly UV light, as evidenced by colour loss and potentially other microstructural damage such as cracks, pinholes and voids, that could explain the increase in the ice adhesion force.

Moreover, we employed a mould with a diameter of 3 mm to facilitate ice removal measurements from the microtextured tracks, which could not accommodate the standard 8 mm mould due to its narrower width. Notably, the microtextured pillars exhibited approximately half the ice adhesion force compared to that of the freshly prepared superhydrophobic coatings. This significant reduction in adhesion strength is elucidated by scanning electron microscope images, which reveal the presence of carbonized perfluoroalkoxy (PFA) residue atop the pillars [32]. This residue effectively inhibits the infiltration of supercooled water between the pillars, enhancing the anti-icing efficiency of the surface. However, after prolonged exposure to Florida's harsh environmental conditions, a notable increase in the ice adhesion force was observed, with the force required to remove ice from the pillars doubling. This change suggests the degradation or removal of the carbonized residue from the pillar surfaces, likely due to environmental wear and exposure. The hard anodized pillars, inherently hydrophilic in nature, are primarily designed to facilitate water transport across the surface rather than enable passive ice removal. Consequently, with the degradation of the carbonized layer, these pillars revert to displaying a higher ice adhesion force, aligning with their intrinsic hydrophilic properties.

This dynamic indicates the crucial role of surface treatments and their integrity in determining the anti-icing performance of engineered surfaces. It underscores the need for robust surface protection strategies that can withstand environmental degradation to maintain their designed functionality, especially in applications where passive ice removal is critical.

10.4 Conclusion remarks

It is important to note that the findings of this study do not aim to define the failure of any particular coating system. Instead, they serve to underscore the importance of evaluating coatings under standardized durability protocols. The results presented herein are intended to enhance the broader conversation surrounding the development of robust icephobic coatings, emphasizing the need for consistent testing methodologies and benchmarks. By doing so, this work contributes to the collective effort of advancing the design and application of durable coatings across diverse industrial contexts.

The focus of this chapter is to conceptualize the term durability in icephobic materials. Here, we made a step towards it by investigated three fluorinated coatings, each prepared using different application methods and thicknesses. Accelerated ageing lab tests were instrumental in identifying the most promising candidates for anti-icing applications. For first time we tested the novel gradient coating under a corrosive environment and other standardized environmental lab-based tests. These tests revealed that while the ultrathin gradient coating did not perform optimally across various environmental conditions, the necessity for an icephobic coating to effectively reduce ice adhesion under near-operational conditions, such as temperature variations and corrosive environments, was clearly demonstrated. Despite similar thicknesses, the superhydrophobic and PFA coatings exhibited significantly different performances in icephobicity and long-term stability. Remarkably, after 12 months of exposure in Florida, both coatings required the same force to remove ice from their surfaces. However, the microtextured pillars, unlike the other coatings, showed no signs of damage or breakage during the ice adhesion tests, highlighting their robustness. Despite similar thicknesses, the superhydrophobic and PFA coatings exhibited significantly different performances in icephobicity and long-term stability.

In industrial applications, it is often necessary to employ both passive (e.g. coatings) and active (e.g. heating systems) strategies to effectively prevent ice accretion. A heating system is mainly used by all industries, and maintaining surfaces above freezing points could benefit from the use of a contrast wettability surface as an anti-icing solution. The PFA coating, in particular, demonstrated exceptional stability throughout all tests, including ice adhesion tests, and proved effective due to its functionalized surface, which includes durable microtextured tracks that facilitate water transportation across the surface. Infrared spectroscopy of both coatings reveals a broad hydroxyl band, indicating water uptake, which is a critical factor to avoid during the coating application process. The technique used for the superhydrophobic coating involves a cold spray method, and its subsequent drying process is not conducted in a controlled environment, leading to initial water absorption. In contrast, the PFA powder, which undergoes a baking process, exhibits minimal water uptake in the same IR spectral region, suggesting more effective moisture exclusion during the application phase. A longitudinal analysis, comparing IR spectra from reference samples with those subjected to 1 and 12 months of natural weathering, demonstrates notable changes in the surface chemistry of both coatings. This evolution in chemical properties indicates an alteration likely due to environmental exposure, underscoring the importance of assessing long-term chemical stability in outdoor applications.

The novel gradient polymer, while durable under de-icing cycles as proven by a previous work, failed to withstand critical tests involving corrosion and direct water-jet impact, which are vital for offshore and onshore engineering applications. This led to a significant decrease in wettability and an increase in the ice adhesion force required to remove ice. The gradient polymer, synthesized via iCVD using an organosilicon and a fluoropolymer, addresses the challenge of thermal stress differentials between the substrate and the coating efficiently. While its thin profile is unsuitable for external applications due to its diminished durability under harsh conditions, and inevitable corrosion, it could be tuned for enclosed applications. This adaptation enhances its utility in encapsulated systems, such as electronic devices, where the effect of corrosion will be minimal due to the protection that offers the design of any device.

This work highlights the complexity of designing effective icephobic coatings and the necessity of rigorous, standardized environmental testing to ensure their long-term functionality and reliability. By continuing to refine the materials and methodologies based on specific application needs and environmental conditions, the development of more resilient and effective anti-icing solutions can be achieved.

Notes

All figures, except figures 10.3(a) and (b), and related content presented in this chapter are adapted from the doctoral dissertation of Theodoros Dimitriadis, 'Study on the efficiency of discontinuity-enhanced coatings under near-operational conditions', completed at Graz University of Technology, Institute of Solid State Physics, 2024 [84]. The author retains the right to reuse this material.

The authors declare no competing financial interest.

TD works for FT Technologies UK Ltd while collaborating with universities on SURFICE Marie Skłodowska-Curie project. TD has no employment, financial, or personal conflict of interest with other organizations.

Acknowledgments

TD acknowledges Luca Stendardo for technical assistance with the ice adhesion set-up and helpful discussions regarding its operation. TD gratefully acknowledges Professor Carlo Antonini (SEFI Lab, Department of Materials Science at the University of Milano-Bicocca) for granting access to laboratories and providing the necessary resources to conduct this research. This project has received funding from the European Union's Horizon 2020 research and innovation program under the Marie Skłodowska-Curie Grant Agreement 956703 (SURFICE Smart Surface Design for Efficient Ice Protection and Control).

References

[1] Huang X *et al* 2019 A survey of icephobic coatings and their potential use in a hybrid coating/active ice protection system for aerospace applications *Prog. Aerosp. Sci.* **105** 74–97

[2] Fortin G 2013 Considerations on the use of hydrophobic, superhydrophobic or icephobic coatings as a part of the aircraft ice protection system *Ice Accretion and Icing Technology* (Piscataway, NJ: IEEE) pp 13–23

[3] Golovin K *et al* 2016 Designing durable icephobic surfaces *Sci. Adv.* **2** e1501496

[4] Kreder M J, Alvarenga J, Kim P and Aizenberg J 2016 Design of anti-icing surfaces: smooth, textured or slippery? *Nat. Rev. Mater.* **1** 15003

[5] Kim P *et al* 2012 Liquid-infused nanostructured surfaces with extreme anti-ice and anti-frost performance *ACS Nano* **6** 6569–77

[6] He Z, Zhuo Y, Zhang Z and He J 2021 Design of icephobic surfaces by lowering ice adhesion strength: a mini review *Coatings* **11** 1343

[7] Golovin K, Boban M, Mabry J M and Tuteja A 2017 Designing self-healing superhydrophobic surfaces with exceptional mechanical durability *ACS Appl. Mater. Interfaces* **9** 11212–23
[8] Jamil M I *et al* 2018 Icephobic strategies and materials with superwettability: design principles and mechanism *Langmuir* **34** 15425–44
[9] Azimi Yancheshme A, Allahdini A, Maghsoudi K, Jafari R and Momen G 2020 Potential anti-icing applications of encapsulated phase change material-embedded coatings: a review *J. Energy Storage* **31** 101638
[10] Kulinich S A, Masson D, Du X and Emelyanenko A M 2020 Testing the durability of anti-icing coatings *Ice Adhesion* ed K L Mittal and C-H Choi (New York: Wiley) pp 495–519
[11] Boinovich L B, Emelyanenko A M, Emelyanenko K A and Modin E B 2019 *Modus operandi* of protective and anti-icing mechanisms underlying the design of longstanding outdoor icephobic coatings *ACS Nano* **13** 4335–46
[12] Nistal A, Sierra-Martín B and Fernández-Barbero A 2023 On the durability of icephobic coatings: a review *Materials* **17** 235
[13] Tagliaro I, Cerpelloni A, Nikiforidis V-M, Pillai R and Antonini C 2022 On the development of icephobic surfaces: bridging experiments and simulations *The Surface Wettability Effect on Phase Change* ed M Marengo and J De Coninck (Cham: Springer International) pp 235–72
[14] Hernández Rodríguez G *et al* 2024 Icephobic gradient polymer coatings deposited via iCVD: a novel approach for icing control and mitigation *ACS Appl. Mater. Interfaces* **16** 11901–13
[15] Guo P *et al* 2012 Icephobic/anti-icing properties of micro/nanostructured surfaces *Adv. Mater.* **24** 2642–8
[16] Beemer D L, Wang W and Kota A K 2016 Durable gels with ultra-low adhesion to ice *J. Mater. Chem.* A **4** 18253–8
[17] Yeong Y H, Milionis A, Loth E and Sokhey J 2018 Self-lubricating icephobic elastomer coating (SLIC) for ultralow ice adhesion with enhanced durability *Cold Reg. Sci. Technol.* **148** 29–37
[18] Wang J and Wu M 2024 Study on durable icephobic surfaces modified with phase change oil impregnation *Surf. Coat. Technol.* **481** 130646
[19] Streitberger H-J and Goldschmidt A 2019 *BASF Handbook Basics of Coating Technology* 3rd revised edn (Hannover: Vincentz Network)
[20] Broughton W R and Maxwell A S 2023 Accelerated ageing of polymeric materials *Measurement Good Practice Guide* 103 National Physical Laboratory
[21] Jacques L F E 2000 Accelerated and outdoor/natural exposure testing of coatings *Prog. Polym. Sci.* **25** 1337–62
[22] Shen Y *et al* 2015 Anti-icing potential of superhydrophobic Ti6Al4V surfaces: ice nucleation and growth *Langmuir* **31** 10799–806
[23] Liu B *et al* 2016 Strategies for anti-icing: low surface energy or liquid-infused? *RSC Adv.* **6** 70251–60
[24] Bakhshandeh E, Sobhani S, Jafari R and Momen G 2024 From theory to application: Innovative ice-phobic polyurethane coatings by studying material parameters, mechanical insights, and additives *Surf. Interfaces* **51** 104545
[25] Fan L *et al* 2023 Superhydrophobic epoxy/fluorosilicone/PTFE coatings prepared by one-step spraying for enhanced anti-icing performance *Coatings* **13** 569
[26] Patil A H *et al* 2024 Facile approach for designing icephobic coatings using polymers/silica nanoparticle composites via self-formation of superhydrophobic surfaces *Adv. Mater. Interfaces* **11** 2300919

[27] Ng Y-H, Tay S-W and Hong L 2018 Formation of icephobic surface with micron-scaled hydrophobic heterogeneity on polyurethane aerospace coating *ACS Appl. Mater. Interfaces* **10** 37517–28
[28] Sarshar M A, Song D, Swarctz C, Lee J and Choi C-H 2018 Anti-icing or deicing: icephobicities of superhydrophobic surfaces with hierarchical structures *Langmuir* **34** 13821–7
[29] Zhuo Y, Wang F, Xiao S, He J and Zhang Z 2018 One-step fabrication of bioinspired lubricant-regenerable icephobic slippery liquid-infused porous surfaces *ACS Omega* **3** 10139–44
[30] Maitra T *et al* 2014 On the nanoengineering of superhydrophobic and impalement resistant surface textures below the freezing temperature *Nano Lett.* **14** 172–82
[31] Schröder S, Polonskyi O, Strunskus T and Faupel F 2020 Nanoscale gradient copolymer films via single-step deposition from the vapor phase *Mater. Today* **37** 35–42
[32] Dimitriadis T *et al* 2023 Capillary-driven water transport by contrast wettability-based durable surfaces *ACS Appl. Mater. Interfaces* **15** 27206–13
[33] Claussen K U, Scheibel T, Schmidt H and Giesa R 2012 Polymer gradient materials: can nature teach us new tricks? *Macromol. Mater. Eng.* **297** 938–57
[34] Yoo Y, You J B, Choi W and Im S G 2013 A stacked polymer film for robust superhydrophobic fabrics *Polym. Chem.* **4** 1664
[35] Gleason K K 2015 *CVD Polymers: Fabrication of Organic Surfaces and Devices* (Weinheim: Wiley)
[36] Gupta M and Gleason K K 2006 Initiated chemical vapor deposition of poly(1H,1H,2H,2H-perfluorodecyl acrylate) thin films *Langmuir* **22** 10047–52
[37] Coclite A M 2013 Smart surfaces by initiated chemical vapor deposition *Surf. Innov.* **1** 6–14
[38] Lambourne R and Strivens T A 1999 *Paint and Surface Coatings* (Oxford: Woodhead)
[39] Wernståhl K M and Carlsson B 1997 Durability assessment of automotive coatings—design and evaluation of accelerated tests *J. Coat. Technol.* **69** 69–75
[40] Malavasi I, Bernagozzi I, Antonini C and Marengo M 2015 Assessing durability of superhydrophobic surfaces *Surf. Innov.* **3** 49–60
[41] Licari J J 2003 *Coating Materials for Electronic Applications: Polymers, Processes, Reliability, Testing* (Norwich, NY: Noyes/William Andrew)
[42] Varshney P, Lomga J, Gupta P K, Mohapatra S S and Kumar A 2018 Durable and regenerable superhydrophobic coatings for aluminium surfaces with excellent self-cleaning and anti-fogging properties *Tribol. Int.* **119** 38–44
[43] Milionis A, Loth E and Bayer I S 2016 Recent advances in the mechanical durability of superhydrophobic materials *Adv. Colloid Interface Sci.* **229** 57–79
[44] Yanagisawa T, Nakajima A, Sakai M, Kameshima Y and Okada K 2009 Preparation and abrasion resistance of transparent super-hydrophobic coating by combining crater-like silica films with acicular boehmite powder *Mater. Sci. Eng.* B **161** 36–9
[45] Li B, Zhang J, Wu L and Wang A 2013 Durable superhydrophobic surfaces prepared by spray coating of polymerized organosilane/attapulgite nanocomposites *ChemPlusChem* **78** 1503–9
[46] Kostal E, Stroj S, Kasemann S, Matylitsky V and Domke M 2018 Fabrication of biomimetic fog-collecting superhydrophilic–superhydrophobic surface micropatterns using femtosecond lasers *Langmuir* **34** 2933–41
[47] Moradi S, Kamal S, Englezos P and Hatzikiriakos S G 2013 Femtosecond laser irradiation of metallic surfaces: effects of laser parameters on superhydrophobicity *Nanotechnology* **24** 415302

[48] Zhang D *et al* 2012 A simple way to achieve pattern-dependent tunable adhesion in superhydrophobic surfaces by a femtosecond laser *ACS Appl. Mater. Interfaces* **4** 4905–12
[49] Sarbada S and Shin Y C 2017 Superhydrophobic contoured surfaces created on metal and polymer using a femtosecond laser *Appl. Surf. Sci.* **405** 465–75
[50] Lin Y *et al* 2018 Durable and robust transparent superhydrophobic glass surfaces fabricated by a femtosecond laser with exceptional water repellency and thermostability *J. Mater. Chem.* A **6** 9049–56
[51] Brassard J-D, Laforte C, Guerin F and Blackburn C 2018 Icephobicity: definition and measurement regarding atmospheric icing *Contamination Mitigating Polymeric Coatings for Extreme Environments* ed C J Wohl and D H Berry (Cham: Springer International) pp 123–43
[52] Kulinich S A, Farhadi S, Nose K and Du X W 2011 Superhydrophobic surfaces: are they really ice-repellent? *Langmuir* **27** 25–9
[53] Jung S *et al* 2011 Are superhydrophobic surfaces best for icephobicity? *Langmuir* **27** 3059–66
[54] Schutzius T M *et al* 2015 Physics of icing and rational design of surfaces with extraordinary icephobicity *Langmuir* **31** 4807–21
[55] Nosonovsky M and Hejazi V 2012 Why superhydrophobic surfaces are not always icephobic *ACS Nano* **6** 8488–91
[56] Zhang H Y *et al* 2019 Compare study between icephobicity and superhydrophobicity *Physica* B **556** 118–30
[57] Farhadi S, Farzaneh M and Kulinich S A 2011 Anti-icing performance of superhydrophobic surfaces *Appl. Surf. Sci.* **257** 6264–9
[58] Jung S, Tiwari M K, Doan N V and Poulikakos D 2012 Mechanism of supercooled droplet freezing on surfaces *Nat. Commun.* **3** 615
[59] Wang Y, Orol D, Owens J, Simpson K and Lee H J 2013 Design and development of anti-icing aluminum surface *Mater. Sci. Appl.* **04** 347–56
[60] Kulinich S A and Farzaneh M 2009 Ice adhesion on super-hydrophobic surfaces *Appl. Surf. Sci.* **255** 8153–7
[61] Meuler A J *et al* 2010 Relationships between water wettability and ice adhesion *ACS Appl. Mater. Interfaces* **2** 3100–10
[62] Baylakoğlu İ *et al* 2021 The detrimental effects of water on electronic devices *E-Prime—Adv. Electr. Eng. Electron. Energy* **1** 100016
[63] Collins D H, Freels J K, Huzurbazar A V, Warr R L and Weaver B P 2013 Accelerated test methods for reliability prediction *J. Qual. Technol.* **45** 244–59
[64] Tagliaro I, Radice V, Nisticò R and Antonini C 2024 Chitosan electrolyte hydrogel with low ice adhesion properties *Colloids Surf. Physicochem. Eng. Asp.* **700** 134695
[65] *The History of Aluminium Industry* https://aluminiumleader.com
[66] Runge J M 2018 A brief history of anodizing aluminum *The Metallurgy of Anodizing Aluminum* (Cham: Springer International) pp 65–148
[67] Jani A M M, Losic D and Voelcker N H 2013 Nanoporous anodic aluminium oxide: advances in surface engineering and emerging applications *Prog. Mater. Sci.* **58** 636–704
[68] Li F, Zhang L and Metzger R M 1998 On the growth of highly ordered pores in anodized aluminum oxide *Chem. Mater.* **10** 2470–80
[69] Bharathidasan T, Kumar S V, Bobji M S, Chakradhar R P S and Basu B J 2014 Effect of wettability and surface roughness on ice-adhesion strength of hydrophilic, hydrophobic and superhydrophobic surfaces *Appl. Surf. Sci.* **314** 241–50

[70] Nguyen T-B, Park S and Lim H 2018 Effects of morphology parameters on anti-icing performance in superhydrophobic surfaces *Appl. Surf. Sci.* **435** 585–91
[71] Ghalmi Z and Farzaneh M 2015 Experimental investigation to evaluate the effect of PTFE nanostructured roughness on ice adhesion strength *Cold Reg. Sci. Technol.* **115** 42–7
[72] Momen G, Jafari R and Farzaneh M 2015 Ice repellency behaviour of superhydrophobic surfaces: effects of atmospheric icing conditions and surface roughness *Appl. Surf. Sci.* **349** 211–8
[73] Maghsoudi K, Vazirinasab E, Momen G and Jafari R 2021 Icephobicity and durability assessment of superhydrophobic surfaces: the role of surface roughness and the ice adhesion measurement technique *J. Mater. Process. Technol.* **288** 116883
[74] Moon H *et al* 2015 Synthesis of ultrathin polymer insulating layers by initiated chemical vapour deposition for low-power soft electronics *Nat. Mater.* **14** 628–35
[75] Schröder S *et al* 2022 Sensing performance of CuO/Cu_2O/ZnO:Fe heterostructure coated with thermally stable ultrathin hydrophobic PV3D3 polymer layer for battery application *Mater. Today Chem.* **23** 100642
[76] Johnson B W and McIntyre R 1996 Analysis of test methods for UV durability predictions of polymer coatings *Prog. Org. Coat.* **27** 95–106
[77] Han D *et al* 2015 Synthesis of fluorinated monomer and formation of hydrophobic surface therefrom *RSC Adv.* **5** 22847–55
[78] Giorgini L, Fragassa C, Zattini G and Pavlovic A 2016 Acid aging effects on surfaces of PTFE gaskets investigated by Fourier transform infrared spectroscopy *Tribol. Indust.* **38** 286–96
[79] Mackie N M, Castner D G and Fisher E R 1998 Characterization of pulsed-plasma-polymerized aromatic films *Langmuir* **14** 1227–35
[80] Bhullar S K, Bedeloglu A and Jun M B G 2014 Characterization and auxetic effect of polytetrafluoroethylene tubular structure *Int. J. Adv. Sci. Eng.* **1** 8–13
[81] Adamčík L, Dzurenda L, Banski A and Kminiak R 2023 Comparison of surface roughness of beech wood after sanding with an eccentric and belt sander *Forests* **15** 45
[82] Karakoç E and Çakır O 2023 Examination of surface roughness values of 6061-T6 aluminum material after machining and after anodizing process *Mater. Today Proc.* **80** 32–9
[83] 2011 *Nanocoatings and Ultra-Thin Films: Technologies and Applications* ed A S H Makhlouf and I Tiginyanu (Oxford: Woodhead)
[84] Dimitriadis T 2025 *PhD Thesis* Graz University of Technology

www.ingramcontent.com/pod-product-compliance
Ingram Content Group UK Ltd.
Pitfield, Milton Keynes, MK11 3LW, UK
UKHW051901100926
8556IPUK00005B/717

9 780750 3600